D1037508

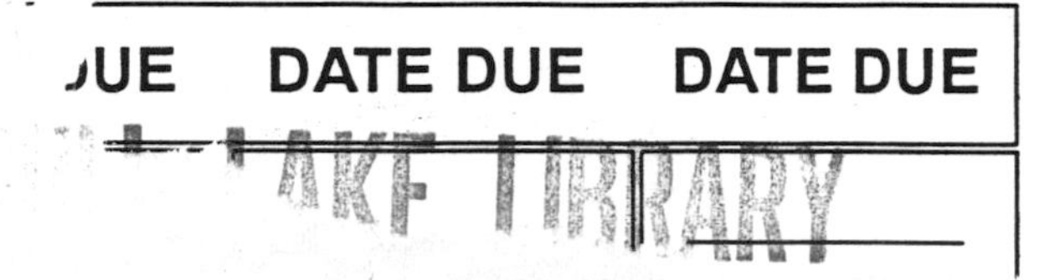
RETURN BOX to remove this checkout from your record.
FINES return on or before date due.
DUE DATE DUE DATE DUE

WEED SCIENCE

WEED SCIENCE

Principles and Practices

THIRD EDITION

Floyd M. Ashton
University of California
Davis, California

Thomas J. Monaco
North Carolina State University
Raleigh, North Carolina

With contributions from
Michael Barrett
University of Kentucky
Lexington, Kentucky

A WILEY-INTERSCIENCE PUBLICATION

JOHN WILEY & SONS, INC.

New York • Chichester • Brisbane • Toronto • Singapore

Library of Congress Cataloging in Publication Data:

Ashton, Floyd M.

Weed science: principles and practices/Floyd M. Ashton, Thomas J. Monaco; with contribution from Michael Barrett.

p. cm.

Rev. ed. of: Weed science/Glenn C. Klingman, Floyd M. Ashton. 2nd ed. c1982.

"A Wiley–Interscience publication."

Includes bibliographical references and index.

ISBN 0-471-60084-9

1. Weeds—Control. 2. Herbicides. I. Monaco, Thomas J. II. Barrett, Michael, 1951- . III. Klingman, Glenn C. Weed science. IV. Title.

SB611.A74 1991 90-24784

632′.58—dc20 CIP

Printed in the United States of America

10 9 8 7 6 5 4 3 2

PREFACE

Weed control is an essential part of crop production. It is expensive and directly increases the price of food. Weeds may poison livestock or seriously slow their weight gains. They cause human allergies such as hay fever and poison ivy. They infest lawns and gardens. Weeds create problems in recreation areas such as golf courses, parks, and fishing and boating areas. They are troublesome along highways and railroads, and in industrial areas and irrigation and drainage systems. Tillage to control weeds may seriously increase soil erosion. Human labor used to control weeds manually is not available for more-productive uses. Weeds affect everyone; therefore, they should be of concern to everybody.

The science of weed control has advanced more during the past 50 years than in the previous 100 centuries. In this book, the old and reliable methods of weed control are integrated with the modern chemical techniques. Together, these practices result in more-effective and less-expensive weed-control programs. These programs help reduce the cost of food and fiber production.

This textbook is designed principally for college classroom instruction in the principles and practices of weed science. However, its detailed coverage of these topics also makes it useful to country agents, farm advisors, extension specialists, herbicide-development personnel, research scientists, and farmers. The first third of this book deals with principles of weed science, the second third with herbicides, and the last third with weed-control practices in specific crops or situations. In total, this book brings together the modern philosophy of weed science and the techniques of weed control.

In recent years, weed scientists and the general public have become increasingly interested in the environmental impact and safety of weed control practices. This is especially true in regard to the use of herbicides. This has led chemical companies to stress the development of new herbicides of low mammalian toxicity that can be used at very low rates. These minimize environmental impact and maximize safe use. There has also been an increase in studies on the biology and ecology of weeds directed toward more effective control methods. An increased emphasis on these topics is provided in the 3rd edition of this book.

The 3rd edition also provides more details on the use of herbicides than previously given. We hope that this will increase its value to those using these chemicals in the field. However, nothing in this book is to be construed as recommending or authorizing the use of any weed-control practice or chemical. A *current* manufacturer's or supplier's label is the final word for the use of herbicides including method and time of application, rates to be used, permissible crops or situation, weeds controlled, and special precautions. *Label recommendations must be followed*—regardless of statements in this book.

Herbicide usage is being continuously revised. The Environmental Protection Agency's policy for the reregistering of pesticides is having a major impact on the continued registration of numerous pesticides and their uses. Many changes can be expected in the future; therefore, check the current registration before use.

Common names of herbicides are usually used throughout the text. However, trade names and chemical names are also given. Trade names have been given as a convenience to the reader, *not* as an endorsement of any one product. Common names and trade names are cross-indexed in the appendix.

We dedicate this book to all children of the world, including our children and grandchildren, in the hope that it contributes to the production of a safe, healthful, and abundant food and fiber supply with a safe environment.

With great appreciation, we wish to acknowledge the contributions of Glenn C. Klingman as sole author of the first edition (1961) and as lead author of the second edition (1975). Lyman J. Noordhoff served as editor of both editions. However, they are in no way responsible for the content of this 3rd edition. We also gratefully acknowledge the assistance of Jerome B. Weber for his review and suggestions regarding Chapter 6, Herbicides and the Soil.

We are grateful to our wives Theo E. Ashton, Jenny S. Monaco, and Adele Barrett for their patience, support, and assistance during the preparation of this manuscript.

FLOYD M. ASHTON
THOMAS J. MONACO
MICHAEL BARRETT

Davis, California
Raleigh, North Carolina
Lexington, Kentucky
March, 1991

CONTENTS

WEED SCIENCE

PART 1
Principles

1 Introduction

Weed control is as old as the growth of food crops. Progressively and with increasing momentum, man has learned to use his bare hands, hand tools, horsepower, tractor power, and herbicides to control weeds. Other approaches that are used to a limited degree and/or are being extensively investigated include biological control by insects and plant disease organisms, herbicide antidotes, breeding of herbicide tolerant and more competitive crops, allelopathy, and genetic engineering.

During thousands of years man has achieved amazing advances. First he substituted a sharp stick and other wooden tools for his fingers. Centuries later he discovered the metal hoe. Then man greatly reduced his labor by harnessing a horse or ox to drag the hoe or plow. In 1731 a major advance was proposed—planting crops *in rows* to permit "horse-hoeing." Jethro Tull explained his idea in *Horse-Hoeing Husbandry*. He was among the first to use the word *weed* with its present spelling and meaning. And less than 200 years later tractors started to replace horses.

All of these methods used brute force to control weeds. However, with the introduction of herbicides and their subsequent extensive use, chemical energy has largely replaced these other methods of weed control in the United States and many other countries (Figure 1-1).

Biological control of weeds by insects and plant disease organisms has had considerable success in a limited number of cases (Andres, 1977; Charudattan and Walker, 1982). The use of herbicide antidotes to protect crop plants has been successful for a few herbicides in specific crops (Hatzios and Hoagland, 1988). The breeding of herbicide tolerant crops has been successful for the use of certain triazine herbicides in the rape seed oil crop (LeBaron and Gressel, 1982). Although allelopathy and use of genetic engineering are in the forefront of research in weed science, a practical application of these approaches has not been realized. Allelopathy is usually defined as any direct or indirect, inhibitory or stimulative, effect by one plant (including microorganisms) on another through the production of a chemical compound(s) (Rice, 1984; Putnam and Tang, 1985). Genetic engineering primarily involves the transfer of the gene(s) responsible for plant tolerance to herbicides from tolerant plants to susceptible crops and the development of more competitive crops. Cell and tissue culture and transfer of recombinant DNA techniques are used (LeBaron and Gressel, 1982).

Improved agricultural technology over the years has contributed greatly to increased food production (Figure 1-2) and a related increase in our standard of

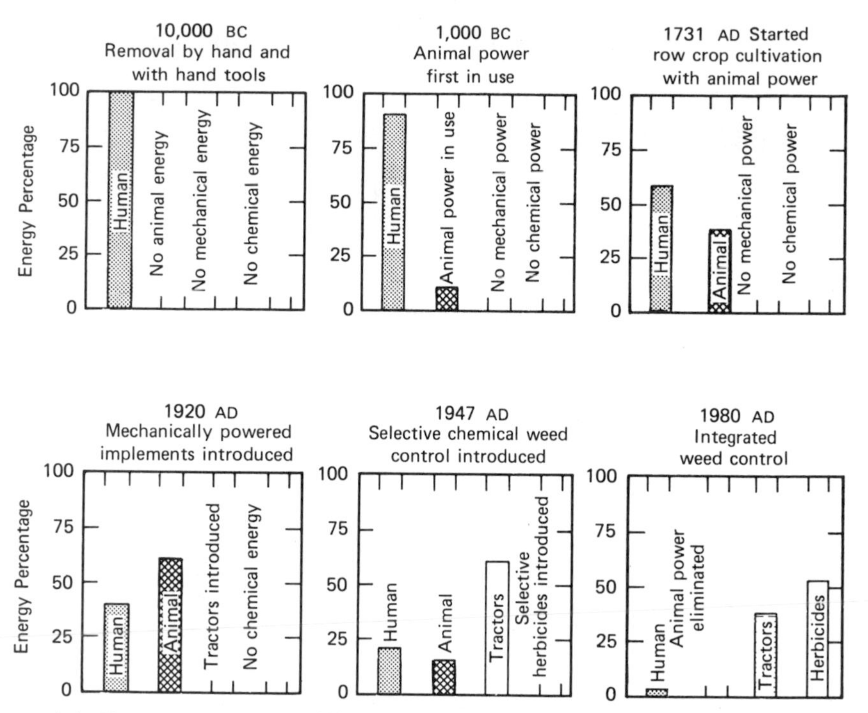

Figure 1-1. Energy sources providing weed control at different times. Data shown for 1920, 1947, and 1980 are for the United States.

living. Advances in weed control practices has been an important part of these gains.

Before 10,000 BC, weeds were removed from crops by hand. One person could hardly feed himself, and starvation was common. During centuries to follow, man slowly improved the uses of hand tools.

By 1000 BC, man used an animal to drag the hoe (as a crude plow), thus reducing human labor, mostly in seedbed preparation. Still, one person could produce only enough food for two people, and starvation continued to be widespread.

After 1731, when growing crops in *rows* with horse-hoeing was introduced, each person could then provide food for four persons.

In 1920 tractors were beginning to be widely used. This new-found power enabled each farmer to produce enough food for eight people.

About 1947 herbicide usage began to be a common practice and at that time one farmer could feed 16 people. During the intervening years many new herbicides have been developed and widely used. Many other improvements in agricultural technology have also contributed to the increase in food production during this period. In 1990, one farmer could feed 75 people. This means that multitudes of the people previously working on the farm have greatly increased our standard of living by providing other goods and services.

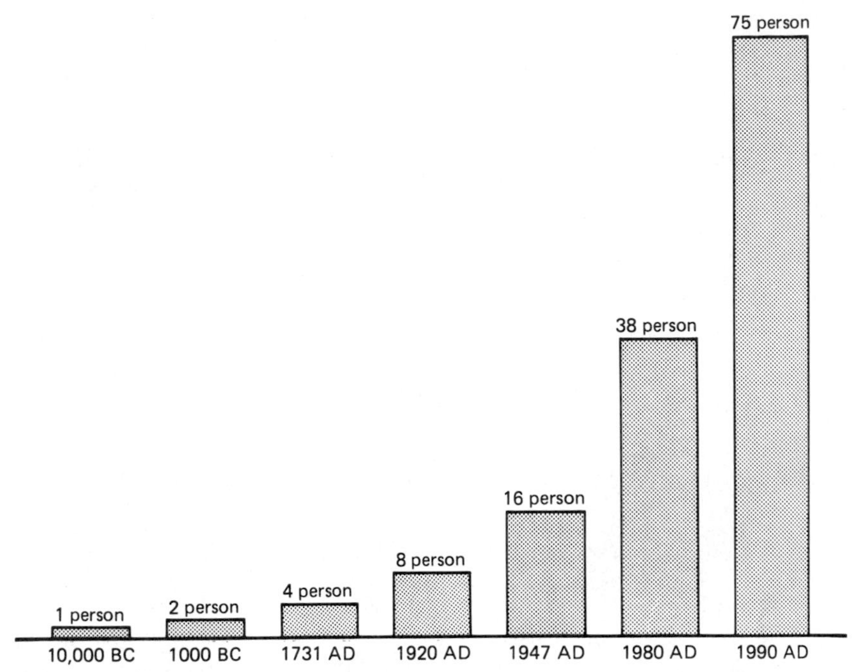

Figure 1-2. Crop energy output per farmer or the number of people fed by one farmer. Data for 1920, 1947, 1980, and 1990 are for the United States.

DEFINITIONS OF A WEED, PESTICIDE, AND HERBICIDE

A weed is a *plant growing where it is not desired*, or a *plant out of place*. Therefore, rye in a wheat field is a weed; so is corn in a peanut field. Weeds encompass all types of undesirable plants—trees, broadleaf plants, grasses, sedges, aquatic plants, and parasitic flowering plants (dodder, mistletoe, witchweed).

Weeds are also classified as *pests*. Other important pests are insects, plant diseases, nematodes, and rodents. A chemical used to control a pest is called a *pesticide*. A chemical used specifically for *weed control* is known as a *herbicide*.

WEED IMPACTS

Weeds should be everybody's business since they affect everyone in one way or another. They not only reduce crop production and increase the cost of agricultural products but they also cause problems for the general public in many other ways, e.g., health, home landscape area, recreational area, and noncrop area maintenance. Specific problems include lower crop and animal yields, less efficient land use, higher costs of insect and plant disease control, poorer quality products, more water management problems, and lower human efficiency.

Figure 1-3. Weeds often reduce the yeild of crops through allelopathic effects. *Left*: Alfalfa growing in weed-free soil. *Center*: Alfalfa watered with filtrate drained from ground quackgrass rhizomes. *Right*: Alfalfa grown in soil containing ground quackgrass rhizomes. (T. Kommendahl, University of Minnesota.)

Lower Plant and Animal Yields

Weed control is an expensive but necessary part of agricultural production, directly affecting the price of food and other agricultural products. However, these would be less abundant and more expensive without modern agricultural and weed science technology. Weeds reduce yields of plant and animal crops. Plant yields are primarily reduced by competition between the weed and the crop for soil water, soil nutrients, light, and carbon dioxide. Certain weeds may also reduce plant yields by releasing allelopathic compounds into the environment (Figure 1-3).

Livestock yields may be reduced by weeds through less pasture or range forage, or by poisonous or toxic plants that cause slower animal growth or death (Figure 1-4).

Less Efficient Land Use

The presence of weeds on a given piece of land can reduce the maximum efficiency of the use of that land in a number of ways. These include increased costs of

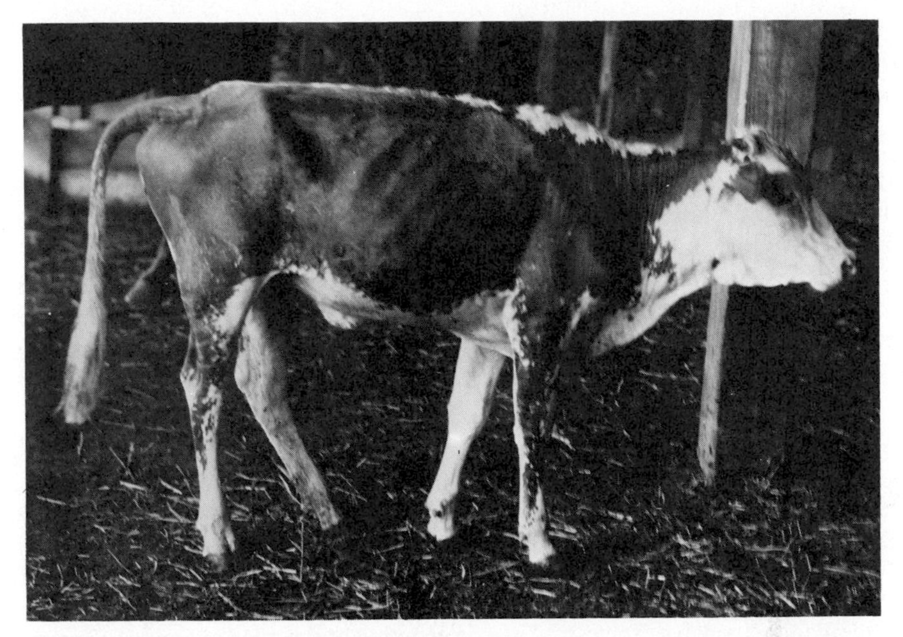

Figure 1-4. Weeds may reduce livestock yields. This animal was fed prostrate spurge. The abnormal stance and digestive disturbances became more pronounced as the spurge feeding continued. (O. E. Sperry, Texas A&M University.)

production and harvest, reforestation, and noncropland maintenance, as well as reduced plant growth, root damage from cultivation, limit the crops that can be grown, and reduce land values.

Higher Insect and Plant Disease Control Costs

Weeds harbor insect and disease organisms that attack crop plants. For example, the carrot weevil and carrot rust-fly may be harbored by the wild carrot, only later to attack the cultivated carrot. Aphids and cabbage root maggots live in mustard, and later attack cabbage, cauliflower, radish, and turnips. Onion thrips live in ragweed and mustards and may later prey on the onion crop. The disease of curly top on sugar beets is carried by insect vectors that live on weeds in wastelands. Many insects overwinter in weedy fields and field borders.

Disease organisms such as black stem rust may use the European barberry, quackgrass, or wild oat as hosts prior to attacking wheat, oats, or barley. Some virus diseases are propagated on members of the weedy nightshades. For example, the virus causing "leaf roll" of potatoes lives on black nightshade. It is thought that aphids carry the virus to potatoes. A three-way harboring and transmission of a mycoplasma disease from weeds to citrus has been discovered in California. Leaf hoppers transmitted the disease organism, citrus stubborn disease (*Spiroplasma citri*), to and from diseased periwinkle and to and from

London rocket (*Sisymbrium irio*). These weedy plants act as a source of the disease organism to infect citrus trees.

Poorer Quality Products

All types of crop products may be reduced in quality. Weed seeds and onion bulblets in grain and seed, weedy trash in hay and cotton, spindly "leaf crops," and scrawny vegetables are a few examples.

Livestock products may be lower priced or unmarketable because of weeds; for example, onion, garlic, or bitterweed flavor in milk, and cocklebur in wool reduce the quality of the products. Poisonous plants may kill animals, slow down their rates of growth, or cause many kinds of abnormalities (Figure 1-4).

More Water Management Problems

Aquatic weeds can be a major problem in irrigation and drainage systems, lakes, ponds, reservoirs, and harbors. They restrict the flow of water (Figure 1-5), interfere with commercial and recreational activities, and may give off undersirable flavors and odors in domestic water supplies. Their control is often difficult and expensive. Terrestrial weeds growing at the edge of aquatic sites can also be a problem. Chapter 23 is devoted to the weed problems and control methods on these sites.

Figure 1-5. In St. Cloud, Florida, The State Mosquito Control Board sprayed ditches with diuron to eliminate weeds and improve drainage. The weed-control program reduced the expenses of mosquito control enough to permit a one-third savings in the total budget. *Left*: One year after hand weeding. *Right*: One year after chemical treatment with diuron. (E. I. du Pont de Nemours and Company.)

Lower Human Efficiency

Weeds have been a plague to man since he gave up a hunter's life. Traveling in developing nations, one may feel that half the world's population work in the fields, stooped, moving slowly, and silently weeding. These people are a part of the great mass of humanity that spends a lifetime simply weeding. Many young people doing such work in Africa, Asia, and Latin America can never attend school; women do not have time to prepare nutritious meals or otherwise care for their families. Modern weed-control methods integrated into the economies and cultures of developing nations provide relief from this arduous chore and give nations the opportunity to improve their standards of living through more productive work.

Weed control constitutes a large share of a farmer's work required to produce a crop. This effort directly affects the cost of crop production and thus the cost of food. It affects all of us, whether we farm or not.

Weeds reduce human efficiency through allergies and poisoning. Hay fever, caused principally by pollen from weeds, alone accounts for tremendous losses in human efficiency every summer and fall. Poison ivy, poison oak, and poison sumac cause losses in terms of time and human suffering; children occasionally die from eating poisonous plants or fruits.

COST OF WEEDS

The cost of weeds to man are much higher than generally recognized. Because weeds are so common and so wide spread, people do not fully appreciate their significance in terms of losses and control costs. Although relatively accurate estimates have been made of losses and control costs on farms in the United States (USDA, 1981; Chandler et al., 1984), other areas of economic impact are much more difficult to estimate. The latter include noncropland, recreational areas, homesite maintenance, and lower human efficiency, as well as many others.

Weeds are common on all 485 million acres of U.S. cropland and almost one billion acres of range and pasture. In the United States, the estimated annual yield loss caused by weeds in 64 crops is 7.5 billion dollars; approximately 85% in field crops, 8% in vegetable crops, 6% in fruit and nut crops, and less than 1% in forage seed crops (Chandler et al., 1984). In Canada, the estimated annual yield loss caused by weeds in 36 crops is 909 million dollars. These values do not include losses due to weeds in pastures, rangeland, and hay crops or livestock poisoning from poisonous weeds. A significant increase in efficiency of food production could result from effective, efficienct, and integrated weed-control strategies.

The cost of controlling weeds in these crops and the losses and control costs in pastures, rangeland, hay crops, and nonfarm sites would greatly increase the above values. It has been estimated that U.S. farmers annually spend 3.6 billion dollars for chemical weed control and 2.6 billion dollars for cultural and other methods of weed control. The total cost of weeds in the United States could approach 15 to 20 billion dollars.

PREVENTION, CONTROL, AND ERADICATION

Prevention

Prevention means stopping a given species from contaminating an area. Prevention is often the most practical means of controlling weeds. This is best accomplished by (1) making sure that new weed seeds are not carried onto the farm in contaminated crop seeds, feed, or on machinery, (2) preventing weeds on the farm from going to seed, and (3) preventing the spread of perennial weeds that reproduce vegetatively.

Control

Control is the process of limiting weed infestations. In crops, the weeds are limited so that there is minimum competition. The degree of control is usually a matter of economics, a balance between the costs involved and the increase in profits due to the control of the weeds. On noncropland, it is often desirable to remove essentially all vegetation for a specific period of time. Weeds are limited to a level that does not allow them to interfere with man's activities.

Most biological control programs using a highly specific insect or plant disease organism as a control agent can be expected to results in adequate control, but not eradication.

Eradication

Eradication is the complete elimination of all living plants including their vegetative propaguies and seeds. Eradication is much more difficult than prevention or control. In general, it is justified only for the elimination of a serious weed on a limited area, e.g., a perennial weed on a small area of a field.

The most difficult part is the elimination of the vegetative propagules and seeds. Seeds of many weeds may remain dormant in the soil for many years and in this dormant state they are not usually killed by standard practices. Vegetative propagules are also often difficult to kill by standard practices. Although most vegetative propagules and many seeds can be killed by soil fumigation, e.g., methyl bromide, some dormant seeds are not killed. This practice is also relatively expensive (Chapter 8). Presistent soil applied herbicides are also used, but they prevent the growth of desirable species for some period of time.

WEED CLASSIFICATION

The plant's life span, season of growth, and its methods of reproduction largely determine the methods needed for control or eradication. In temperate climates there are three principal groups: annuals, biennials, and perennials.

Annuals

Annual plants complete their life cycle in less than 1 year. Normally they are considered easy to control. This is true for any one crop of weeds. However, because of an abundance of dormant seed and fast growth, annuals are very persistent, and they actually cost more to control than perennial weeds. Most common field weeds are annuals. There are two types—summer annuals and winter annuals.

Summer Annuals Summer annuals germinate in the spring, make most of their growth during the summer, and the plants mature and die in the fall. The seeds lie dormant in the soil until the next spring. Summer annuals include such weeds as cockleburs, morning glories, pigweeds, common lambsquarters, common ragweed, crabgrasses, foxtails, and goosegrass. These weeds are most troublesome in summer crops like corn, sorghum, soybeans, cotton, peanuts, tobacco, and many vegetable crops.

Winter Annuals Winter annuals germinate in the fall and winter and usually mature seed in the spring or early summer before the plants die. The seeds often lie dormant in the soil during the summer months. In this group, high soil temperatures (125°F or above) have a tendency to cause seed dormancy—inhibit seed germination.

This group includes such weeds as chickweed, downy brome, hairy chess, cheat, shepherdspurse, field pennycress, corn cockle, cornflower, and henbit. These are most troublesome in winter-growing crops such as winter wheat, winter oats, or winter barley.

Biennials

A biennial plant lives for more than 1 year but not over 2 years. Only a few troublesome weeds fall in this group. Wild carrot, bull thistle, common mullein, and common burdock are examples.

There is confusion between the biennials and winter annual group, because the winter annual group normally lives during two calendar years and during two seasons.

Perennials

Perennials live for more than 2 years and may live almost indefinitely. Most reproduce by seed and many are able to spread vegetatively. They are classified according to their method of reproduction as *simple* and *creeping*.

Simple Perennials Simple perennials spread by seed. They have no natural means of spreading vegetatively. However, if injured or cut, the cut pieces may produce new plants. For example, a dandelion or dock root cut in half longitudinally may produce two plants. The roots are usually fleshy and may grow very large. Examples include common dandelion, dock, buckhorn plantain, broadleaf plantain, and pokeweed.

Creeping Perennials Creeping perennials reproduce by creeping roots, creeping aboveground stems (stolons), or creeping belowground stems (rhizomes). In addition, they may reproduce by seed. Examples include red sorrel, perennial sow thistle, field bindweed, wild strawberry, mouseear chickweed, ground ivy, bermudagrass, johnsongrass, quackgrass, and Canada thistle.

Some weeds maintain themselves and propagate by means of tubers, which are modified rhizomes adapted for food storage. Nutsedge (nutgrass) and Jerusalem artichoke are examples.

Once a field is infested, creeping perennials are probably the most difficult group to control. Cultivators and plows often drag pieces about the field. Herbicides applied and mixed into the soil may reduce the chances of establishment of such plant pieces. Continuous and repeated cultivations, repeated mowing for 1 or 2 years, or persistent herbicides are often necessary for control. Cultivation in combination with herbicides is proving effective on some weeds. An eradication program requires the killing of seedlings as well as dormant seeds in the soil.

SUGGESTED ADDITIONAL READING

Andres, L. A., 1977, The biological control of weeds, in J. D. Fryer and S. Matsunaka, Eds., *Integrated Control of Weeds*, pp. 153–174, Univ. of Tokyo Press, Tokyo, Japan.

Auld, B. A., K. M. Menz, and C. A. Tisdell, 1987, *Weed Control Economics*, Academic Press, New York.

Chandler, J. M., A. S. Hamill, and A. G. Thomas, 1984, *Crop Losses Due to Weeds in Canada and the United States*, Weed Science Society of America, Champaign, IL.

Charudattan, R., and H. L. Walker, 1982, *Biological Control of Weeds with Plant Pathogens*, Wiley, New York.

Hatzios, K. K., and R. E. Hoagland, 1988, *Crop Safeners for Herbicides*, Academic Press, New York.

Herbicide Handbock, 1989, Weed Science Society of America, Champaign, IL.

LeBaron, H. M., and J. Gressel, 1982, *Herbicide Resistance in Plants*, Wiley, New York.

Putnam, A. R., and C.-S. Tang, Eds., 1985, *The Science of Allelopathy*, Wiley, New York.

Rice, E. L., 1984, *Allelopathy*, Academic Press, New York.

USDA, 1981, *The Economics of Agricultural Pest Control*, an Annotated Bibliography, 1960–80, Econ. and Stat. Ser., Bib. and Lit. of Agric. No. 14.

Weed Research, Periodical, European Weed Research Society, Wageningen, Netherlands.

Weed Science, Periodical, Weed Science Society of America, Champaign, IL.

Weed Technology, Periodical, Weed Science Soceity of America, Champaign, IL.

2 Weed Biology and Ecology

The *biology* of weeds is concerned with their establishment, growth, and reproduction. The *ecology* of weeds is concerned with the development of a single species within a *population* and the development of all populations within a *community* on a given site. The numerous factors of the environment have a pronounced influence on all of these processes and systems. The environment and the living community are considered to be an *ecosystem.*

Heredity and environment are the master factors governing life. Heredity determines what an organism becomes by controlling life form, growth potential, method of reproduction, length of life, and so on. The environment largely determines the extent to which these life processes proceed.

Knowledge of weed biology and use of environmental management makes it possible to shift plant populations and communities in a desired direction. Cultivation in crop fields makes the environment favorable to the crop plants and unfavorable to competing weeds. Proper grazing and/or fertilization management shifts the environment in pastures and range areas to maximize the production of desirable species by reducing yield-reducing weeds (see Chapter 21). The several mechanical methods used to remove undesirable species from forest lands are environmental management tools (see Chapter 22). The use of herbicides can also be considered to be an environmental management tool.

ENVIRONMENTAL FACTORS

The environmental factors that need to be considered in relation to weed biology, ecology, and control include climatic, physiographic, and biotic aspects.

Climatic

Climatic factors include the following:

- Light (intensity, quality, duration including photoperiod)
- Temperature (extremes, average, frost-free period)
- Water (amount, percolation, runoff, evaporation)
- Wind (velocity, duration)
- Atmosphere (CO_2, O_2, humidity, toxic substances)

Physiographic

Physiographic factors include the following:

Edaphic (soil factors including pH, fertility, texture, structure, organic matter, CO_2, O_2, water drainage)
Topographic (altitude, slope, exposure to the sun)

Biotic

Biotic factors include the following:

Plants (competition, released toxins or stimulants, diseases, parasitism, soil flora)
Animals (insects, grazing animals, soil fauna, man)

Many of the most common weeds have a broad tolerance to environmental conditions. In fact, that is a major reason why they are so common and troublesome. For example, common weeds such as lambsquarters, common chickweed, and shepherdspurse grow on almost all types of soils. Rarer species such as saltgrass, halogeton, and alkali heath are usually found only on alkali soils.

Similar environmental requirements of certain weeds and selected crop species produce some rather common crop–weed associations. Some examples are mustard in small grain, barnyardgrass in tomatoes, burning nettle in lettuce, common chickweed in celery, and pigweeds in sugar beets.

COMPETITION

Crop–weed competition generally infers an inhibition of crop growth by weeds. However, more technically, competition is one of several types of *interference* among species or populations. There are both positive and negative interferences. *Competition* can be considered to be a negative interference that induces decreased growth of two species of plants because of an insufficient supply of some necessary factor, e.g., water, mineral element. *Amensalism* is another type of negative interference and can be defined as the inhibition of one species by another. However, in contrast to competition, which involves the removal of a resource, amensalism involves the addition of something to the environment. Allelopathy is a type of amensalism and will be discussed in the next section. These terms are discussed in greater detail in several books cited at the end of this chapter, including Harper (1977), Radosevich and Holt (1984), and Rice (1984).

Weeds are considered to primarily compete with crops for soil nutrients, soil moisture, light, and carbon dioxide. Weeds are naturally strong competitors, otherwise they would not be weeds. Some of the factors that contribute to the strong competitive nature of weeds include high seed production leading to high plant density, rapid germination, very rapid early growth, and a long duration (life cycle).

Weed competition early in the season usually reduces crop yields far more than late-season weed growth. Although late-season weed growth may not seriously reduce yields, if often makes harvesting difficult, reduces crop quality, reinfests the land with weed seeds, and may favor the overwintering of insect and disease pests.

Zimdahl's review (1980) cited numerous experiments on the effect of both density and duration of weeds on yield of many crops. For example, (1) in regard to density, Weatherspoon and Schweizer (1971) reported that kochia plants at 0.1, 0.2, 0.5, and 1.0 per foot of row reduced yields of sugar beets, 26, 44, 67, and 79%, respectively; (2) in regard to duration, Dawson (1965) reported that maintaining sugar beets weed free for 0, 4, 8, or 12 weeks resulted in yields of 20, 46, 60, and 100%, respectively.

A considerable amount of the recent and current research on weed–crop competition involves computer modeling. Rejmánek et al. (1989) discuss this approach and give many references to this subject in their recent paper entitled "Weed–Crop Competition: Experimental Design and Models for Data Analysis."

ALLELOPATHY

Allelopathy refers to chemical interactions among plants (microbes and higher plants) including stimulatory as well as inhibitory influences (Molisch, 1937; Rice, 1984; Putnam and Tang, 1986). However, inhibitory effects of weeds on crop yields is the main interest in this phenomenon to weed science. These may originate from a direct release of the toxin(s) from the living plant or be released from decaying plant litter, residues, or root tissues. From these nonliving plant tissues, microorganisms have been implicated in the release of the toxin or modification of nontoxic compounds to toxic compounds.

Many weed species, perhaps as many as 90, may interfere with plant growth through allelopathic mechanisms (Putnam and Tang, 1986). Perennial weeds including quackgrass, johnsongrass, and nutsedges have often been implicated.

Rice (1984) classified allelopathic agents into 14 chemical categories (plus a miscellaneous) that are either secondary compounds (not involved in basic metabolism) or are associated with the shikimic acid and acetate pathways. Most of the allelopathic compounds that have been isolated and identified have one or more rings and many have quite complicated chemical structures.

The unequivocal proof that a postulated allelopathic phenomenon is not some other type of interference is quite difficult. It has been suggested that a specific protocol similar to the established procedure for proof of disease (Koch's postulates) be followed (Fuerst and Putnam, 1983; Putnam and Tang, 1986). They list a four-step sequence of studies that should be conducted.

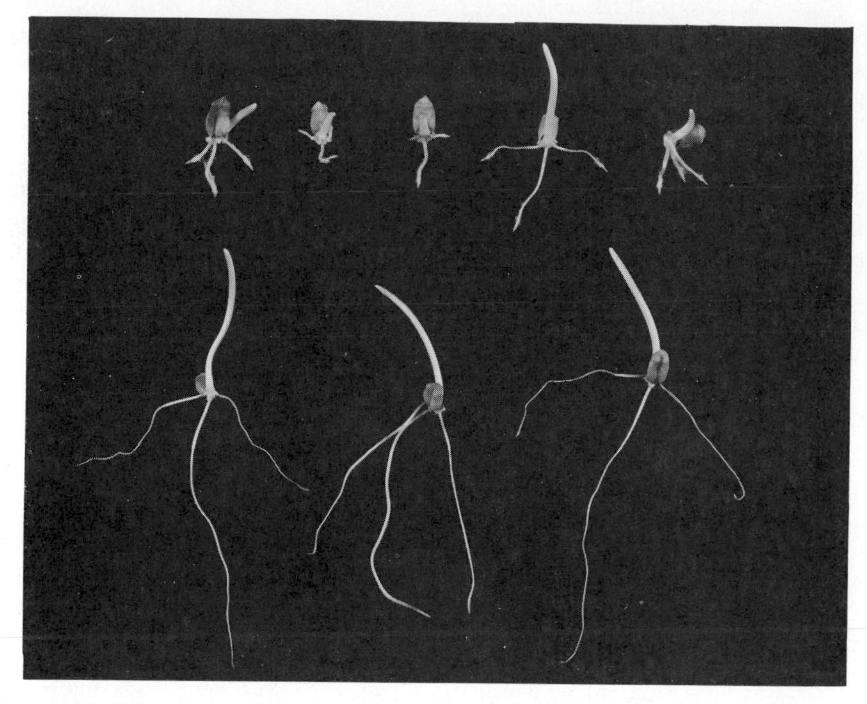

Figure 2-1. Allelopathic effects. *Upper row*: Stunted wheat seedlings caused by a water leachate from quackgrass roots and rhizomes. *Lower row*: Normal wheat seedlings grown in tap water. (T. Kommedahl, University of Minnesota.)

Putnam and Tang (1986) suggest that "allelopathy is now a maturing scientific discipline" and that the future of allelopathy will probably be involved in ecosystem management, pest management, and the development of novel agricultural chemicals.

REPRODUCTION OF WEEDS

Weeds multiply and reproduce by both sexual and asexual (vegetative) means. Sexual reproduction requires fertilization of the egg by sperm. This usually proceeds via the pollination of a flower, which subsequently produces seed. The viable seed then has the potential of producing a new plant. Asexual reproduction involves the development of a new plant from a vegetative organ such as stems, roots, leaves, or modifications of these basic organs. These include underground stems (rhizomes), aboveground stems (stolons), tubers, corns, bulbs, and bulblets.

Although several of the above environmental factors influence both vegetative and seed reproduction in a general way, the influence of day-length or photoperiod can be quite specific. These terms refer to the relative lengths of day and night. In many plants, flowering and/or development of certain vegetative reproduction organs are controlled by the photoperiod. This may control the development of propagules of a given species on a particular site and limit its geographic distribution.

The discussion of the several aspects of the reproduction of weeds from seeds is followed by a brief coverage of the reproduction of weeds by vegetative means in the last section of this chapter.

SEED DISSEMINATION

Seeds in general have no method of movement; therefore, they must depend on other forces for dissemination. Regardless of this fact, they are excellent travelers. The spread of seeds, plus their ability to remain viable in the soil for many years (dormant), poses on the most complex problems of weed control. This fact makes "eradication" nearly impossible for many seed-producing weeds.

Weed seeds are scattered by (1) crop seed, grain feed, hay, and straw; (2) wind; (3) water; (4) animals, including man; (5) machinery; and (6) weed screenings.

Crop Seed, Grain Feed, Hay, and Straw

Weeds are probably more widely spread through crop seeds, grain feed, hay, and straw than by other means.

Studies of wheat, oats, and barley seed sown by farmers in six North Central states reveal the seriousness of the problem. Researchers took seed directly from drill boxes and analyzed for weed seeds. About 8% of the samples contained primary noxious weeds, and about 45% contained secondary noxious weed seeds. About 80% of the seed had gone through a "recleaning" operation. Much of the recleaning was of limited benefit, because only part of the weed seeds was removed.

In the above study, *certified*, *registered*, and *foundation* seeds were free of primary noxious weeds and contained only a small percentage of the common weeds (Furrer, 1954).

Farmers often feel that a low percentage of weed seeds on the seed label means that the few weeds present are of little importance. This may be a serious mistake (Table 2-1). One dodder plant may easily spread to occupy one square rod during one season; thus only 0.001% dodder seed is enough to completely infest a legume crop the first year!

The prevalence of weed seed in legume seed is clearly shown by data collected by official state seed analysts. In 3643 samples, weed seeds averaged from 0.10 to 0.38% by weight. These figures indicate that certain weed species can completely infest a field the first year, despite an extremely low percentage of weed seeds present in the crop seed. These percentages are often below legal tolerances, which make it necessary to state their presence on the seed label.

Weeds are commonly spread through grain feed, hay, and straw. Where straw

TABLE 2-1. Field Dodder[1] and Its Rate of Planting in Contaminated Legume Seed Sown at the Rate of 20 lb/acre

Dodder Seed by Weight (%)	No. of Dodder Seeds (per lb of legume seed)	No. of Dodder Seeds Sown	
		Per acre	Per square rod
0.001	8	160	1
0.010	80	1,600	10
0.025	200	4,000	25
0.050	400	8,000	50
0.100	800	16,000	100
0.250	2,000	40,000	250

[1] There are 550,000 to 800,000 dodder seeds per pound.

is used for mulching, it is important that the straw be free of viable weed seeds as well as grain seeds. The grain seed in the straw may prove to be a weed under such circumstances. Most of the grain seed will germinate and die if the straw is kept moist for 30 days with temperatures favorable to germination.

As shown later in this chapter, an appreciable number of weed seeds in grain and hay are viable after passing through the alimentary canal of the animal. If the manure is allowed to become "well rotted," the weed seeds will be killed.

Wind

Weed seeds have many special adaptations that help them spread. Some are equipped with parachutelike structures (pappus) or cottonlike coverings that make the seed float in the wind. Common dandelion, sowthistle, Canada thistle, wild lettuce, some asters, and milkweeds are examples (Figure 2-2).

Water

Weed seed may move with surface water runoff, in natural streams and rivers, in irrigation and drainage canals, and in irrigation water from ponds.

Some seeds have special structures to help them float in water. For example, curly dock has small pontoon arrangements on the winged seed covering. Other seeds are carried in moving water or along the river bottom. Flooded areas from river overflows are nearly always heavily infested with weeds.

Irrigation water is a particularly important means of scattering seed. In 156 weed–seed catches in Colorado, in three irrigation ditches, 81 different weed species were found. In a 24-hour period, several million seeds passed in a 12-foot ditch.

Scientists have found great variation in the length of time that seeds remain viable in fresh water. Results in Table 2-2 are based on weed seed suspended in bags (luminite screen sewn with nylon thread) at 12- and 48-inch depths in a freshwater canal at Prosser, Washington. Water could circulate freely within the

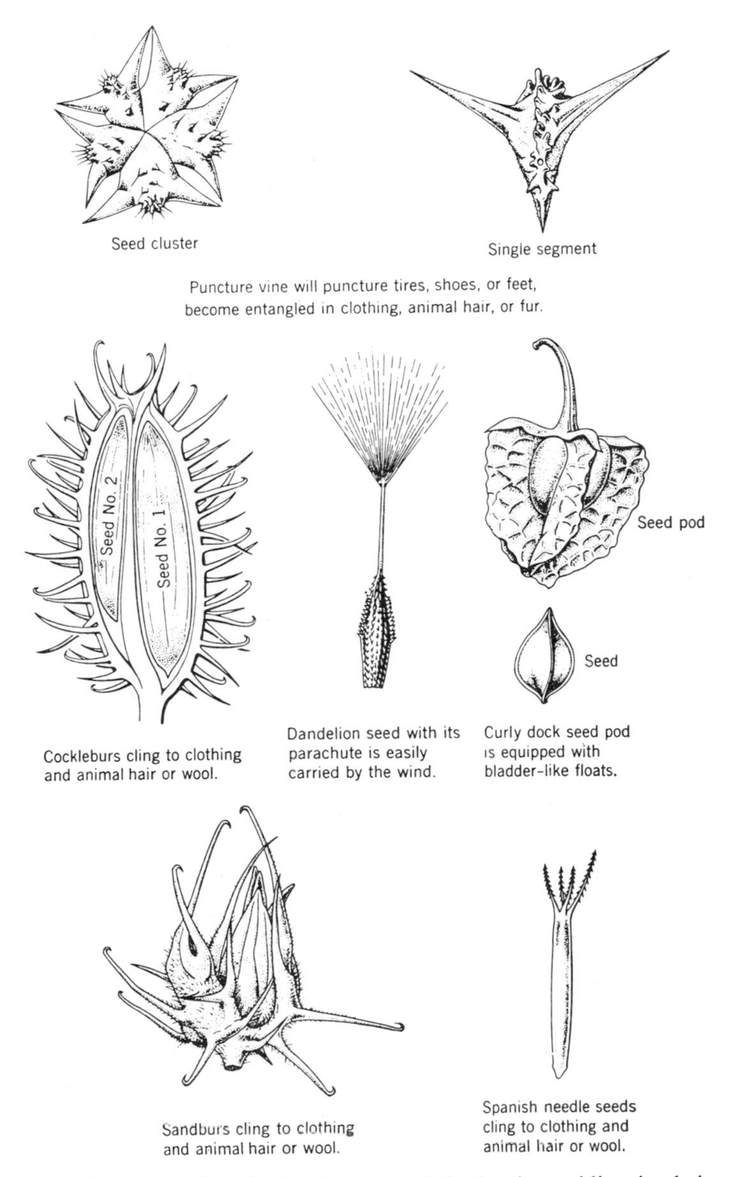

Figure 2-2. Examples of seeds that have special adaptions aiding in their spread.

bags. The bags were removed periodically over a 5-year period and germination counts were made.

There was little or no variation in germination between the 12- and 48-inch depths, but there was considerable difference in viability among species. Clearly, some seeds can be stored in fresh water for 3 to 5 years and still germinate. In some cases storage in water tended to break dormancy and increased the percent

TABLE 2-2. Germination of Weed Seed after Storage in Fresh Water

Field bindweed	After 54 months, 55% germinated
Canada thistle	After 36 months, about 50% germinated After 54 months, none germinated
Russian knapweed	After 30 months, 14% of seed still sound; none germinated after 5 years
Redroot pigweed	After 33 months, 9% still sprouted
Quackgrass	None sprouted after 27 months
Barnyardgrass	After 3 months less than 1% germinated; none germinated after 12 months
Halogeton	After 3 months less than 1% germinated; none germinated after 12 months
Hoary cress	After 2 months germination dropped to 5% or less, and dropped to zero after 19 months

germination; this was especially true after 2 to 4 months in the water (Bruns and Rasmussen, 1958).

Seed production from plants along irrigation ditches, drainage ditches, and reservoirs is probable the major source of weed seed contamination of irrigation waters. Every effort should be made to keep the banks of these areas free of seed producing plants.

Animals

Animals, including man, are responsible for scattering many seeds. They may carry the seed on their feet, clinging to their fur or clothes, or internally (ingested seed).

Many seeds have specially adapted barbs, hooks, spines, and twisted awns that cling to the fur or fleece of animals or to man's clothing. Sandburs, cockleburs, sticktights, and beggarticks are examples. Others may imbed themselves in the animal's mouth, causing sores. Examples are wild barley, downy brome, and various needlegrasses. Other seeds have cottonlike lint or similar structures that help the seed cling to fur or clothes. Annual bluegrass and bermudagrass seed will stick to the fur of rabbits and dogs. Mistletoe seeds are sticky and become attached to the feet of birds. Birds often carry away fleshy fruits containing seeds for food.

Many weed seeds pass through the digestive tracts of animals and remain viable. Often weed seedlings are found germinating in animal droppings. Table 2-3 shows results of feeding weed seeds to different kinds of livestock and washing the seed free from the feces. The figures are given as a percentage of viable seeds recovered.

Machinery

Machinery can easily carry weed seeds, rhizomes, and stolons. Weed seeds are often spread by harvesting equipment, especially combines. Cultivation

TABLE 2-3. Percentage of Viable Seeds Passed by Animals Based on Total Number of Seeds Fed

	Percentage of Viable Seeds Passed by					
Kind of Seeds	Calves	Horses	Sheep	Hogs	Chickens	Average
Field bindweed	22.3	6.2	9.0	21.0	0.0	11.7
Sweetclover	13.7	14.9	5.4	16.1	0.0	10.0
Virginia pepperweed	5.4	19.8	8.4	3.1	0.0	7.3
Velvetleaf	11.3	4.6	5.7	10.3	1.2	6.6
Smooth dock	4.5	6.5	7.4	2.2	0.0	4.1
Pennsylvania smartweed	0.3	0.4	2.3	0.0	0.0	0.6
Average	9.6	8.7	6.4	8.8	0.2	6.7

From Harmon and Keim (1934).

equipment, tractors, and tractor tires often carry dirt that may include weed seeds. Also, cultivation equipment may drag rhizomes and stolons, dropping them later to start new infestations.

Weed Screenings

Most weed seeds have a reasonably high feed value. Common ragweed seed, when chemically analyzed, had 20.0% crude protein, 15.7% crude fat, 18% nitrogen-free extract, and 4.83% ash. Because of their relative cheapness, weed-seed screenings are often included in livestock feeds (see Tables 3-5 and 21-1).

Over the years, considerable attention has been given to the problem of destroying the viability of weed seeds in screenings. Seeds are usually finely ground, or soaked in water and then cooked. Fine grinding with the hammer mill has reduced to a minimum the hazard of scattering live weed seeds.

Pelleting destroyed the seed's viability when high temperatures were used. Fish solubles with a high protein content were added to screenings after heating to about 200°F. The mixture was pressed into pellets. Germination tests indicated that the seeds were no longer viable with that amount of heat.

NUMBER AND PERSISTENCY OF WEED SEEDS

Number in the Soil

Some species produce a remarkable number of living seeds per acre. Wild poppies seriously infested the Rothamsted (England) Experiment Station in 1930 with an estimated 113 million poppy seeds per acre.

In Minnesota, weed-seed counts at four different locations on 24 different plots showed from 98 to 3068 viable weed seeds per square foot of soil, 6 inches deep. Converted to an acre basis, this is between 4.3 million and 133 million seed per acre in the upper 6 inches.

Soil samples were taken on 10 plantations in Louisiana heavily infested with johnsongrass. The average number of viable johnsongrass seeds per acre was 1,657,195. A sugar cane crop was grown for 3 years without permitting the addition of new johnsongrass seed; after 3 years the johnsongrass seed population in the upper 2.5 inches of soil dropped to 1.3% of the original number.

Number Produced by Plants

The persistence of annual and biennial weeds depends mainly on their ability to reinfest the soil. The first infestation of most perennial species depends on seed. Obviously, if we could control the production of seed, we could eventually eliminate many species.

One scientist, reporting on 245 species, found that the number of weed seeds produced by one plant ranges from 140 for leafy spurge to nearly a quarter of a million seeds for common mullein. Of these, 24 species have been selected and are listed in Table 2-4. Only plump, well-developed seeds were counted from well-developed plants growing with comparatively little competition. Those 24 species averaged 25,688 seeds per plant. A second report listed 263 species. One witchweed plant may produce as many as one-half million seeds.

Age of Seed and Viability

The length of time that a seed is capable of producing a seedling varies widely with different kinds of seeds and with different conditions. Certain seeds keep their viability for many years, whereas others die within a few weeks after maturing if they do not find a suitable environment for germination.

A short-lived species, silver maple, has seeds with about 58% moisture when shed, and they will germinate immediately. However, when moisture content drops to 30 to 34%, they die. In nature this occurs within a few weeks.

Lotus (*Nelumbo nucifera*) seeds found in a lake bed in Manchuria were still viable after approximately 1000 years. The seeds were 1040 ± 210 years old, as measured by a radioactive carbon dating technique. The seeds had been covered in deep mud and very cold water.

To determine the longevity of seed, Duvel placed 107 species in porous clay pots and buried them 8, 22, and 42 inches deep in the soil. He removed samples at various intervals and left others undisturbed. His findings, reported by Toole and Brown (1946), were as follows:

After 1 year, seed of 71 species germinated
After 6 years, seed of 68 species germinated
After 10 years, seed of 68 species germinated
After 20 years, seed of 57 species germinated

TABLE 2-4. Number of Seeds Produced per Plant, Number of Seeds per Pound, and Weight of 1000 Seeds

Common Name	Number of Seeds (per plant)	Number of Seeds (per lb[1])	Weight of 1000 Seeds (g)
Barnyardgrass	7,160[2,3]	324,286	1.40
Buckwheat, wild	11,900	64,857	7.0
Charlock	2,700	238,947	1.9
Dock, curly	29,500	324,286	1.4
Dodder, field	16,000[3]	585,806	0.77
Kochia	14,600	534,118	0.85
Lambsquarters	72,450	648,570	0.70
Medic, black	2,350	378,333	1.2
Mullein	223,200	5,044,444	0.09
Mustard, black	13,400[4]	267,059	1.7
Nutsedge, yellow	2,420[2]	2,389,474	0.19
Oat, wild	250[2]	25,913	17.52
Pigweed, redroot	117,400[2]	1,194,737	0.38
Plantain, broadleaf	36,150	2,270,000	0.20
Primrose, evening	118,500	1,375,757	0.33
Purslane	52,300	3,492,308	0.13
Ragweed, common	3,380[2]	114,937	3.95
Sandbur	1,110[2]	67,259	6.75
Shepherdspurse	38,500[2,3]	4,729,166	0.10
Smartweed, Pennsylvania	3,140	126,111	3.6
Spurge, leafy	140[4]	129,714	3.5
Stinkgrass	82,100[2,3]	6,053,333	0.07
Sunflower, common	7,200[2,3]	69,050	6.57
Thistle, Canada	680[2,3]	288,254	1.57

[1] Calculated from the weight of 1000 seeds.
[2] Many immature seeds also present.
[3] Many seeds shattered.
[4] Yield of one main stem.
From Stevens (1932).

After 30 years, seed of 44 species germinated
After 38 years, seed of 36 species germinated

After 38 years:
- 91% of jimsonweed seed germinated
- 48% of mullein seed germinated
- 38% of velvetleaf seed germinated
- 17% of eveningprimrose seed germinated
- 7% of lambsquarters seed germinated
- 1% of green foxtail seed germinated
- 1% of curly dock seed germinated

The actual percentage of germination is not as important as the fact of survival. Programs to eradicate plants with long seed dormancy are doomed to failure until techniques are developed to break (100%) seed dormancy in soil.

The three depths did not greatly influence the data; however, seed at the 42-inch depth lived slightly longer. The 8-inch depth is far below the ideal depth for most seeds to germinate. If one sample had been buried 1 inch deep or less, greater differences caused by depth would have been likely.

In another test started in 1879, Beal mixed seeds of 20 different weed species with sand and buried the mixtures in uncorked bottles, with the opening tipped downward. After 20 years, 11 of the species were still alive. After 40 years, 9 species were still alive. These 9 were redroot pigweed, prostrate pigweed, common ragweed, black mustard, Virginia pepperweed, eveningprimrose, broadleaf plantain, purslane, and curly dock.

After 50 years, moth mullein germinated for the first time. Therefore, the seeds were still alive at 40 years but did not germinate because they were dormant. After 70 years, moth mullein had a germination rate of 72%, eveningprimrose 17%, and curly dock a few percent. After 80 years, moth mullein still had a germination rate of 70 to 80%, but with the other two species only a few seeds germinated. It appeared that these two species were nearing the end of their survival period.

Thus many weed seeds retain their viability for 40 years and longer when buried deep in the soil. Scientists believe that much of the longevity depends on the seeds being buried deep, with a reduced oxygen supply available. If brought to the surface with other conditions favorable, many of the seeds will germinate. It is evident that repeated cultivation for several years without opportunity of reinfestation effectively reduces the weed–seed population in a soil.

Depending on climatic conditions, abandoned cultivated crop land will usually return to a forest or grassland vegetation. If cultivated after 30 to 100 years, the original field weeds immediately appear. The weed seeds have remained dormant while buried deep in the soil.

GERMINATION AND DORMANCY OF SEEDS

Germination includes several steps that result in the quiescent embryo changing to a metabolically active embryo as it increases in size and emerges from the seed. It is associated with an uptake of water and oxygen, use of stored food, and, normally, release of carbon dioxide. For a seed to germinate, it must have an environment favorable for this process. This includes an adequate, but not excessive, supply of water, a suitable temperature and composition of gases (O_2/CO_2 ratio) in the atmosphere, and light for certain seeds. Specific requirements for seed germination differ for various species. Although these factors are optimal, a seed may not germinate because of some kind of dormancy.

Dormancy is a type of resting stage for the seed. Dormancy may determine the time of year when a seed germinates, or it may delay germination for years and thus guarantee the viability of the seed in later years. Five environmental factors affect seed dormancy: *temperature, moisture, oxygen, light, and the presence of inhibitors* including allelopathic effects. Other factors directly related to the seed and its dormancy include impermeable seed coats (to water, oxygen, or both), mechanically resistant seed coats, immature embryos, and afterripening.

Temperature

The temperature that favors seed germination varies with each species. There is a *minimum* temperature below which germination will not occur, a *maximum* temperature above which germination will not occur, and an *optimum*, or ideal, temperature when seeds germinate quickest. Thus some seeds germinate only in rather cool soils, whereas others do so only in warm soils.

The temperature requirements for most crop seeds are well established, and farmers recognize these requirements and plant accordingly. Cotton, for example, requires relatively high temperatures for germination, whereas the small grains will germinate at relatively cool temperatures.

Russian pigweed seed has germinated in ice and on frozen soil. Wild oat may germinate at temperatures of 35°F. Comparatively low temperatures (between 40 and 60°F) are necessary before certain winter annuals will germinate. High temperatures may cause a secondary type of seed dormancy, especially with some winter annual weeds. Wormseed mustard was introduced to secondary dormancy by temperatures of 86°F, and many summer annuals require temperatures of 65 to 95°F to germinate. Alternating temperatures are often better than a constant temperature for seed germination.

When redroot pigweed seed (a summer annual) was placed in germinators at 68°F, some seed remained dormant for more than 6 years. The seeds could be induced to germinate at any time in three ways: (1) by raising the temperature to 95°F, rubbing with the hand, and replacing at 68°F, (2) by partial desiccation, or (3) by alternating the temperatures.

Temperature alone does not completely explain the periodicity of seed germination. Often the seeds have another form of dormancy that temporarily stops germination. This may be a survival mechanism to keep the plant from germinating immediately upon maturity in a season not suited to the plant.

Seeds may lie dormant for as little as several weeks or as much as several years. For example, cheat and hairy chess have a primary dormancy of 4 to 5 weeks after maturity. During this period they will germinate only if subjected to low temperatures (59°F or below). However, if the seeds are stored for 4 to 5 weeks, germination will then readily occur at temperatures of 68 to 77°F.

Moisture

Germination is normally a period of rapid expansion and high rates of metabolism or cell activity. Much of the expansion is simply an increase in water, expanding cell walls. If water content of the seed is reduced, the activity of enzymes—and consequently metabolism—slows down. The amount of moisture contained in seeds may determine their respiratory rate. During germination the seed respires at a very rapid rate. Many seeds cannot maintain this high rate of respiration until they reach a moisture content of 14% or more. Thus in dry soils the seed remains dormant.

Dry seeds can tolerate severe conditions; some have been kept in boiling water for short intervals without injury and others in liquid air (– 310°F). When moist enough for germination, the same seeds may be killed by cold temperatures of 30°F or warm temperatures of 105°F.

Oxygen

In addition to the right temperature and sufficient moisture, germination depends on oxygen. Aerobic respiration requires more free oxygen than anaerobic respiration; thus some seeds start germination under anerobic conditions, then shift to aerobic respiration when the seed coat ruptures.

The percentage of oxygen found in the soil varies widely, depending on soil porosity, depth in the soil, and organisms in the soil that use oxygen and release CO_2 (microorganisms, roots, etc.). The percentage of oxygen in the soil is usually inversely proportional to the percentage of CO_2. In swampy rice land there may be less than 1% oxygen in the soil atmosphere; in freshly green-manured land, 6 to 8%; and where corn is growing rapidly, 8 to 9%, compared with about 21% in a normal atmosphere. The percentage of carbon dioxide may range from 5 to 15% under such conditions, compared with 0.03% for normal air.

Different species vary considerably in the amount of oxygen needed for seeds to germinate. Wheat seed germinated well when the replenished oxygen supply of the soil was 3.0 mg/m^2/hr or more. It failed to germinate when the rate was below 1.5 mg. Rice seed germinated at 0.5 mg. Broadleaf cattail and some other acquatic plants germinate better at low oxygen concentrations than with normal air.

The effect of different oxygen concentrations on the seeds of field bindweed, leafy spurge, hoary cress, and horsenettle was determined. Oxygen concentrations of 5% produced little to no germination of hoary cress, horsenettle, and leafy spurge. At 10% oxygen and below, these three weeds germinated at a rate far below normal, whereas bindweed was reduced somewhat. The highest percentage germination was found at about 21% oxygen (normal air) for leafy spurge and hoary cress, but the best germination for horsenettle was at 36% oxygen and for field bindweed 53%.

Wild oat and charlock germination can be greatly suppressed by reducing the oxygen supply by soil compaction. Cultivation increased sixfold the number of wild oat that germinated and the number of charlock twofold, compared with the compacted plots. Cultivation increased soil aeration and thus increased the content of oxygen in the soil atmosphere.

Excess water in the soil cuts down seed germination of most plant species. Researchers believe that lower germination is related to a smaller supply of oxygen in waterlogged soils, rather than merely excess water.

Many small-seeded weed seeds germinate only in the upper 1 to 2 inches of soil, mostly in the upper 1 inch. A limited number, however, germinate below 2 inches. In sandy soils, seeds germinate deeper than in clay soils as a result of better aeration or better oxygen supply in the sands. Some seeds buried deep in the soil do not germinate but lie dormant for many years. When brought to the surface, they germinate promptly. Aeration, involving increased oxygen supply, is probably responsible.

By using herbicides, successive crops of weed seedlings may be killed without disturbing the soil. Few to no viable weed seeds may remain in the upper soil layer. The soil surface may then remain relatively free of weeds. Repeated treatments to kill annual weeds before they produce seed is especially useful in areas where the soil is not disturbed by cultivation, such as in some perennial crops, permanent sod, or turf areas.

Light

Some kinds of seed germinate best in light, others in darkness, and others germinate readily in either light or darkness. Among several hundred species in which the role of light has been investigated, about half require light for maximum germination. The length of day and quality (color) of light also have influence. Here is a brief review of the electromagnetic spectrum (color of wavelengths).

Name	Wavelength (Å)
X ray	10–150
Ultraviolet	Below 4000
Visible spectrum	
Violet	4000–4240
Blue	4240–4912
Green	4912–5750
Yellow	5750–5850
Orange	5850–6470
Red	6470–7000
Infrared	Greater than 7000

Germination of lettuce seed was promoted by radiation at 6600 Å (red light) and inhibited at 7300 Å (infrared light). A later study with lettuce showed that the inhibitory effect of infrared light could be reversed by red light. Regardless of the number of alternating periods of red and infrared light to which the seeds were exposed, the final type of light determined the percentage germination. For example, when the final light was infrared, germination was about 50%; but when the final light was red, germination reached almost 100%. Without light, germination fell to about 8%.

Here are a few examples of species and the light requirement of their seeds for germination.

Germination Favored by Light		Germination Favored by Darkness	Germination in Either Light or Darkness	
Bluegrass	Dock	Onion	Salsify	Wheat
Tobacco	Primrose	Lily	Bean	Rush
Mullein	Buttercup	Jimsonweed	Clover	Toadflax

These effects vary from species to species. In some seeds, the light requirement can be replaced by afterripening in dry storage, alternating the daily temperature, higher temperatures, and treatment in potassium nitrate or gibberellic acid solutions.

Seed Coat Impermeable to Water, Oxygen

Seed coats may be waterproof and may prevent the seed from absorbing water. Such seed will not germinate even if soil moisture is plentiful. Seed that fails to germinate because of a waterproof seed coat is called *hard seed.* Hard seed is common in annual morningglory, lespedeza, clovers, alfalfa, and vetch. Researchers believe that many weed species have hard seeds.

Some seed coats are impermeable to oxygen but not water. Cocklebur has two seeds per fruit, one set slightly below the other. The lower seed usually germinates during the first spring and the upper remains dormant until the next year. Both can be made to germinate immediately by breaking the seed coats, or by simply increasing the oxygen supply. Seeds of ragweed, several grasses, and lettuce also show this type of dormancy.

As with waterproof seed coats (hard seeds), anything that breaks the seed coat—scarification, acids, soil microorganisms—will break this type of dormancy. Under laboratory conditions, oxygen dormancy can usually be broken by increasing the oxygen supply. Cultivation of the soil often has a similar effect by increasing the oxygen level in the upper layer of soil.

Mechanically Resistant Seed Coats

A tough seed coat may forcibly enclose the embryo and prevent germination. While the seed absorbs oxygen and water, it builds pressures in excess of 1000 psi. The seed will quickly germinate if the seed coat is removed. Pigweed, wild mustard, shepherdspurse, and pepperweed have this type of dormancy.

As long as the seed coat remains moist, it remains tough and leathery—for as long as 50 years. Any factor that weakens the seed coat will help break dormancy. Drying at temperatures of 110°F or mechanical or chemical injury to the seed coat may break this type of dormancy.

Immature Embryos

The outside of the seed may appear fully developed, but it may have an immature embryo that needs more growth before the seed can germinate. Therefore, the seed appears dormant, although the embryo is slowly growing and developing. Seeds of orchids, holly, smartweed, and bulrush show this type of dormancy.

Afterripening

In some species the embryos appear completely developed, but the seed will not germinate although the seed coat has been carefully removed to permit easy absorption of water and oxygen. Light and darkness have no effect. In this case germination occurs normally after a period of *afterripening*. Occasionally cool temperatures for several months will end this type of dormancy. Afterripening is especially common in the grass, mustard, smartweed, rose, and pink families.

Afterripening is a physiological change of a complex physicochemical nature. Although the exact processes are not completely understood, they may be associated with changes in the storage materials present, substances promoting germination may appear, or substances inhibiting germination may disappear.

GERMINATION AS AFFECTED BY CERTAIN CONDITIONS

Burning

Burning fields after weed seeds have matured gives only partial and erratic destruction of seeds. Seeds that lie on the ground may readily escape, whereas those held on the plants often burn completely. The degree of weed–seed destruction depends largely on the intensity of the heat, and this in turn on dryness and the amount of litter and debris to be burned.

The aftereffects of burning are usually pronounced, especially after forest fires. Weed species absent, for the most part, from the area for many years may suddenly appear after the fire and dominate other vegetation.

Data and experience clearly show that moderate heat may end seed dormancy. Five other factors that normally follow a fire may also terminate dormancy; these factors are (1) greater alternation of temperature in the upper soil layers between day and night, (2) more light is reaching the surface soil, (3) removal of litter, (4) removal of competition from other plants, and (5) removal of plants previously living in the area that had soil-inhibiting substances that prevented seed germination. With removal of those plants, those substances are no longer present. This is probably an alleopathic effect.

Cutting (Stage of Maturity)

Cutting weeds to prevent seed production is a common recommendation. The practice is important in agricultural croplands, in turf, and in hay crops.

TABLE 2-5. Germination of Weed Seeds Cut at Various Stages of Maturity

	Percent Germinated		
Weed Seeds	Cut When Dead Ripe	Cut When in Flower	Cut in Bud Stage
Groundsel, ragwort	72	80	0
Sowthistle, common	100	100	0
Groundsel, common	90	35	0
Sea aster	90	86	0
Dandelion	91	0	0
Catsear, spotted	90	0	0
Canada thistle	38	0	0

In South Dakota, Canada thistle and perennial sowthistle heads were removed from the plant and dried at daily intervals after the flowers opened. The experiment was continued for 3 years. Perennial sowthistle heads harvested 3 days after blooming had 0.0% viable seed; 6 days after blooming they had an average of 6% viable seed; and 8 days after blooming, 65% viable seed.

Canada thistle harvested 6 days after blooming had an average of 0.03% viable seed; 8 days after blooming, 6.7% viable seed; and 11 days after blooming, 73% viable seed.

Another study compared removal of the heads as described above with the effect of cutting the entire plant and leaving the heads on the plant during drying. Results of the two methods were similar; thus little seed development takes place after either type of cutting.

In summary, either mowing in or before the bud stage prevents viable seed production. If seeds reach medium ripeness, probably a large percentage of viable seeds will be produced. In the case of many species, cutting after that time does little or no good in preventing viable seed production. With other species, mowing is not effective because some of the heads are short and missed by the mower, and these produce seed.

Storage in Silage

Many seeds lose their viability in silage. Many weed seeds lose their germinating power 10 to 20 days after being placed in silage. Other reports indicate, however, that some seeds will germinate after being in the silo for periods up to 4 years. These variations are possibly a result of differences in the silage as to moisture content, temperature, and amount of organic acids produced (Tildesley, 1936–1937).

Storage in Manure

When manure is spread fresh from the stable, viable weed seeds are usually spread with it. But if manure is stored, heating and decomposition will begin, and, in time, the weed seeds are destroyed.

TABLE 2-6. Effect of Length of Time in Cow Manure on the Viability of Various Weed Seeds

Kind of Seeds	Percent Viability before Burial	Percent Viability after Storage			
		1 Month	2 Month	3 Month	4 Month
Velvetleaf	52.5	2.0	0.0	0.0	0.0
Field bindweed	84.0	4.0	22.0	1.0	0.0
Sweetclover	68.0	22.0	4.0	0.0	0.0
Pepperweed	34.5	0.0	0.0	0.0	0.0
Smooth dock	86.0	0.0	0.0	0.0	0.0
Smartweed	0.5	0.0	0.0	0.0	0.0
Cocklebur	60.0	0.0	0.0	0.0	0.0
Puncturevine	52.0	0.0	0.0	0.0	0.0

From Harmon and Keim (1934).

In Table 2-6 only three weeds showed any viability after a 1-month storage. All seeds were destroyed at the end of 4 months.

Stoker et al. (1934) found complete destruction of hoary cress and Russian knapweed seeds in moist, compacted chicken manure at the end of 1 month. However, seed of field bindweed was still viable at the end of 4 months.

These tests were conducted during the summer in cow or chicken manure. Horse and mule manure tends to heat, whereas cow or chicken manure does not. Decomposition of the weed seed would be more rapid at the higher temperature. If the manure is frozen or cold, the seed would live longer.

If the edges or outside of the manure pile dry out, decomposition slows down and viable weed seeds would likely persist. Therefore, manure should be turned occasionally to kill all seeds. Well-rotted manure is free of viable seeds.

DISSEMINATION BY RHIZOMES, STOLONS, TUBERS, ROOTS, BULBS, AND BULBLETS

Although weeds reproduce and spread most widely by means of seeds, they also multiply by vegetative, or asexual, methods. Rhizomes, stolons, tubers, roots, bulbs, and bulblets are all vegetative, or asexual, methods of reproduction. Many perennial weeds classed as "serious" reproduce vegetatively, and most of these also reproduce by means of seeds. Vegetative organs occasionally have a short period of dormancy.

Most plants spread slowly by vegetative means alone. Without help from man and his cultivation equipment, weeds such as quackgrass, field bindweed, johnsongrass, bermudagrass, nutsedge, and Canada thistle would spread vegetatively less than 10 ft/year. However, the rhizomes, stolons, roots, and tubers are dragged about the field with soil-tillage equipment. Wherever these

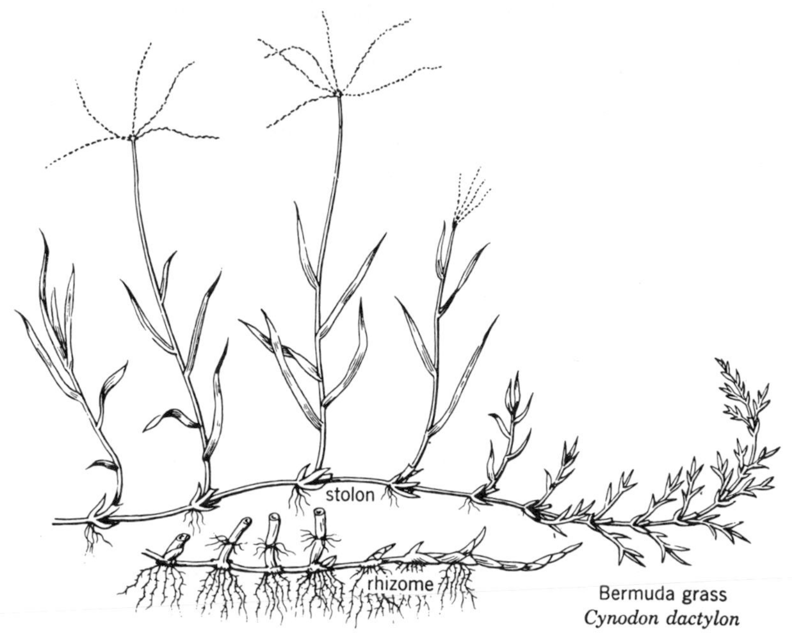

Figure 2-3. Bermudagrass reproduces by seeds, stolons, and rhizomes. (North Carolina State University.)

plant pieces drop, a new infestation is likely. Disc-type cultivation equipment is less likely to drag the plant parts than are shovels, sweeps, and plows.

Repeated tillage will kill most plants possessing rhizomes, stolons, roots, and tubers. If cut off and in *dry soils*, the vegetative parts may quickly dry and die, preventing new growth.

In most soils, the cut vegetative parts quickly take root and establish new plants. Under such conditions, repeated tillage may exhaust the underground food reserves. Most such weeds are killed through *carbohydrate starvation* by repeated tillage for 1 to 2 years.

Some chemicals mixed into the soil will retard the development of new roots, especially after the plant has been cut off. With a combination treatment of an effective herbicide plus repeated cultivation, many serious perennial weeds may be controlled in a short time. For example, johnsongrass can be controlled by trifluralin plus repeated cultivation.

SUGGESTED ADDITIONAL READING

Baker, H. G., 1974, *The Evolution of Weeds*, Annual Review of Ecology and Systematics, Annual Reviews, Palo Alto, CA.

Bruns, V. F., and L. W. Rasmussen, 1958, *Weeds*, **6**, 42.

Dawson, J. H., 1965, *Weeds*, **13**, 245.

Fuerst, E. P. and A. R. Putnam, 1983, *J. Chem. Ecol.*, **9**, 937.

Furrer, J. D., 1954, *North Central Weed Cont. Conf. Proc.* **11**, 26.

Gill, N. T., 1938, *Ann. Appl. Biol.*, **25**, 447.

Harmon, G. W., and F. D. Keim, 1934, *J. Am. Soc. Agron.*, **26**, 762.

Harper, J. L., 1977, *The Population Biology of Plants*, Academic Press, London.

Hill, T. A., 1977, *The Biology of Weeds*, Edward Arnold, Ltd., London.

Holzner, W., and M. Numata, Eds., 1982, *Biology and Ecology of Weeds*, Dr. W. Junk, The Hague, Netherlands.

King, L. J., 1966, *Weeds of the World: Biology and Control*, Interscience, New York.

Mayer, A. M., and A. Poljakoff-Mayber, 1975, *The Germination of Seeds*, Macmillan, New York.

Molisch, H., 1937, *Der Einfluss einer Planze auf die andere-Allelopathie*, Fischer, Jena.

Putnam, A. R., and W. B. Duke, 1978, *Annu. Rev. Phytopathol.*, **16**, 431.

Putnam, A. R., and C.-S. Tang, Eds., 1986, *The Science of Allelopathy*, Wiley, New York.

Radosevich, S. R., and J. S. Holt, 1984, *Weed Ecology: Implications for Vegetation Management*, Wiley, New York.

Rejmánek, M., G. R. Robinson, and E. Rejmánková, 1989, *Weed Sci.*, **37**, 276.

Rice, E. L., 1984, *Allelopathy*, Academic Press, Orlando, FL.

Stevens, O. A., 1932, *Am. J. Bot.*, **19**, 784.

Stoker, G. L., D. C. Tingey, and R. J. Evans, 1934, *J. Am. Soc. Agron.*, **26**, 600.

Thompson, A. E., Ed., 1985, *The Chemistry of Allelopathy*, ACS Symposium Series No. 268, American Chemical Society, Washington, D.C.

Tildesley, W. T., 1936–1937, *Sci. Agric.* **17**, 492.

Toole, E. H., and E. Brown, 1946, *J. Agric. Res.* **72**, 201.

Weatherspoon, D. M., and E. E. Schweizer, 1971, *Weed Sci.*, **19**, 125.

Zimdahl, R. L. 1980, *Weed-Crop Competition*, International Plant Protection Center, University of Oregon, Corvallis.

3 Weed Management Practices

Weed control is essential for successful crop production. Although attacks on crops by insects and disease can be more dramatic, weeds potentially reduce crop yields every year. Weed populations in a field are relatively constant from year to year, whereas insect and disease outbreaks can be sporadic. Farmers and vegetation managers can *plan* a weed control strategy based on prior knowledge of the weeds to expect. This chapter outlines some of the factors and practices to consider when developing a weed management plan.

LEVELS OF WEED MANAGEMENT

Weed management practices are commonly grouped into three levels: prevention, control, and eradication.

Prevention

Prevention means stopping a new weed from invading an area or limiting weed build-up in a field. Prevention is practiced by (1) not planting crop seed contaminated by weed seeds, (2) not carrying weed seeds or vegetative propagules into an area with machinery, contaminated manure, or irrigation water, (3) not allowing weeds to go to seed, and (4) stopping the spread of vegetatively reproducing perennial weeds. Prevention can be a cost-effective and practical way to control weeds. This is particularly true for discouraging outbreaks of new weed problems. Unfortunately, perfect (100%) weed control is needed to prevent seed production by a general weed population in a field. This can be very difficult and uneconomical to achieve. Limiting weed seed production is a desirable goal but total prevention of weeds from seeding is usually used only for isolated occurrences of new weeds.

Control

Control is a degree of weed management that decreases weed populations to noninterfering levels. Weeds are reduced in crops to numbers that do not reduce yield or interfere with harvest operations. The level of weed suppression desired is usually balanced between the economic gains and the actual cost of the weed control practices. Control is the normal target for weeds in farm crops and most other weed management situations.

Eradication

Eradication is the complete elimination of a weed species from an area. This includes both live plants and reproductive parts, and both seeds and vegetative reproductive structures. The weeds currently present are eliminated and the weed does not reappear in succeeding years. Thus eradication differs from perfect (100%) control in dealing not only with the present weed problem, but also in addressing future weed problems. Eradication programs must (1) eliminate living weeds and (2) remove reproductive structures, particularly seeds, from the soil. Direct eradication of weed seed and reproductive structures in soil is difficult to achieve. This is particularly true on a large acreage basis. The long life of many weed seeds also makes eradication a difficult goal.

On the farm, eradication is most feasibly applied to the elimination of the first occurrence of a new weed species. This is accomplished before any seeds are produced. Elimination may not be possible after seed production due to the persistence of the seeds in the soil.

Eradication of weeds from smaller areas where high value horticultural or ornamental plants will be grown is a worthwhile consideration. This is particularly true given the limited weed management options available for these crops. Eradication commonly employs a soil fumigation to rid the soil of weed seeds and vegetative parts. Insect and disease control can be an additional benefit of fumigation.

WEED MANAGEMENT PRACTICES

Achievement of the desired level of weed suppression requires the use of specific weed management practices. Although these are discussed as separate strategies, it must be emphasized that the most effective and economical weed control plan will almost always employ several approaches to reduce the weed problem. Each component contributes to the overall level of weed control. Omitting or reducing the control achieved from one or more components increases the level of control needed from the remaining weed control practices. Integration of weed management practices is discussed later in this chapter. Four general areas of weed control tactics are (1) mechanical, (2) cultural, (3) biological, and (4) chemical.

MECHANICAL PRACTICES

Tillage, handweeding, mowing, mulches, burning, and flooding are considered mechanical weed control methods. *Primary tillage* is the initial ground breaking in preparation for crop production. *Secondary tillage* is additional tillage performed to smooth and level the ground prior to planting.

Tillage systems are often defined by the type of primary tillage practiced. *Conventional tillage* uses a moldboard plow for the primary tillage. This is

TABLE 3-1. Effect of Tillage Systems on Soil Pulverization, Inversion, and Plant Residue Cover

Tillage System	Tillage Operation	Soil Pulverization Rank[1]	Inversion Rank[1]	Residue Cover (%) After Corn	After Soybeans	After Small Grain
Conventional	Moldboard plow, disk twice	1	1	5	2	5
Conservation	Chisel plow,[2] field cultivate	2	2	30	10	30
No-tillage	—	3	3	80	60	80

[1] 1 = highest; 3 = lowest.
[2] Chisel points.

Adapted from Griffith, D. R., J. V. Mannering, and J. C. Box, 1986, Soil and moisture management with reduced tillage, pp. 19–57, in M. A. Sprague and G. B. Triplett, Eds., *No-Tillage and Surface-Tillage Agriculture*, Wiley, New York.

followed by additional secondary tillage with harrows to level the ground prior to planting. *No-tillage* is at the opposite extreme. This system uses herbicides to kill existing vegetation followed by planting into the undisturbed seedbed. A fluted coulter on no-till planters opens a narrow slit ahead of the planter unit to allow seed placement into the soil.

Conservation tillage, *reduced tillage*, and *minimum tillage* are all terms commonly used to describe primary tillage systems between the extremes of conventional and no-tillage, although, the no-tillage system also fits these categories. Conservation tillage focuses on the savings in soil and water losses gained by using tillage alternatives to conventional tillage. Reduced tillage decreases the tillage operations compared to the conventional system, often by employing specialized planting techniques. Minimum tillage uses the minimum amount of tillage operations required to prepare the crop seedbed. Excessive tillage is eliminated.

Conventional, reduced, and no-till tillage systems can also be characterized by the amount of soil pulverization, inversion, and plant residue cover associated with each tillage (Table 3-1).

The objective of the tillage operations is to prepare a crop seedbed and weed control effects are secondary. However, the tillage operations both directly and indirectly impact weed management. Tillage kills perennial weeds by physically damaging the vegetative reproductive parts that can accelerate microbial attacks on the plants. Tillage operations can also leave reproductive organs on the soil surface exposing them to freezing and/or drying conditions. Reductions in tillage

TABLE 3-2. Effect of Tillage System on Perennial Weed Populations[1]

Tillage Treatment	Weeds per Acre		
	Hemp Dogbane	Common Milkweed	Canada Thistle
Moldboard plow	880	0	0
Chisel plow	925	3	3
No-till	1850	34	5

[1] Population counts taken after 4 years of treatment with the tillages. The counts are the average of those from continuous corn and corn–soybean rotation cropping patterns.

From Fawcett, R. S., 1982, Can you control weeds with reduced tillage? Proc. 34th Iowa Fertilizer and Agricultural Chemical Dealers Conference.

can increase perennial weeds (Table 3-2). Both simple and creeping perennial weed populations can be higher with reduced tillage.

Tillage, especially soil inversion with a moldboard plow, buries weed seed and places the seed in an unfavorable environment for germination. The seed burial can reduce the weed population the year after heavy seed production by uncontrolled weeds. However, the reservoir of dormant buried weed seeds serves as a source of continuing weed problems for the future. Plant residue levels (Table 3-1) present in various tillage systems also affect weed control. The plant residue intercepts the soil applied herbicides and reduces the herbicide reaching the soil (Figure 3-1). Herbicides on the plant residue may be washed to the soil with subsequent rainfall. However, herbicides that are degraded by sunlight or are volatile may be lost from the plant residue before rainfall occurs. Even with

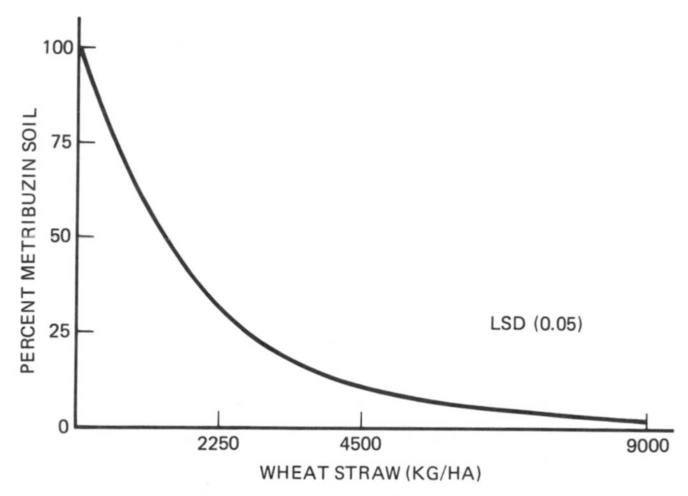

Figure 3-1. Influence of wheat straw on the amount of metribuzin reaching the soil surface. (Banks, P. A., and E. L. Robinson, 1982, *Weed Sci.* **30**, 164.)

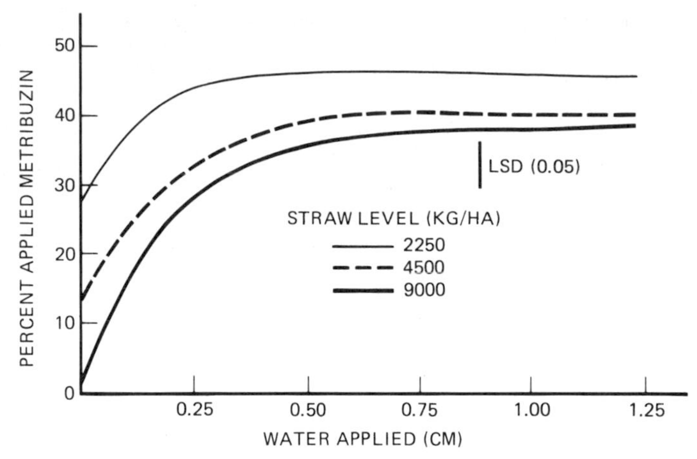

Figure 3-2. Metribuzin reaching the soil surface as influenced by wheat straw and irrigation. (Banks, P. A., and E. L. Robinson, 1982, *Weed Sci.* **30**, 164.)

rain, herbicides can remain trapped on the plant residues (Figure 3-2). Weed control can suffer due to reduced herbicide availability in these cases.

Secondary tillage can also contribute to weed suppression. A large portion of the potential weed population can germinate and emerge before crop planting if planting is delayed. Reworking the ground with harrows just before planting will kill these weeds. These weeds could also be controlled with herbicides. Herbicide use may be mandated if the weeds become too large for effective control with tillage. Killing the initial flush of weed seedlings can make further weed control much easier to achieve. Unfortunately, delaying planting may not be practical for farmers with large acreages to plant. Delayed planting can also decrease potential crop yield. However, should the opportunity to control this first set of weed seedlings present itself, the weed control benefits can be very significant.

Row Crop Cultivation

The primary objective of cultivation is to control weeds while primary and secondary tillage are aimed at preparing a suitable crop seedbed. One of the most important and primary reasons for growing crops in rows was to allow the passage of cultivation equipment pulled by draft animals. Row widths were dictated by the minimum distance needed for the draft animal. Herbicide use and machine cultivation have allowed reductions in the row widths necessary for crop production. Cultivation for weed control was used extensively in the past and still represents an excellent and economical method for weed control.

Other benefits for cultivation are occasionally claimed, such as increased soil aeration, the breaking of surface crusts, and increased rainfall penetration. The majority of research has demonstrated that the yield gains from cultivation are due to the weed control obtained. Controlling weeds without soil disturbance (with very shallow scraping, clipping, or herbicide use) gave the same crop yield as treatments in which weeds were eliminated and the soil was disturbed. However, exceptions can occur on soils that become hard and dry. Research in North Carolina showed a slight yield advantage from cultivation, beyond weed control (G.C. Klingman, unpublished) on a soil that becomes hard and cracked when dry. Cultivation evidently enhanced rainfall penetration in this case. There was no benefit from more than one cultivation. The advantage of cultivation was also not detected in years with adequate moisture or on more porous soils.

Cultivation kills weeds by breaking the contact between the roots and soil, by cutting the tops from the roots, and by burial of the aerial growing plants and leaves with soil. These actions can lead to weed desiccation and depletion of plant food reserves.

Best results from cultivation are obtained with small (<2.5 inches) weeds. Larger weeds are difficult to bury and have sufficient roots to escape total separation from the soil. Seedlings of perennial weeds are easily controlled by cultivation but older plants can escape the disturbance. Cultivation equipment can also be clogged by the larger weeds. Effective cultivation needs dry soil both at the surface and below the depth of cultivation. Dry soil promotes desiccation of the uprooted weeds. Proper soil moisture for working the ground will also avoid damage to soil structure. Cultivation while the soil is too wet will simply transplant weeds, especially the vegetative reproduction organs of perennial weeds. The same problem can occur if rainfall occurs soon after cultivation. Ample moisture in the soil will promote weed survival after cultivation. The inhibition of root growth by some herbicides, such as the dinitroanilines, can increase the weed control from cultivation.

The criteria for optimal weed size and soil moisture are two limitations to the use of cultivation for weed control. These can be especially critical if cultivation is used as the sole means of weed control. Untimely rain that delays the use of cultivation can result in large uncontrollable weeds.

Cultivation is often used as a complement to herbicides. The cultivation will control weeds escaping the herbicide and extend the longevity of the weed suppression. There is a danger of cultivation following application of a herbicide to soils. The cultivator can bring untreated soil from beneath the layer of herbicide-treated soil to the surface. Weed seeds in this untreated soil can germinate and escape the herbicide. Shallow cultivator operation will help prevent this problem and also avoids "root pruning." Crop roots can be damaged by deep cultivation close to the row late in the season. This is avoided by early and shallow cultivation.

Using herbicides with cultivation helps overcome another deficiency of relying only on cultivation for the weed control program. The herbicide can control weeds directly in the crop row. These weeds are generally inaccessible to the mechanical action of cultivation. One exception to this is controlling small weeds with a rotary hoe in crops such as corn and soybean. The rapidly moving rotary hoe uproots small weed seedlings in the rows of small soybean or corn seedlings.

The shallow operation of the rotary hoe is sufficient to uproot small weed seedlings while deeper rooted and stronger soybean and corn survive. However, use of a rotary hoe over the row should be avoided if the corn or soybean are very turgid as breakage of the plants might occur. A slightly wilted crop condition is preferred over a fully turgid status to avoid crop injury.

Although cultivation is not considered highly effective for weed control within the crop row, some measure of weed suppression in the row can be obtained by burying small weed seedlings with soil thrown into the row by the cultivation. Careful cultivator adjustment and operation are needed for maximum burial.

Reduced tillage systems can limit the use of cultivation. Soil in these tillages is often hard and restricts penetration of the cultivation tool. Plant residues on the soil surface in reduced tillage can interfere with cultivation by clogging the cultivator. Special cultivation tools have been developed to help overcome these problems in reduced tillage. However, reductions in tillage generally mean a higher reliance on herbicides and other means of weed control than in conventional tillage.

In summary, cultivation is an excellent component of a weed control program. It can provide effective weed control very economically.

Hand Weeding

Hand pulling and hoeing go back to the very beginnings of agriculture when man favored one plant over another. They are also the most selective and sure way of removing weeds from a crop. The major disadvantages of these methods are the time, monetary, and energy expenses they require (Table 3-3). Arguably, a major social benefit of modern weed control methods is the release of workers from the drudgery of manual weed control. In the United States today, the cost of manual weed removal restricts its use to roguing fields of a few new weeds before they become established or on a few acres of high value crops. The high cost of labor makes hand weeding uneconomical for most weed control situations.

TABLE 3-3. Yield, Profit, and Man-Hour Requirement Comparisons for Weed Control Practices[1]

Method of Weed Control	Corn Yield (bushels)	Profit from Weed Control ($/acre)	Man-Hours Required (hr/acre)
None	54	—	0
Cultivation	81	61.23	0.57
Herbicide	90	78.49	0.05
Hand labor	92	−65.90	60.00

[1] Based on 1976 dollar values and six experiments in Minnesota.

Adapted from Nalewja, J. D., 1984, Do herbicides consume or produce energy?, in *Energy Use and Production in Agriculture*, CAST Rep. **9**, 23–25.

Mowing

Mowing can effectively prevent seed formation on tall annual and perennial weeds, deplete food reserves of perennial weed vegetative reproductive organs, and favor competitive crops adapted to mowing. Unfortunately, mowing can favor weeds that grow and reproduce below the cutting height. Repeated mowing can cause a shift in the dominant biotype of a weed species from an upright growing form to a more prostrate form.

Effective prevention of seed formation by mowing requires cutting the weeds before flower formation. Pollination, fertilization, and production of viable weed seed occur so soon following flower appearance in a number of weed species that mowing flowering weeds is often only a cosmetic solution.

Mowing kills existing shoot growth but mowed plants can produce additional flushes of shoot material. The previously dormant lateral buds may start to grow, with more new stems developing. This may appear to thicken the stand; however, this is desirable if you mow this new growth repeatedly. The new stems grow at the expense of the below-ground stored food, and repeated cutting hastens food depletion and death of the plant.

Some annual weeds sprout new stems below the mower cut. This growth can often be controlled by cutting rather high at the first mowing and enough lower with the second mowing to cut off the sprouted stems. By the second mowing, the stem is often hard and woody and cannot develop new sprouts below the cut. This procedure is effective on bitter sneezeweed, horseweed, and many other weeds.

A single mowing will often not prevent seed production. The new stems produced below the initial cut will flower and form seeds. Two or three mowings will be needed to ensure prevention of seed formation.

Repeated mowing not only prevents seed production of perennial weeds, but may also starve the underground parts. Cutting the leaves removes the food (photosynthate)–producing organs. Second, the regrowth stimulated by the cutting (see above) draws on the stored food reserves. Mowing must be often and start early in the season to be successful. The plant must not be allowed to replenish its underground food supply. Even following these guidelines, it may take 2 or more years of this treatment to completely kill a perennial weed stand. A small amount of reproductive organs left after 1 year of mowing could easily reestablish the weed problem.

The best time to begin mowing is usually when the underground root reserves are at a low level, between full leaf development and the flower appearance (Figure 3-3).

These principles of timing for mowing can also be applied to the use of herbicides for perennial weed control.

Mowing for harvest and maintenance of hay, pasture, turf, and cover crops help eliminate tall growing weeds. Crop plants grown in these situations are adapted to mowing and the cutting favors them over nonadapted weeds. As is generally the case, combination of mowing with a competitive crop is more effective for weed suppression than either element alone. Weeds that are also

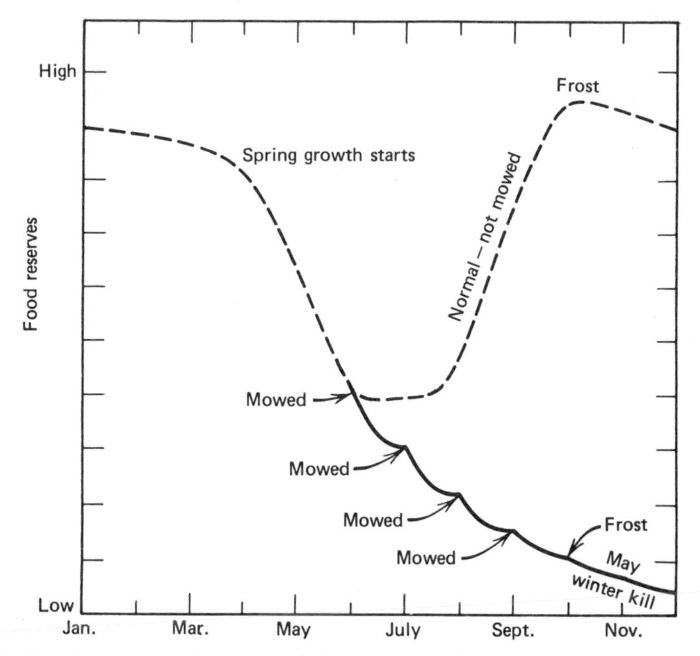

Figure 3-3. Food reserves of a perennial unmowed plant compared with reserves of a repeatedly mowed plant.

adapted to mowing, forming rosettes, mats, or close growing sods, will not be controlled by the cutting. This is why mowing does not kill established plants of weeds such as common dandelion, buckhorn plantain, violets, bermudagrass, crabgrass, and goosegrass.

Mulches

Mulches stop weed growth by restricting the penetration of sunlight to the soil surface. Weed seedlings emerging from the soil are killed through starvation from lack of photosynthesis. Light-promoted weed seed germination may also be inhibited under a mulch. Vigorous perennial weeds are not well controlled by mulching. These weeds have sufficient stored food reserves to allow them to emerge through the mulch despite the lack of photosynthesis.

Mulches can either be nonliving or living material of enough density to restrict light penetration. Nonliving materials used for mulches include wood chips, bark chips, sawdust, straw, grass clippings, paper, black plastic, and stones. The cost of these materials and the labor intensity of their use have restricted application of mulching to high value and low acreage crops. Their use is also usually restricted to established plants.

Use of plants suppressed or killed with herbicides prior to crop emergence as a mulch may extend mulching as a weed control technique to large acreage crops.

Crownvetch (*Coronilla varia* L.) has been tested for use as a living mulch in corn. Herbicides are used to suppress, but not kill, the crownvetch to avoid competition with the corn. Crownvetch recovers later in the season and the living mulch is maintained. The dense cover provided by the crownvetch helps reduce weed growth. Use of legumes such as crownvetch as living mulches also has potential for supplying some of the nitrogen requirement of the companion crop. The mulch also protects the soil from erosion. A major problem with living mulches is potential competition with the companion crop. Insufficient mulch suppression with the herbicide can cause crop yield losses from competition with the mulch. A fine line is usually drawn between the quantity of herbicide needed to manage the mulch but not to kill the mulch plants.

Another mulch system for larger acreage that overcomes this problem is use of a killed (by herbicides) cover crop of cereal grains or other grasses as the mulch. The cover crop is planted in the fall and then killed in the spring. Wheat, winter rye, and ryegrass have shown promise for use as mulches. One of the disadvantages of using a mulch for weed control is that the soil covering can restrict use of other weed control methods. All weeds are unlikely to be controlled by the mulch. Mulch can interfere with cultivation, mowing, and herbicide application. It was shown mulch reduces the herbicide reaching the soil (Figure 3-1). Thick layers of organic mulches around landscape plants can also encourage rots at the base of the plant stems. Potential benefits of mulches beyond weed suppression include soil moisture conservation, lowered soil temperatures (higher under black plastic mulch), protection of the soil from erosion, and added organic matter to the soil. The relative benefits or detriments of these must be evaluated for each case. Wetter and cooler soil can delay spring planting but moisture conservation may help avoid later plant drought stress. Warmer soil and a more uniform moisture supply with black plastic mulch can improve growth and yield of tomatoes.

Burning

Fire can be used to remove undesirable plants from ditch banks, roadsides, and other waste areas, to remove undesirable underbrush and broadleaf species in conifer forests, and for annual weed control in some row crops. Burning must be repeated at frequent intervals if it is to control most perennial weeds. In alfalfa, weeds and some insects can be controlled by burning.

In waste areas, if the vegetation is green, a preliminary searing will usually dry the plants enough so that they will burn by their own heat 10 to 14 days later.

Proper burning techniques can favor conifer trees over hardwood species in forestry. This *controlled burning* can also remove undesirable underbrush if done at regular intervals. Controlled burning reduces the hazard of uncontrolled forest fires.

Flaming has been used most successfully for selective weed control in cotton. Special burners are used to direct the flame at the base of the cotton plants. The hard woody cotton stem escapes injury but young weed seedlings are killed. Two passes are normally done a few days apart for best results. Proper adjustment and speed of operation are essential to avoid crop injury. Flaming is similar to row cultivation or use of a directed spray herbicide application (see section on Chemical Methods) in requiring a size difference between the crop and weed for effective weed suppression and crop safety. The increased fuel costs for flaming have diminished the economy of the technique for weed control.

Flooding

Flooding has been used to control weeds in rice fields. Water-saturated soil limits oxygen availability. Whereas rice can grow under the flooded conditions, many weeds cannot. Aquatic plants also tolerate the flooded conditions and are not controlled. Perennial weeds can also be controlled by prolonged (3 to 8 weeks) flooding. Flooding is used to control established perennials such as silverleaf nightshade, camelthorn, and Russian knapweed in the Western United States but the expense of creating dikes and maintaining the water level for the prolonged periods prevents wider use.

Flooding has limitations for further use in weed control in rice production. Further reliance on flooding can increase rice production costs and reduce yields. Perennial reproductive structures that become dormant due to flooding will not be controlled by the treatment. Flooding will have little effect on weed seeds in the soil.

CULTURAL PRACTICES

Crop selection, rotation, variety selection, planting date, plant population and spacing, plus fertility and irrigation are all cultural practices that affect weed management.

Crop Selection

Selection of a crop determines strategies for the subsequent battle with weeds. Crop selection will determine the level of weed control needed for efficient crop production and, in many cases, which weeds will be most troublesome. The crop grown will also determine which weed management options are available to the farmer. These include the cultural, mechanical, and biological techniques for weed control plus the herbicides that can be safely and legally employed to manage weed infestations. The potential monetary return from a particular crop will also determine the economics of weed control practices. It is very important to select a crop that allows selection of weed control strategies that can manage the weeds present. Unfortunately, crop selection can rarely be made solely from a weed management perspective. Climate, soil adaptability, history, market availability, and the potential economic return are all factors that must be considered by a farmer in deciding on a crop.

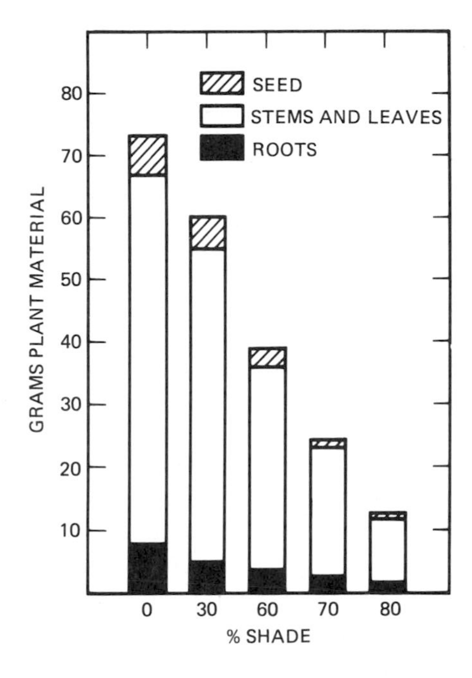

Figure 3-4. Effect of shade on weight of roots, stems, leaves, and seeds of giant foxtail. (Knake, E. L., 1972, *Weed Sci.* **20**, 588.)

Generally, selection of the crop with the highest growth rate and final height are important because they reflect the ability of a crop to grow above weeds. Closure of the crop canopy over weeds decreases sunlight and directly limits weed growth. This was demonstrated for yellow nutsedge and giant foxtail (Figure 3-4). Crop species differ in their interception of sunlight.

Practices such as narrow rows that promote early closure of the crop canopy help maximize the effect of crop competition. Conversely, conditions such as poor crop growth, planter skips, or wide row spacing that delay canopy closure will make weed control, especially late in the season, more difficult. An example is the increased difficulty in controlling grass weeds in fields of corn grown for seed compared to corn grown for feed. The inbreds grown for seed production grow slowly; their final height is much less than hybrid corn, and they are planted in wide rows to allow passage of crews for detassling and other hand labor operations.

The greatest amount of weed suppression due to crop competition occurs when a dense perennial sod, turf, or hay crop is grown. The heavy perennial growth, combined with cutting for harvest, can greatly reduce weed seedling establishment, seed production, and perennial growth. A perennial crop will rarely eliminate a weed from an area due to survival of dormant weed seed in the soil. However, raising a perennial crop or other dense cover crop, especially for several years in succession, can reduce a problem weed population to manageable levels. Examples of crops useful for weed suppression include alfalfa, buckwheat, sudangrass, and densely planted small grains.

Crop Rotation

Crop rotations help prevent the build-up of weeds adapted to a particular cropping system. Certain weeds are more common in some crops than others. Pigweed, lambsquarters, common ragweed, velvetleaf, cocklebur, foxtail species, and crabgrass are found in summer cultivated crops such as corn. Mustards, wild oat, wild garlic, chickweed, and henbit are associated with fall sown small grains. Pastures often contain perennial weeds such as ironweed and thistles. Changing crops changes the cultural conditions (planting date, crop competition, fertility, etc.) that a weed must tolerate. Rotating crops also often means a different set of herbicides will be used. The overall success of a crop rotation in managing weeds depends on the ability to control the weeds in each crop grown in the rotation. Rotation will prevent a weed species from becoming dominant in a field but will also maintain a diversity of weed species present in the same area.

Crop rotation historically was very important for managing weed problems. Today, rotation is used more for managing diseases and insects than weeds. Availability of abundant fertilizer and diverse herbicides made it possible to minimize the need for crop rotations for weed control. There remain sound reasons to practice crop rotation for general crop production and pest management purposes. As an example, corn grown in rotation with soybean consistently yields more than corn planted after corn. In addition, there is negligible danger of corn rootworm damage to corn planted after soybean but corn grown after corn must often be treated with an insecticide to avoid damage. Rotation is not an option with long-term perenial crops such as orchards, tree plantings, and perennial forages. Rotation can also require additional farmer knowledge and equipment to manage the various rotational crops.

Some of the benefits of rotation can be retained in monoculture cropping systems by selection of a variety of herbicides, especially differing in mode of action and use of cultivation. The herbicide diversity and cultivation will help prevent the development of tolerant weed populations adapted to both an unchanging herbicide program and crop.

Crop Varieties

Development of new higher yielding crop varieties is generally done under conditions of minimal weed, insect, and disease interference. Normal variety development schemes yield little information on the differential competitive ability of cultivars. However, more vigorous, faster growing, and taller crop varieties are likely to be better competitors. Differential competitive ability among varieties has been amply demonstrated as shown in Table 3-4. Ennis (1976) estimated selection of competitive soybean cultivars could provide up to 80% control of selected weeds in the crop. The majority of weeds discussed were suppressed less than 50%. Suppression of less than 50% would not be sufficient to eliminate competitive yield losses but would contribute to overall control when other weed control practices were also used. Generally, cultivars with the highest

TABLE 3-4. Yield Reductions in Selected Soybean Varieties due to Johnsongrass or Cocklebur Competition

	Competing Weed	
Soybean Variety	Johnsongrass (% soybean yield	Cocklebur reduction)
Davis	34	56
Lee	41	67
Semmes	23	53
Bragg	24	57
Jackson	30	67
Hardee	23	26

From McWhorter, C. G., and E. E. Hartwig, 1972, Competition of Johnsongrass and Cocklebur with six soybean varieties. *Weed Sci.* **20**, 56–59.

yield potential are selected by growers and weed control is implemented to allow expression of the yield potential.

Planting Date

The trend in crop production is for earlier planting to increase yields. The longer exposure to sunlight with earlier planting is primarily responsible for the higher yields associated with this practice. Early planting can establish adapted crops before weeds and provide the crop with a competitive edge.

There are weed control disadvantages from early planting. Early planting means soil-applied herbicides may need to persist longer in the environment for the most effective weed control. It also eliminates cultivation just before later planting, which often destroys the first flush of germinating weed seedlings.

Unfortunately, any advantages gained for weed control by delayed planting are often outweighed by decreased crop yield potential. The cost of using delayed planting as a weed control strategy must be weighed for each crop–weed–environment situation.

Plant Population and Spacing

Historically, crops were planted in rows spaced wide enough to allow passage of draft animals pulling cultivation equipment to control weeds. Development of herbicides removed this constraint and allowed adoption of narrow row production systems that can produce higher crop yields. The narrower rows can also produce less weed growth, especially if herbicides are relied on for weed control (Figure 3-5). This advantage may be lost if cultivation is needed to control weeds as cultivation is more difficult in the narrower rows. The weed control advantages for narrow rows stem from the more rapid establishment of a closed

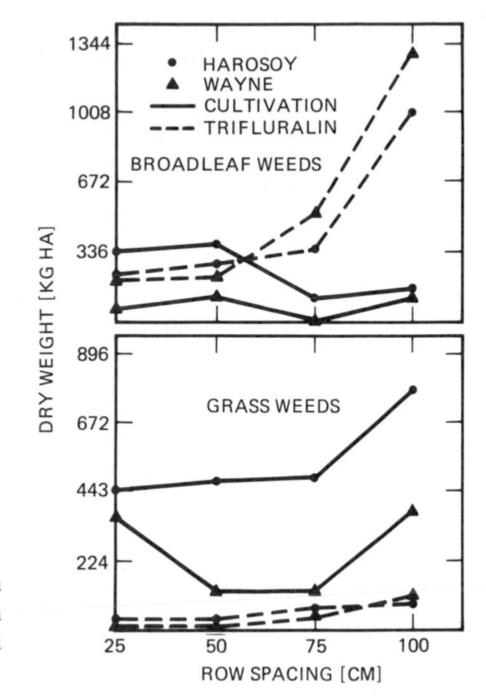

Figure 3-5. Effects of row spacing, soybean variety, and weed control method on weight of weeds. (Wax, L. M., and J. W. Pendleton, 1968, *Weed Sci.* **6**, 462.)

crop canopy and resultant reduction in light reaching the soil surface to support weed growth.

Fertility and Irrigation

Nitrogen is generally the nutrient of greatest concern in weed competition. Increasing the nitrogen supply can increase crop yields, have no effect, or reduce crop yields when weeds are present. Final crop yield with increased nitrogen and weed competition is rarely as great as yield without weed competition. Extra fertilizer is not generally an efficient way to avoid crop losses from weed competition. Some weeds use more fertilizer than needed for growth. These "luxury consumers" may actually benefit from the fertilizer more than the crop (Table 3-5).

Similar to fertilizer, added moisture, through irrigation can overcome a portion of crop yield losses due to weed competition but full crop yield potential will unlikely be reached (Table 3-6). Also like fertilizer, additional water can favor weeds because many weeds have high water requirements, water used per plant weight.

BIOLOGICAL CONTROL

Biological control of weeds uses the insects and diseases that naturally attack weeds, the competitive ability of crops, and herbivores to reduce or eliminate the

TABLE 3-5. Chemical Composition of Corn and Corn Weeds in September

		Mean Percentage Composition, Air-Dry Basis				
Plant	Growth Stage	Nitrogen	Phosphorus	Potassium	Calcium	Magnesium
Corn	Total plant, early milk stage	1.20	0.21	1.19	0.18	0.15
Pigweed	Green plants, 25–50% seeds ripe	2.61	0.40	3.86	1.63	0.44
Lambsquarters	Green plants, 25–50% seeds ripe	2.59	0.37	4.34	1.46	0.54
Smartweed	Green plants, 25–50% seeds ripe	1.81	0.31	2.77	0.88	0.56
Purslane	Lush plants, seeds partly ripe	2.39	0.30	7.31	1.51	0.64
Galinsoga	Lush plants, seeds partly ripe	2.70	0.34	4.81	2.41	0.50
Ragweed	Green plants, seeds partly ripe	2.43	0.32	3.06	1.38	0.29
Crabgrass	After bloom, seeds not ripe	2.00	0.36	3.48	0.27	0.54
Average for weeds		2.36	0.34	4.23	1.36	0.50

From Vengris, J., et al., *Agron. J.* 1953, **45**, 213.

TABLE 3-6. Effect of Quackgrass Interference on Yield of Irrigated and Nonirrigated Soybeans

Soybean treatment		
Irrigation	Quackgrass	Soybean yield (ton/acre)
No	No	1.3
Yes	No	1.2
No	Yes	0.7
Yes	Yes	1.0

From Young, F. L., D. L. Wyse, and R. L. Jones, 1983, Effect of irrigation or quackgrass (*Agropyron repens*) interference in soybeans. *Weed Sci.*, **31**, 720–727.

potential detrimental effects of weed populations. Using competitive crops for weed management was discussed under Cultural Control. Although there is much public awareness about the concept of biological control, currently it is only a limited alternative to other weed control methods. The major successes for biological weed control have occurred in relatively undisturbed areas such as rangeland or aquatic weed infestations.

There are several reasons why biological weed management is not better adapted to many weed control situations. Biological agents generally attack only a single weed species. Cropland almost always contains a complex of weed species. A particular crop can have two dozen weeds commonly associated with its culture. Other tactics will be needed to control the weeds the biological agent does not attack. A biological control program may be best suited for management of a single weed species that is poorly controlled by the other available management techniques. Unfortunately, the profitability of a biological control agent that attacks only a single weed species, perhaps only a problem in one crop, may be low. Commercial companies can be unwilling to undertake the costs of discovery, research, development, production, and marketing for such a product. The majority of the work and expense required for discovery and research on biological weed control is done by the federal government and universities.

The action of biological agents on weeds can be slow. Reductions in weed growth may be too slow to avoid crop yield losses from weed competition. This is primarily a problem when the objective of the control strategy is to quickly *reduce* a weed population below an economic threshold. A longer term objective can be establishment of an ecological balance between the biological control agent and the weed. This is not constrained by the time factor. The long-term goal is to eventually *prevent* the weed population from exceeding a threshold. Long-term objectives are easier to adopt in rangeland or aquatic situations.

The ecological balance achieved with biological control is, in effect, a permanent solution to the weed problem. This is an obvious benefit. However, a long-term balance between the biological control agent and the weed population means some portion of the weed population will continue to survive. Survival of some weeds is necessary to ensure survival of the biological control agent. As was shown in Chapter 2, the presence of even a relatively few weeds in an annual crop may be enough to cause significant yield loss.

Perennial crops, rangeland, and aquatic areas represent relatively stable ecosystems. This stability will allow build-up of biocontrol control agent populations and improve the effectiveness of the biocontrol strategy. Annual crop production is the opposite of stability. Populations of biocontrol agents must survive annual tillage operations, cultivations, crop rotations, and pesticide applications.

Economics favors biocontrol of weeds in rangelands and aquatic areas. These are often large acreages with difficult accessibility, on which little can be spent on weed control technology. Crop production, in contrast, represents a comparatively high economic investment for which the potential economic return from weed control promotes greater use of active weed management strategies.

Biological control of weeds is often more easily applied and suited for management of a single aggressive perennial weed species that is introduced and invades rangeland or aquatic areas.

Criteria for Biological Control Agents

The most important criteria for any potential biological control agent is that it attacks only the target weed species and no other plants. Fear of accidentally introducing new pests has diminished as our knowledge of biological control agents has grown. Plant feeding insects are often very specific in their feeding preferences. The danger of insects switching feeding preferences seems remote.

Although a successful biological control agent will attack only one plant, it should be remembered that just as definitions of a weed can differ, different opinions on the worth of a plant can exist. Some weeds were imported as ornamentals and are still used this way. Weeds can be important sources of forage in one situation and a weed in others. Total economic impact of biological control agents must be evaluated before their introduction. It is difficult to reverse or contain biological control programs following their release.

An introduced biological control agent should reproduce quickly to build the population fast enough to effect weed control. The biological control agent should also be adapted to the new environment and be rid of its own parasites, predators, and diseases. This freedom will help maintain the population of the agent at high levels.

Types of Biological Weed Control Agents

Insects Plant-attacking insects are currently the most widely used biological control agents for weed control. They have a specific host range, can be mobile (which promotes their dispersion), and can destroy both vegetative and reproductive portions of weeds. Insect attacks can also predispose weeds to attack by other factors, such as diseases. Action of several biocontrol agents on a weed is often more effective than attack by only one agent.

The outstanding example of biological weed control is the prickly pear or cactus (*Opuntia* spp.) in Australia. The prickly pear was originally planted for ornamental purposes, but it spread rapidly from 1839 to 1925, covering 60 million acres and threatening much of the cultivated land. In seeking control measures, research scientists found insects that attacked the prickly pear and no other plants. A moth borer (*Cactoblastis cactorum*) from Argentina was the most effective. It tunneled through the stems, underground bulbs, and roots. The damage was even more effective after several bacterial root organisms were accidentally introduced into the wounds caused by the insect.

The moth borer was released in 1926 and by 1931 had multiplied to such numbers that nearly all the prickly pears were destroyed by midseason. With little or no food supply, many of the moth borers starved. Several years followed when prickly pear increased, with a later increase in the moth borer. Several "waves" of each type of growth occurred before an equilibrium or "balance of nature" was reached. For this reason, biological methods control rather than eradicate. This is especially true of weed species reproducing from seeds that may lie dormant in the soil for years.

There are other cases of effective biological control. In the western United States, St. Johnswort (*Hypericum perforatum*), a poisonous range weed, is being controlled by leaf-eating beetles (*Chrysolina* spp.) (see Figure 3-6). In Hawaii, pamakani (*Eupatorium adenophorum*), a range weed, is being controlled by a stem gallfly (*Procecidochares utilis*), and in Fiji, curse (*Clidemia hirta*), another range weed, is being controlled by a shoot-feeding thrip (*Liothrips urichi*).

Other more recent biological weed-control research has indicated possible control of musk thistle by a weevil (*Rhinobyllus conicus*), purple nutsedge by a moth (*Bactra verutana*), alligatorweed by three different insects, and puncturevine by seed weevils (*Microlarinus lareynii*).

Pathogens Two approaches are utilized for employing plant pathogens, primarily fungi, for biological control of weeds. The *classical* approach aims at establishing a self-maintaining plant pathogen innoculum level. This technique has widest application to the traditional areas of biological weed control: rangeland, aquatic sites, and noncropland. The alternative approach to use of plant pathogens relies on repeated application of high levels of disease innoculum to control weed problems. This is called the *mycoherbicide* approach. The mycoherbicide technique currently offers the best opportunity for extension of biological weed control into nontraditional disturbed areas. Mycoherbicides are being used commercially in citrus groves and rice fields.

An example of the classical approach is the use of a rust fungus (*Puccina chondrilla*) to control rush skeletonweed (*Chondrilla juncea* L.) in areas of the Western United States. This fungus was previously successful in reducing rush skeletonweed populations in Australia. The specificity of this biocontrol agent is

Figure 3-6. St. Johnswort or Klamath weed control by *Chrysolina quadrigemina* (beetles) at Blocksburg, California. *Top*: Photograph taken in 1946. The foreground shows weeds in heavy flower, whereas the rest of the field has just been killed by beetles. *Middle*: Portion of the same location in 1949 when heavy cover of grass had developed. *Bottom*: Photograph taken in 1966 showing the degree of control that has persisted since 1949. Similar results were reported throughout the state. (C. B. Huffaker, University of California, Berkeley.)

illustrated by the difficulty in introducing it to the United States. Initial attempts to infect rush skeletonweed in the United States with rust from Australia failed, but rust from Italy infected and spread through California and Oregon. This strain from Italy did not establish in Idaho. A second strain from Italy established on southern, but not northern, biotypes of rush skeletonweed in Idaho. Additional screening on infected U.S. plants was needed to find a rust that would infect rush skeletonweed in Washington State.

Spores of the rust need 6 hours of dew during the dark for infection to be successful. This requires aerial spreading of the spores in the arid Western United States for quick dispersal in the brief favorable environmental conditions for infection.

An example of a mycoherbicide is a preparation of the fungus *Colleotrichum gloesporiodes* trade named Collego® used to control northern jointvetch (*Aeschynomene virginica*) in rice. Collego® is used like a herbicide applied to the leaves of the target weed. It is sold as a two-component formulation of dry fungal spores plus a liquid spore rehydrating agent that is mixed before use. Care must be taken to avoid fungicide use 1 week before and 3 weeks after Collego® application. In addition, aerial application equipment is cleaned prior to Collego® application with a suspended charcoal solution to ensure removal of any contaminating pesticides from the sprayer. These precautions illustrate the difficulty of integrating a mycoherbicide into crops that require extensive fungicide use.

Herbivores Grazing animals such as geese, goats, sheep, and cattle have been used to selectively control weeds in crops, pastures, and noncropland. The White Chinese breed of geese and others have found specialized use for control of seedling grass weeds in a number of broadleaf row crops, particularly cotton. They have also been used in orchards, nurseries, and some perennial crops. Additional control measures are needed to manage broadleaf weeds where the geese are used. Sheep and goats can be used to improve grazing conditions for cattle. They can remove weedy annual broadleaf species and brush, which allows improved rangegrass growth for cattle feed. There is little food competition between goats and cattle if sufficient shrubs and broadleaf plants are available for the goats. Sheep and goats can be used to help avoid *Senecio* poisoning in cattle. About 20 times as much *Senecio* on a body weight basis is needed to poison sheep and goats as for a cow. Sheep and goats will readily graze *Senecio* and stocking sheep and goats with cattle will help prevent cattle poisoning.

A number of possible animal management costs and efforts can be associated with the use of grazing animals for weed control. These include feeding for a balanced diet, fencing, herding, sheltering, protection from predators, and general care.

Aquatic weed eating animals and fish are promising for weed management in aquatic sites. The manatee or sea cow consumes large amounts of vegetation, but its use is limited to tropical areas.

Waterfowl such as geese, ducks, and swans can be used for removing floating

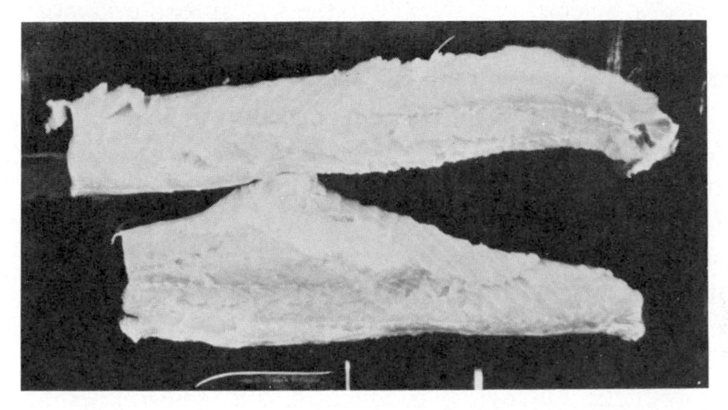

Figure 3-7. *Top*: The Chinese grass carp, or white amur, grows rapidly by grazing on water weeds. *Bottom*: A valuable by-product of this activity is a highly nutritious fish flesh. (Richard R. Yeo, USDA, SEA, ARS and LeRoy Holms, Madison, WI.)

and submersed plants from small ponds. Unfortunately, waterfowl can interfere with use of the pond due to their aggressiveness and their deterioration of the water quality and bank stability. Waterfowl must also be protected from predators.

Plant-eating fish offer one of the more promising biological approaches to aquatic weed control. These are imported fish that feed exclusively on vegetation. Young *Tilapia* spp. can feed on phytoplankton and unicelluar algae and older *Tilapia* also feed on larger plants. *Tilapia* are sensitive to cold temperatures and die below 46 to 48°F. Annual stocking of *Tilapia* has been used for aquatic weed management in irrigation ditches of Southern California.

The grass carp or white amur (*Ctenophryngodon idella*) (Figure 3-7) can survive in water below an ice cover. Grass carp has a life span of several years. It feeds exclusively on filamentous algae, Chara, weeds below the water surface, and duckweed. The fish consumes its weight in vegetation each day and can grow to 20 to 50 lb. As with any introduced species, there is some concern with the competition of grass carp with desirable fish species and other environmental impacts. Grass carp does not appear to reproduce naturally in the United States. Thus, its population will not crowd out desirable game fish. Commercial hatcheries use hormonal treatments and special environment conditions to raise fish for release. The grass carp also does not roil the sediments and cloud water as does the common carp. Although a number of states outlawed the importation of the grass carp, this fish is gaining wider acceptance and use.

CHEMICAL WEED CONTROL

Use of chemicals that selectively kill weeds in crops is a integral part of many modern weed management systems. The specific *pesticides* for controlling weeds are called *herbicides.* Selectivity is the key to the widespread utility of herbicides. Bases for herbicide selectivity between plants are discussed in Chapter 5. Chemicals were used in ancient times to control unwanted vegetation. The Roman Army would salt the fields of their enemies, preventing growth of all plants, both crops and weeds. Weeds were controlled selectively in the early twentieth century using inorganic salts such as sulfuric acid, but selectivity was limited to a few crops and required special care in use of the chemical.

Modern selective herbicides were borne from the discovery during the Second World War of the herbicidal properties of synthetic plant growth regulators. The Weed Science Society of America listed over 250 chemicals for herbicide use in 1985. These chemicals are offered to consumers over the world in thousands of trade named products. Discovery, production, and sale of herbicides are a multibillion dollar worldwide industry. Because of the importance of herbicides to modern agriculture, a large portion of this book is devoted to the characteristics use, environmental behavior, and toxicology (health considerations) of these chemicals.

Herbicide Classification

Herbicides can be grouped into various categories for convenient discussion. These include chemical similarity, mode of action (how they kill plants), herbicide movement within plants (mobile versus immobile), selectivity (selective versus nonselective), and application and use patterns. This book utilizes chemical similarity by grouping herbicides into chemical "families." Further discussion of specific herbicides is found in Chapters 8–14.

Adoption of Herbicides

The widespread adoption of herbicides by farmers is linked to clear advantages for herbicide use in several areas. Herbicide use increased largely at the expense of cultivation for weed control. However, it can result in higher crop yields than reliance on cultivation alone (Table 3-7). It is worth reemphasizing that the best and most efficient weed management scheme will utilize all available weed control tactics as appropriate.

Herbicides can control weeds beyond the reach of the cultivator. Weeds

TABLE 3-7. Average Cost Relationships in Weed Control in a 3-Year Experiment in Mississippi

	Cotton Yield per Acre			Values per Acre		
Treatment	Lint (lb)	Seed (lb)	Energy (kcal)	Cost of Weed Control ($)	Received for Lint and Seed ($)	Received above Weed Control Cost ($)
Four herbicides, no cultivation	553	764	2,739.722	26.11	364.95	338.84
Three herbicides, two cultivations	487	673	2,413.144	23.07	348.67	325.60
No herbicides, five cultivations	158	218	782.144	10.00	104.26	94.26
No herbicides, five cultivations, hand hoeing (75 hr/acre)	529	731	2,621.168	261.25	349.12	87.87

From Dowler, C. C., and E. W. Hauser, 1975, Weed control systems in cotton in Tifton loamy sand soil. *Weed Sci.* **23**, 40–42.

directly within the crop row, in closely seeded (drilled) crops such as small grains, and in notillage can be managed with herbicides but not with cultivation. Control of weeds that directly compete with crop plants in the row is especially important to minimize crop yield losses.

Herbicides, compared to cultivation, help reduce the labor and time needed for effective weed management. These reductions can directly lead to increased economic return for a farmer. Reduced time and labor requirements for weed control also allow farming of larger acreages despite a shrinkage of available farm labor.

Herbicides contribute to higher crop yields in ways other than improved weed control. Crop yield is increased for some crops, such as corn, by early planting. Herbicides allow planting into soils too wet for a final tillage to kill any emerged weeds. Crops can also be planted before the soil warms enough to promote weed seed germination and seedling growth. Herbicides applied at planting or later will control weeds as they germinate without needing to wait for a tillage operation to kill them after they emerge.

Herbicides are the reason reduced tillage systems, particularly no-tillage, are now feasible. Without herbicides primary tillage, secondary tillage, and cultivation would still be needed to suppress weed populations. Herbicides are available for many crops to kill existing vegetation before planting, to keep weeds from becoming a problem later, and to control any escaping weeds.

Herbicides eliminated or greatly reduced the need for hand weeding in many weed control situations. This frees labor from the drudgery and hard work of hand pulling or hoeing weeds. Freedom from manual weed control allows workers to seek better employment elsewhere. This is an advantage where there is work available for workers released from weed control but has been criticized by some when employment is not available for displaced workers. As with the introduction of any new technology, herbicide use can bring sociological as well as agricultural change. This is more of a concern in less developed countries of the world where the labor supply exceeds the available work opportunities.

Herbicides have increased the choices available to farmers. Selections of crops, rotations, and tillages are not as dependant on the impact of these choices on the existing weed population. Growing crops in closely spaced rows for higher crop yields and ease of mechanical harvest is aided by effective herbicide programs. Given the availability of effective herbicide programs, a farmer's choice of production system variables is not limited by weed problems.

Herbicide use has resulted in energy-efficient and economic weed control (Tables 3-3 and 3-7). Farmers have adopted herbicides for weed control because the chemicals can increase profit, weed control efficacy, and production flexibility and reduce time and labor requirements for weed management.

Herbicide use is not without its potential problems. Farmers who use herbicides need to be concerned about possible crop injury. Selectivity can be reduced under adverse environmental conditions (see Chapter 5) or be marginal for some herbicides even with good growing conditions. There is also the danger of injury to nontarget plants in adjacent fields and, in some cases, areas far removed from herbicide applications. Herbicides can potentially move off target as volatile gases, in water running off a field, and attached to dust particles or sediments in runoff water. There have even been cases in which an aircraft flying to treat a crop field leaked herbicide over a wide area of its flight path. Concern must also be given to potential environmental damage associated with herbicide use. Herbicide residues in soil can restrict or prevent rotational crop growth. Groundwater (water in the water table) and surface water contamination by herbicides must also be assessed. The effects of herbicides on other lifeforms in the environment, such as fish, animals, birds, invertebrates, and microorganisms, should be evaluated.

The danger of direct human toxicity from herbicide exposure should also be of paramount concern. This involves not only danger of immediate (acute) effects but the implications from low term (chronic) exposure. These occur through direct exposure of herbicide production workers and applicators to high amounts of chemical and possible indirect exposure of others to very low levels of herbicides in food and drinking water. These can all be controversial and emotional subjects for discussion. However, they should not generally obscure the benefits that farmers have realized from herbicide use. Herbicide environmental impacts and safety requirements are further discussed in Chapter 4.

Many of the worst problems that have occurred with herbicides could have been avoided through their proper use. Herbicide users have a responsibility to select the right herbicide for the weed problem and crop, handle and store the herbicide with respect, and correctly apply the chemical. There are plentiful aids available from the herbicide manufacturers and their representatives plus literature from the Cooperative Extension Service and others to help with proper herbicide use. Above all, it is imperative to read and follow the herbicide product label directions.

Time of Herbicide Application

One of the major distinguishing characteristics between different herbicide programs is the time the chemicals are applied. These timings are defined both with respect to the stage of weed and crop growth. In the broadest sense, herbicides can be applied either directly to the soil (soil active) or directly to the foliage of weeds (foliar applied). Some herbicides are effective with only one of these types of applications while other herbicides can be applied either way. More specific application timings are *preplanting*, *preemergence*, and *postemergence*. Examples of these are illustrated in Table 3-8.

Preplanting Preplanting treatments are made anytime before crop planting. *Soil fumigations* and *preplow*, *early preplant*, plus *preplant incorporation* treatments are all examples of preplanting applications. Soil fumigation places a nonselective herbicide into the soil to eliminate many existing weed seeds and vegetative reproduction structures. The soil fumigant, such as methyl bromide, must dissipate from the soil before planting for crop safety. Preplow treatments are applied to the soil prior to primary tillage for seedbed preparation.

TABLE 3-8. Examples of Preplant, Preemergence, or Postemergence Herbicide Use Defined by the Crop, Weed, or Both

Application Type	Crop Stage	Weed Stage	Example
Preplant	Preplant	Pre	Early preplant application of atrazine in corn to control annual weeds
Preplant	Preplant	Post	Application of 2,4-D before soybean planting in no-tillage to control perennial weeds
Preplant incorporated	Preplant	Pre	EPTC application in corn or trifluralin application in soybeans to control annual weeds
Preemergence	Pre	Pre	Alachlor in corn or soybeans to control annual grass weeds
Postemergence	Pre	Post	Paraquat or glyphosate after planting in no-tillage but before corn emergence to control existing weeds
Postemergence	Post	Post	Sethoxydim use in established alfalfa to control grass weeds
Postemergence	Post	Pre	Diphenamid use after transplanting tobacco to control germinating weeds

Early preplant applications generally use herbicides that persist in soils and that are applied to no-tillage fields 2 or more weeks before planting. These early applications may be done before any weeds emerge before planting. The residual herbicide prevents early weed growth and can reduce the need for herbicide control of existing weeds at planting. Firmer soil in no-tillage fields than in tilled fields allows sprayer passage early in the season without the threat of getting stuck in muddy tilled fields. Early treatment of the soil surface also increases the likelihood of rainfall occurring to move the herbicide into the soil before weed emergence. Soil applied herbicides must be moved into the soil to be active on weeds. Early preplant treatments can increase the reliability of herbicide treatments. Good weed control relying on herbicides is essential for success in no-tillage. Early preplant treatments also let farmers and other herbicide applicators begin their work early in the season. This can be a big advantage when many acres must be treated.

There are some disadvantages to early preplant programs. Early herbicide application can reduce the period of effective weed control after planting. Herbicides in the soil generally only last for a few weeks for weed control (see Chapter 6 for discussion of herbicide fate in soils). Additional herbicide treatments at planting or later may be necessary following early preplant applications to ensure adequate long-term weed management.

Preplant incorporation is the most common form of preplant herbicide application. Discussion of equipment for this purpose is in Chapter 7. Incorporation (mixing) of herbicides into the soil before planting can offer several advantages. These include less reliance than with surface application for rainfall to move the herbicide into the soil plus improved control of some weeds that germinate deep in the soil and some perennial weeds. Applying the herbicide before planting helps ensure good weed control during seedling growth of the crop.

Disadvantages of preplant incorporation include the monetary cost of the tillage operation, possible soil drying and erosion losses due to the tillage, and the potential for improper incorporation (too deep or streaked in the soil) causing reduced weed control. Higher herbicide rates can be needed to offset the herbicide dilution in the soil. However, some herbicides must be incorporated to stop losses from herbicide volatility (gaseous loss) or ultraviolet degradation that would otherwise happen if the chemicals remained on the soil surface. Incorporation can also be difficult to coordinate with aerial or contract (custom sprayed) ground application of herbicides. It is not possible to use herbicide incorporation in no-tillage systems and it may be contrary to the objectives of reduced tillage programs.

Preemergence Preemergence treatments made shortly after crop planting but before weeds emerge are a very common way to use soil applied herbicides. There are several benefits of this type of application. Herbicides are often more concentrated after preemergence application in the upper soil layers than where the chemical was mechanically mixed into the soils. The higher herbicide concentration can produce better control of shallow germinating weed seedlings. Longer residual control is also possible as preemergence applied herbicides are not as subject to leaching (downward movement in the soil with water) below the weed seed germination depth as incorporated herbicides. Preemergence herbicide applications can be made with the planting equipment avoiding an extra trip across the field. Weeds are controlled early in crop growth, minimizing competitive effects, and preemergence applications are suitable for a variety of tillage practices. Aerial and contract applications are easy to arrange with preemergence treatments. Greater crop safety can occur with preemergence applied herbicides due to the spatial separation of the herbicide-treated soil layer from the crop seed. Preemergence herbicide applications can be made to soil that would be too wet for effective incorporation.

The most severe limitation for preemergence herbicide treatments is the requirement for rainfall (or irrigation water) to move the chemical into the soil to achieve weed control. Delay in rainfall can result in loss of weed control. Preemergence applications may not be feasible in arid areas with limited access to irrigation water. The majority of preemergence herbicides are most effective against recently germinated weeds or small weed seedlings. Application soon after crop planting is necessary for effective weed control. Unfortunately, this may slow the planting operation. Finally, although preemergence applications are often safer on crop plants, high rainfall can move a concentrated band of pesticide from the soil surface into the crop root zone and produce crop injury.

Postemergence Postemergence applications are made after emergence of the specified crop and/or weed species. Postemergence application is the only herbicide application strategy that is not strongly influenced by the soil environment. Both preplant and preemergence application rates must be adjusted for soil texture (relative proportions of sand, silt, and clay) and soil organic matter content. Lack of soil influence allows postemergence applications to be used in areas such as high organic matter soils, where soil applications would be totally ineffective. The high rates of soil-applied herbicides needed on organic soils can favor the use of postemergence herbicides. A second advantage of postemergence applications is that they are made after the weed problem appears. This can help avoid unneeded preventative treatments or allow only infested parts of the field to be treated. Postemergence treatments do not take time during planting and make aerial and contract ground applications very feasible. Many postemergence herbicides have little soil activity, removing any threat of injury to rotational crops.

The major disadvantage for postemergence applications is the often limited time over which herbicides can be effectively and safely applied. There can be restrictions on both the size of weeds effectively controlled and crop size for selectivity. The critical period for weed control can be lengthened, in some cases, by increasing the herbicide rate. Of course, this incurs extra costs. The optimum conditions for weed treatment and crop safety must coincide for maximum herbicide effectiveness. Environmental conditions (weather too hot, cold, dry, wet, or wet soil) can impact or delay postemergence applications and prevent full control. The amount of area some farmers must spray can prevent optimum application timing for postemergence control on all fields. Finally, delay in controlling the weeds increases the potential for yield losses due to weed competition.

Another limitation to use of postemergence herbicides is the relatively limited spectrum of weeds controlled by many of these herbicides. Often, postemergence herbicides are effective in controlling only broadleaf (dicotyledonous) weeds or grass (monocotyledonous) weeds. More than one herbicide is often required for control of the total weed population present in a field. Beyond the general group of weeds controlled (broadleaf or grasses) by any one postemergence herbicide, the response of specific weed species will vary widely to a particular herbicide. This may also require that more than one herbicide be used for control of the entire weed population. Herbicides used postemergence can have less selectivity than soil-applied herbicides. Also, with foliar active herbicides there is greater danger of spray drift harming nontarget plants removed from the treatment area.

Specialized postemergence applications include *directed applications* and *lay-by applications.* Directed applications achieve selectivity by specialized application equipment that allows minimal contact of the spray solution with sensitive crop parts. Directed spray operations used contact-type herbicides (such as paraquat) to control weeds within an established crop. The spray is directed to the base of the crop plant to avoid contact with the crop foliage and prevent crop injury (Figure 3-8). Systemic herbicides are also used postemergence directed, for example 2,4-D or dicamba use in corn, when translocation of the herbicide is limited in the crop and directing the spray avoids actual contact with the sensitive crop parts. In the case of 2,4-D or dicamba in corn, protection of the growing point (apical meristem) of corn is the important factor.

Selectivity can also be gained by treating only weeds growing above the crop without contacting the crop below. Systemic herbicides, such as glyphosate, are used for this treatment. Specialized equipment (See Chapter 7) is employed to

Figure 3-8. A directed spray. Nozzles direct spray across the row, killing small weeds. The cotton stem is tolerant of aromatic oil, but the leaves are easily killed. (Delta Branch Experiment Station, Stoneville, MS.)

accomplish this. Unfortunately, allowing the weeds to remain in a field until they overtop the crop means significant competition has occurred prior to the treatment. However, treatment of perennials growing above the crop can reduce the weed population in a field. These applications can also help prevent harvest problems from weed infestations. Over-the-top herbicide treatments also cost little to apply.

A lay-by application is made with or following the last cultivation before it is impossible to move equipment through the field because of the crop size. Lay-by applications are often soil treatments intended to extend the period of residual weed control.

Area of Applications

Herbicides are applied *broadcast*, as a *band*, and as a *spot* treatment. Broadcast treatment or blanket application is uniform application to an entire area.

Band treatment usually means treating a narrow strip commonly directly over or in the crop row. The space between the rows is not chemically treated, but is usually cultivated for weed control. This method reduces the chemical cost, because the treated band is often one-third of the total area with comparable savings in chemical cost. In addition, when the chemical has a long period of residual soil toxicity (remains toxic in the soil for a long time), the smaller total quantity of the chemical reduces the residual danger to the succeeding crop.

Spot treatment is treatment of a restricted area, usually to control an infestation of weed species requiring special treatment. Soil sterilant treatments or nonselective herbicides (sacrificing any crop present) are often used on small areas of serious perennial weeds to prevent their spread.

DEVELOPING A WEED MANAGEMENT PROGRAM

Although this chapter discusses weed management tactics separately, designing a weed control program involves more than simply selecting weed control techniques. A weed management *program* integrates the weed management *tactics* into a long term *strategy* for dealing with weeds in a field. Both the short- and long-term impacts of the weed control system on the weed population should be weighed. In addition, environmental, cultural, economic, and management factors discussed below need to be considered in planning a weed control strategy.

Identify the Problem

All successful weed control programs begin with correct identification of the existing weed problem. It is impossible to judge the potential impact of various weed control tactics on the weed population without knowing the weed species present. The number of different weeds infesting a crop is relatively limited so proper weed species identification is not an impossible task. Check edges of the field and other areas of poor weed control to establish a list of weed species present. This can be done both early in the growing season and at harvest. Keeping good field records will aid not only the weed control program but the total crop production scheme. Records can include field maps, which show areas of weed infestations and note abundance of individual weeds. New weeds can be located and be considered for eradication efforts. The field records help prioritize the weed control needs with respect to the most economically damaging and troublesome weeds. It may be possible to treat only certain parts of a field with obvious savings in weed control costs. Other important records are notes on the yearly success of a weed management program. These will identify both tactics to retain and areas for change.

Crop Selection

The crop selected with impact both the available weed control tactics and the profitability of their use. However, crop selection will be primarily based on potential profit, history of crops previously grown and suitability of a crop to an area, rather than weed control considerations. Despite this, it is imperative to establish that effective weed management tools are available for the chosen crop. Attempts to grow crops for which it is impossible to control the existing weed population are almost certain to fail. The selected crop will determine weed control inputs available for cultural (row spacing, planting date), mechanical (primary, secondary, and cultivation tillages), biological, and chemical (herbicides available) weed control practices. Crop value will also set limits on the cost of weed management that can economically be justified.

Crop Rotation

Not only will the crop grown in the present year impact weed management options, but so will the future cropping intentions (rotations) for the same area. Farmers can use a crop for which good control tactics are available to reduce a weed population to low levels. This can allow later choice of a crop for which good weed control methods are not available. For example, johnsongrass, a serious perennial grass weed, is well controlled in soybeans. The high level of johnsongrass control achieved in soybeans allows subsequent production of a corn crop, in which effective johnsongrass control is difficult, in the same field. Other crop rotations, such as planting a alfalfa sod, can reduce weed populations through their competitive nature.

Crop rotation plans can also very strongly influence appropriate herbicide choices. Many herbicides can persist in the soil. This makes it unsafe to plant crops that are not tolerant of the herbicide(s) until enough time passes for the herbicide(s) to dissipate. More specific examples of "replanting" restrictions following herbicide use are discussed in Chapter 4.

Tillage Selection

The tillage, especially primary tillage, used will partially determine short- and long-term weed problems and suitable herbicide and other weed management options. Growers are adopting reduced tillage systems both by personal choice and because of government regulation. Reduced tillage can increase perennial and annual weed problems, limit cultivation use, and decrease herbicide options. This is especially true of herbicides that must be soil incorporated and that, therefore, cannot be used in no-tillage. Farmers will need to monitor weed populations to assess whether more intensive tillage is periodically warranted.

Herbicide Selection

Once the decision to use herbicides as part of the weed management program is made, several points should be considered before deciding which herbicide(s) to use.

1. Will the herbicide(s) adequately control the weed species present? This is one reason why correct knowledge of the weeds present is so important. Extension service information, herbicide labels, sales literature, and retailers can help identify the best herbicide choice(s). A number of states are also developing computer programs to aid in the herbicide selection process. It is also good to consider whether two or more herbicides applied separately or as a tank mixture are needed to adequately control the weeds.
2. Is the crop sufficiently tolerant of the herbicide? Herbicides are not generally marketed unless the crop will tolerate a herbicide application rate twice that needed to control susceptible weeds. This is called a 2 × safety factor. However, crop tolerance of various herbicides does differ. This is especially true under unfavorable environmental conditions for crop growth or if a high herbicide rate is used for hard to control weeds. The tolerance can also vary among crop varieties. It is necessary to rely on both past herbicide use experience and published information to assess the threat of yield reducing crop injury versus the benefit of herbicide control of weeds.
3. What are the crop rotation plans? As discussed under crop rotation above, it is important not to use a herbicide that will not allow the desired crop rotation. This limitation is primarily due to herbicide residues in the soil. Only when the weed problem cannot be managed in another way should use of herbicide that will limit cropping sequence flexibility be considered.
4. What is the danger of damage to nontarget plants? Movement of herbicides off the application site can potentially injure adjacent and far removed crops and other plants. Volatile (as a gas) herbicide movement is the prime danger but transfer with dust, soil, and water to untreated areas can also occur. This danger should be assessed when selecting particular herbicides and herbicide formulations. Formulation can have a large impact on volatility as discussed in

Chapter 7. Use of volatile herbicides is discouraged or outlawed to prevent damage to sensitive nontarget crops. For example, use of volatile 2,4-D formulations is outlawed in some states during the time from tobacco planting until harvest. Tobacco is very sensitive to damage from 2,4-D vapors.

5. Is the soil suited for the herbicide choice? Soil organic matter, clay content, and pH can all affect the toxicity and persistence of herbicides. Herbicides applied to soils low in organic matter and/or clay may be too toxic and damage crops. Alternatively, herbicides in these soils may leach (wash downward in the soil) too quickly for an effective length of weed control. The opposite extreme can occur on soils with high organic matter and/or clay where the herbicide is not available for weed control. These soils can require impractically high rates of soil-applied herbicides or reliance on foliar-applied (postemergence) herbicides for weed control.

Soil pH can also limit herbicide choices. Some herbicides such as the triazines are very quickly degraded under low soil pH while both triazines and sulfonylureas can be too persistent with high soil pH.

The herbicide label will outline herbicide–soil considerations and should be consulted for information.

6. Are there other environmental dangers from the herbicide use? There are both high-risk herbicides and high-risk areas of the United States for potential groundwater contamination (see Chapter 4). Farmers in these areas need to practice caution in their herbicide choices. Care should also be exercised if contamination of surface water with a herbicide is likely. Selection of herbicides should be restricted to those that are neither prohibited from application near water nor extremely toxic to aquatic organism.

7. Is the herbicide economic to use? It is assumed herbicide users will always weigh whether the potential increased economic return warrants herbicide use. This can be difficult to predict precisely for a soil application but models are being developed for decisions on the economics of postemergence herbicide use.

Management Level

Overriding all other decisions is the management level practiced on a farm. Is there enough time, equipment, and attention available to execute the planned program? Selection of weed management tactics that require very timely implementation over many acres is not realistic if neither the time nor equipment is available to treat the large area. Postemergence herbicide use generally requires more attention to developing weed problems, determination of the correct weed size to treat, and proper application techniques than do many soil-applied herbicides. Reliance on cultivation for weed control is similar to postemergence herbicide use. Soil-applied herbicides may be the best choice for farmers who cannot give the attention needed for postemergence herbicide use.

It is very important to match the weed management program to the time, skills, and equipment available to carry out the plan.

SUGGESTED ADDITIONAL READING

Altieri, M. A., and M. Liebman, 1988, *Weed Management in Agroecosystems*, CRC Press, Boca Raton, FL.

Anderson, W. P., 1983, *Weed Science: Principles*, 2nd Ed., West Publishing, St. Paul, MN.

Andres, L. A., 1981, Insects in the biological control of weeds, in D. Pimental, ed., *Handbook of Pest Management in Agriculture*, Vol. II, pp. 337–344, CRC Press, Boca Raton, FL.

Charudattan, R., and H. L. Walker, 1982, *Biological Control of Weeds with Plant Pathogens*, Wiley, New York.

Ennis, W. B. Jr., 1976, *Proc. World Soybean Conf.* 1975, pp. 375–386.

Hay, A. M., and D. C. Herzong, 1985, *Biological Control in Agricultural IPM Systems*, Academic Press, San Diego, CA.

Microbiological Control of Weeds Symposium. *Weed Science* **34**: Supplement 1. Weed Science Society of America, Champaign, IL.

Nalewaja, J. D., 1984, Do herbicides consume or produce energy? in *Energy Use and Production in Agriculture*, pp. 23–25, CAST Report 99. Cast, Ames, IA.

Papavizas, G. C., 1981, *Biological Control in Crop Production*, Allanheld, Osmun, Granada.

Ross, M. A., and C. A. Lembi, 1985, *Applied Weed Science*, Burgess Pub., Edina, MN.

Scrifes, C. J., 1981, Selective grazing as a weed control method, in D. Pimental, ed., *Handbook of Pest Management in Agriculture*, Vol. II, pp. 377–384, CRC Press, Boca Raton, FL.

Sprague, M. A., and G. B. Triplett, 1986, *No-Tillage and Surface-Tillage Agriculture*, Wiley, New York.

Vengris, J., M. Drake, W. B. Colby, and J. Patt, 1953, *Agron. J.* **45**, 213.

4 Herbicide Registration and Environmental Impact

REGISTRATION

Herbicides and other pesticides must be registered with the Environmental Protection Agency (EPA) before they are distributed or sold in the United States. To obtain the registration for a given product, the applicant must provide extensive data to show that it is effective and safe.

At the federal level, the Federal Insecticide, Fungicide, and Rodenticide Act (FIFRA) and several section of the Food, Drug, and Cosmetic Act (FDCA) provide the authority to regulate pesticides. States also require registration. When state registrations differ from federal registrations, they usually involve *special local needs* or *emergency* pest problems. State registrations may also contain more restrictive requirements that reflect differences in pest control needs or environmental conditions, e.g., lower rates. However, state registrations must be consistent with federal requirements, see 24(c) Section and Section 18 below.

In general, the FDCA involves the establishment of tolerances of pesticides in food and the FIFRA involves the control of pesticides. The regulations for the enforcement of FIFRA are administered by the EPA. These regulations are to assure that health and environmental concerns are balanced against the benefits of use and that the product is useful and poses no undue environmental or health hazard and has an appropriate label.

Pesticides are classified as general-use or restricted-use pesticides. General-use pesticides are relatively safe to use, whereas restricted-use pesticides are more toxic to humans and require a warning label, precautionary safety handling procedures, and a special permit for their use. The use of restricted-use pesticides requires that they be applied by a certified applicator and mixers, loaders, and applicators wear protective clothing.

Other Registration Procedures

In addition to general-use and restricted-use pesticide registrations. Experimental Use Permit (EUP) and special registration procedures [24(c) Section and Section 18] are provided for in the federal law for unique circumstances.

Experimental Use Permits Experimental Use Permits (EUP) are often used by companies during the final stages of the development of a pesticide to expand their data base on field use before full registration. The EUP allows them to use or sell a *limited* amount of the product in specific geographical areas. A food crop must be destroyed if no food tolerance has been established. Essentially all of the data required for full registration of the product must be presented to EPA to obtain an EUP.

24(c) Section The 24(c) Section of the federal regulation provides for state registration of pesticides for *special local need* by the designated state regulatory agency responsible for state registration. It allows for the minor use of a particular pesticide when this use is not registered on the label. No pesticide that has been canceled or suspended may be used under 24(c) registration. EPA must be informed of this action within 10 days. The specific use can continue and becomes part of the label *for that state only* if EPA does not object within 90 days.

Section 18 Section 18 of the federal regulation provides *emergency* exemption from registration of a pesticide. It may allow the use of an unregistered pesticide under emergency conditions. Such an emergency could be a serious unexpected pest problem that has no alternate method of control available, and significant economic or health problems would occur if the pest was not controlled. The pesticide is usually registered for other uses or approaching registration. A Section 18 exemption may be granted by EPA on a request by the designated state regulatory agency responsible for state registration. Section 18 exemptions are only for temporary use.

Information Required

Extensive information on the effectiveness and safety of a new pesticide must be provided before it can be registered for use. During the last two decades, an increasing emphasis has been placed on the safety aspects, especially fate in the environment and environmental impact. The research involved to provide this information is a major cost in the development of a pesticide. All costs, especially more stringent registration requirement, of putting a new herbicide on the market have increased markedly during the past several years. The cost of development of a new herbicide in a new area of chemistry is considered to range from 20 to 50 million dollars and usually requires a period of 6 to 10 years. These increasing costs partly explain the reason that many companies have discontinued the development of pesticides.

Although the type and amount of information required for federal regulation of a given pesticide are not limited, the data required at least includes the following: efficacy, general chemistry, environmental chemistry, crop residues, toxicology, fate in the environment, and environmental impact.

Efficacy involves the demonstration that the product is effective for the stated purpose in the field. *General chemistry* includes composition and analytical methods for the technical and formulated products and environmental and crop residues. *Environmental chemistry* includes field stability, rate of degradation, and degradation compounds formed. *Crop residues* include an analysis of food crops and animal-feed crops for the absence or amount present of the present compound applied and degradation products, where applicable. If there are detectable residues, a safe tolerance level must be established. Depending on the type of study conducted, EPA tolerance values are 100 to 1000 times less than the no-effect level in laboratory animals (Mitich, 1989). Expanded sections on toxicology and environmental impact including fate in the environment are covered later in this chapter.

Toxicology *Toxicology* involves the determination of toxicity values of the parent compound and its formulations from experimental animals. These are very helpful in assessing the potential toxic effects on humans and other animals. These studies include acute toxicity, subacute toxicity, and chronic toxicity. Toxicity is the inherent capacity of a known amount of a substance to produce injury or death. In addition to low-dosage and potential-exposure levels, high-dosage levels are also used to establish an *effect*. In addition to usual experimental laboratory animals, toxicological data from soil microorganisms and wildlife including birds, fish, and other aquatic organisms may also be required. The toxicology of degradation products may also be required. First-aid and diagnostic information is also required.

A biological system's response to chemical exposure is dose related. At a low dosage rate, the chemical may serve as a stimulant; at a moderate dosage rate, the chemical may have no effect; whereas at a high dosage rate, the chemical may act as a serious biological depressant. At high rates of application, the animal biological system may completely fail in its normal response. Thus high dosage levels may cause biological responses completely different from those caused by low levels of chemical exposure. There may be no biological relationship between dosage levels required to produce an *effect* and exposure levels experienced in normal usage.

"Safety in use" is a function of toxicity and exposure (exposure level and length of time of exposure). It is a total estimate of the hazards of use. Thus "safety in use" usually indicates safety or hazard better than toxicity data alone.

Toxicology Terms. LD is *lethal dose* and LD_{50} is the dose that will kill 50% of a population of test animals. LC is *lethal concentration* and LC_{50} is the concentration that will kill 50% of the animals. *Acute oral* refers to a single dose taken by mouth or ingested. *Acute dermal* and *skin effects* refer to a single dose applied directly to the skin. *Inhalation* refers to exposure through breathing or inhaling. *Eye effects* refers to a single dose applied directly to the eye. Acute oral LD_{50} and acute dermal LD_{50} values are expressed in terms of milligrams of the substance per kilogram (mg/kg) of body weight of the test animal. LC_{50} values are expressed in terms of milligrams of the substance, as a mist or dust, per liter (mg/liter) of air.

A toxicity category is assigned to every pesticide according to the criteria in Table 4-1. All labels must contain a Signal Word and state. "Keep Out of the

TABLE 4-1. Toxicity Categories for Pesticides

	Hazard Indicators				
Category	Oral LD_{50} (mg/kg)	Inhalation LC_{50} (mg/liter)	Dermal LD_{50} (mg/kg)	Eye Effects	Skin Effects
I	50 or less	0.2 or less	200 or less	Corrosive, corneal opacity, not reversible within 7 days	Corrosive
II	51 to 500	0.21 to 2.0	201 to 2000	Corneal opacity, reversible within 7 days, irritation persisting for 7 days	Severe irritation at 72 hours
III	501 to 5000	2.1 to 20	2001 to 20,000	No corneal opacity, irritation reversible within 7 days	Moderate irritation at 72 hours
IV	>5000	>20	$>20,000$	No irritation	Mild or slight irritation at 72 hours

Reach of Children." The Signal Word for category I is DANGER POISON (fatal), category II is WARNING (may be fatal), and category III and category IV is CAUTION.

Most herbicides (>90%) have a relatively low toxicity to higher animals and are in category III or IV. Those most toxic and in category II (based of acute oral LD_{50}) include bromoxynil, copper sulfate, cyanazine, diallate, difenzoquat, diquat, endothall (amine), and paraquat. No herbicides are in category I.

Label

The label must show trade name, registrant name, active ingredient (name and amount), inactive ingredient (amount), use classification (general or restricted), net weight or measure of contents, EPA registration number, registration number of formulation plant, directions for use, a signal word, and a warning or precautionary statement. The signal word and warning and precautionary statements are mainly concerned with toxicological, environmental, physical, and/or chemical hazards. The registered label is a legal document that permits the applicant to distribute and sell the product. Furthermore, no one can use or recommend the use of the product in a manner that is not consistent with the label instructions.

Minor Crops

A minor crop is one of limited acreage. Minor crops include most horticultural and speciality crops, e.g., vegetable crops, small fruits, tree fruit and nut crops, and ornamentals.

Since the market is small and the costs of development and registration of a pesticide for use on a minor crop can exceed the potential financial return, chemical companies are reluctant to develop a pesticide specifically for a minor crop. Therefore, most herbicides used in minor crops were previously developed for a major crop. This may result in minor crops being somewhat less tolerant to herbicides than major crops. Herbicides must be used with greater care in minor crops. Normally at least a twofold safety factor is desired; twice as much herbicide is required to induce crop injury as is needed for weed control. An additional fact that reduces a company's interest in registering herbicides for minor crops is the risk/profit ratio. Since minor crops are high-value crops, the liability from crop injury can be high relative to the profit that can be made on the limited amount of chemical sold.

Even though these problems exist, chemical companies do obtain registration for their herbicides on minor crops, although, usually only when they have been previously registered for use on a major crop. Much of the information used for registration of a pesticide on a major crop can also be used for registration on a minor crop, e.g., chemistry, toxicology, environmental impact.

Another approach to the registration of pesticides involves Interregional Research Project No. (IR-4). This project coordinates and supports federal and state research directed toward the development of the information required for the registration of a pesticide for minor crops and other minor uses by EPA. It started as a small USDA regional project but has increased greatly in size, funding, and use in recent years. IR-4 is supported by the federal government but state involvement is also an integral part of the total program. Once the need is established, a coordinated program is established between federal and state research personnel to develop the required efficacy and crop residue data. Chemical company information on other required aspects such as toxicology and environmental impact from existing registrations of the pesticide also are used. All of this information is assembled into a petition that IR-4 submits to EPA. On EPA's approval, the use becomes legal and is registered.

ENVIRONMENTAL IMPACT

The environment is infinitely complex and includes the totality of the land, air, and water that surround us including their interactions. Many climatic, edaphic, biotic, and social factors influence an ecological community and its organisms. These factors ultimately determine an organism's welfare and survival. Natural disasters such as hurricanes, floods, and earthquakes have tremendous environmental impacts on humans and other organisms. However, increasing interest and emphasis are being given to *human activities* as they relate to an altered environment. These activities include almost everything from the construction of a shopping center to the use of pesticides. Many of these require an *environmental impact statement* that must be approved by some level of government before the activity can proceed.

Weeds per se, nonchemical weed control methods, and many standard agricultural practices have environmental impact. The impact of weeds may be beneficial or detrimental. The beneficial effects can include reduced soil erosion and silting of streams and increased wildlife habitat and natural beauty. When a plant normally considered to be a weed has a beneficial effect, it probably should not be considered to be a weed. The detrimental effects include interference with human activities and a reduction in the aesthetics of the environment. Cultural practices for land preparation and weed control may contribute to environmental degradation by increasing wind and water erosion of soil. For example, the plowing and planting of large areas of the Great Plains not suited to conventional farming contributed to the degraded environment and resulting human suffering of the dust bowl era.

The physical and molecular fate of herbicides in the environment is essential information for determining their environmental impact. The fate of herbicides in plants (Chapter 5) and soil (Chapter 6) is discussed in greater detail later. Obviously, a herbicide cannot have an environmental impact unless it enters the environment. Herbicides enter the environment on application and almost always have some kind of impact. The major impact is usually a beneficial biotic response, controlling the target weed species without any detrimental effects.

However, we are primarily interested here in discussing the *potential* hazards of herbicides in the environment.

In general, herbicides have presented little or no problems on the intended target area. However, problems may arise when extensive movement of a herbicide away from the target site occurs. These are usually caused by movement of the herbicide from the application site due to drift (spray or volatility), leaching, or soil erosion by wind or water runoff.

Environmental concerns involving herbicides are mainly related to human health. However, other major concerns include damage to off-site vegetation, including natural vegetation, and detrimental effects on wildlife.

Man's Health

A report from the Council for Agricultural Science and Technology (CAST, 1987), Health Issues Related to Chemicals in the Environment: A Scientific Perspective, is recommended reading for this topic. It includes discussions of chemical effects and their significance, risk of exposure to chemicals, and chemicals in groundwater and drinking water. It was prepared by university, federal, state, and industry personnel.

Most herbicides are relative nontoxic to humans because their action at the molecular level is usually at a site that is not common to higher animals, e.g., photosynthesis or plant-specific enzymes. Furthermore, all chemicals developed for herbicide use in recent years have a low mammalian toxicity. The few older herbicides with higher mammalian toxicity, category II, are either being rapidly phase out or are classified as restricted pesticides and require special handling.

However, all chemicals, synthetic and natural, are toxic and should be handled with due caution. Even aspirin and table salt (sodium chloride) have significant oral LD_{50} values, 1.2 and 3.3 g/kg, respectively, and a toxicity category of III.

The greatest health hazards of herbicides is to people that handle or are otherwise exposed to large quantities, e.g., industrial manufacturing, formulation, and distribution personnel and those involved in applying them in the field: applicators, mixers, loaders, and aircraft flagmen. To the best of our knowledge, herbicide-induced injury to farm workers has not occurred from entering treated areas or handling a commodity from a treated area.

Indirect exposure of the general public to low levels of a herbicide could occur by the ingestion of contaminated food or water. In general, these indirect exposures to herbicides present little hazard because the levels of exposure and mammalian toxicity are low. Absence or safe levels of herbicides in food or animal feed is assured by residue analysis of these products as established by the registration procedures.

Pesticides in Natural Waters An adequate water supply is one of the nation's most precious resources and it must be maintained safe for all plants and animals, including humans (Messersmith, 1988). A great deal of attention has been given to pesticides in natural waters and numerous publications have dealt with this topic including *Pesticides in Soil and Water* (Guenzi, 1974), *Ecology of Pesticides* (Brown, 1978). *The Use and Significance of Pesticides in the Environment* (McEwen and Stephenson, 1979), *Health Issues Related to Chemicals in the Environment: A Scientific Perspective* (CAST, 1987), and *Groundwater Contamination by Herbicides* (Messersmith, 1988).

The CAST (1987) report includes a discussion of the Safe Drinking Water and Toxic Enforcement Act of 1986. The purpose of this Act is to protect the public health against possible harmful effects of chemicals in drinking water. It requires EPA to specify the contaminants that may have any adverse effect on public health and control their concentrations within safe levels.

The great interest in this topic stems from the fact that *trace* amounts of pesticides have been detected in both surface and groundwater. *Surface water* includes stream, rivers, and lakes and *groundwater* is the water that occurs in the earth below the water table (Figure 6-2). The pesticides most likely to become contaminants of natural waters are those used in large quantities. When detected, even these are usually found in concentrations below 1 part per billion (CAST, 1987). However, occasionally they have be found at concentrations up to 20 to 50 parts per billion.

Surface Water. The surveillance of surface water for pesticides has been conducted for many years (Guenzi, 1974). In 1957, the Public Health Service established surveillance stations on the major rivers and the Great Lakes. During the 1960s, the Department of Interior implemented a program for continuous monitoring of major streams. State programs have also been established. More recently, the EPA has become involved.

Water runoff from pesticide-treated land is the major source of pesticide contamination of surface waters. However, some surface water contamination may occur by lateral movement through shallow groundwater. Local contamination can occur from pest control procedures, e.g., mosquito and aquatic weed, where the pesticide is applied directly to surface water and when water retention procedures are inadequate.

Groundwater. Within the past few years traces of agricultural chemicals have been detected nation wide in groundwater (Mitich, 1989). These include fertilizers, insecticides, and herbicides. The potential for groundwater contamination has been studied by Cohen et al. (1984) and the related issues and problems of groundwater quality reported by the National Research Council (1986). Other relevant publications are cited in the introduction to this section, Pesticides in Natural Waters.

The depth of groundwater and the time required for surface water to reach the groundwater pool varies with climatic and geologic conditions. Depending on these conditions, the upper boundary of groundwater may range from a few feet to hundreds of feet below the soil surface. The time for surface water to reach these depths may range from a few days to centuries. Since pesticides are generally bound and/or degraded as they past through the soil profile with water, shallow groundwater has a greater potential for pesticide contamination than deep groundwater.

Groundwater contamination is of particular concern because water from this source is pumped for domestic, agricultural, and industrial use. Leaching of pesticides with water from pesticide-treated land is the major source of pesticide contamination of groundwater (Figure 6-2). Factors affecting the leaching of herbicides through soil are discussed in Chapter 6.

Mitich (1989) states, "trace amounts of pesticides (in ground water) do not necessarily constitute a health hazard. Despite media claims to the contrary, health effects of pesticides are documented carefully, and in their calculations, EPA leaves a generous safety factor."

Off-Site Vegetation

The major source of an adverse effect of herbicides to off-site vegetation is from drift, spray or volatility. When it does occur, it is usually near the area treated and could have been prevented by a more careful application or use of a nonvolatile formulation (Chapter 7). However, under unusual climatic and/or topographic conditions drift injury to off-site vegetation may occasionally occur some distance away. These include atmospheric inversion layers in valleys and alternating land–sea air flows. Leaching of relatively persistent herbicides into the rooting zone of trees outside the treated area has caused tree injury.

Wildlife

The major adverse effects of herbicides to wildlife are indirect since herbicides are relatively nontoxic to higher animals. These indirect adverse effects are primarily related to the removal of vegetation that provide food and habitat for wildlife. However, in some cases the vegetative shifts induced by herbicides can be beneficial to certain species. Deer populations often increase when native grasses and forbs replace heavy brush stands removed by herbicides. Brown (1978), Guenzi (1974), and McEwen and Stephenson (1979) provide considerable information on the effects of herbicides on wildlife including terrestrial and aquatic vertebrate and invertebrate fauna, insects, and microorganisms.

SELECTED ADDITIONAL READING

Brown, A. W A., 1978 *Ecology of Pesticides*, Wiley, New York.

CAST, 1987, *Health Issues Related to Chemicals in the Environment: A Scientific Perspective*. Council for Agricultural Science and Technology, Ames. IA.

Cohen, S. Z., R. F. Carsel, S. M. Creeger, and G. G. Enfield, 1984. Potential for pesticide contamination of ground water resulting from agricultural uses, in F. F. Krueger and J. N. Seiker, Eds., *Treatment and Dispersal of Pesticide Wastes*, pp. 297–325, American Chemical Society, Washington, D.C.

Guenzi, W. D., Ed., 1974, *Pesticides in Soil and Water*, Soil Science Society of America Inc., Madison, WI.

McEwen, F. L., and G. R. Stephenson, 1979. *The Use and Significance of Pesticides in the Environment*, Wiley, New York.

Messersmith, C. G., Ed., 1988, Symposium on Ground Water Contamination by Herbicides, *Weed Tech.* **2**, 206.

Mitich, L. W., 1989, *Agribusiness Fieldman*, p. 6, May.

National Research Council, 1986, *Pesticides and Groundwater Quality: Issues and Problems in Four States*, National Academy Press, Washington, D.C.

5 Herbicides and the Plant

When a herbicide comes in contact with a plant, its action is influenced by the morphology and anatomy of the plant as well as numerous physiological and biochemical processes that occur within the plant. These processes include (1) absorption, (2) translocation, (3) molecular fate of the herbicide in the plant, and (4) effect of the herbicide on plant metabolism. The interaction of these plant factors with the herbicide determines the effect of a specific herbicide on a given plant species. When one plant species is more tolerant to the chemical than another plant species, the chemical is considered to be *selective*. The first four topics will be discussed, followed by a discussion of selectivity.

The life processes of plants are many and varied; they are complex and delicately balanced. Disturb one of these processes, even only slightly, and a chain of events may be initiated that changes the plant's growth and development. Minor changes in the environment may also result in major changes in the life processes of a plant. For example, many perennial plants remain dormant below ground all winter. When the soil temperature increases a few degrees in the spring, a complex series of reactions is initiated that results in the beginning of another annual cycle.

There are different concepts of the terms *mode of action* and *mechanism of action* of herbicides within the scientific community. However, the National Academy of Science book entitled "Weed Control" (Anon., 1968) stated "The term 'mode of action' refers to the entire sequence of events from introduction of a herbicide into the environment to the death of plants. 'Mechanism of action' refers to the primary biochemical or biophysical lesion leading to death."

The following terms will also be used repeatedly in the subsequent discussion of how herbicides kill plants: (1) *herbicide*—a chemical that kills or inhibits growth of plants, (2) *contact herbicide*—a herbicide that causes injury only to tissue to which it is applied, (3) *mobile herbicide*—a herbicide that moves or translocates in a plant, (4) symplast—total living protoplasmic continuum of a plant; it is continuous throughout the plant and there are no islands of living cells; the phloem is a component of the symplast and long distance symplastic translocation is via the phloem and, (5) *apoplast*—total nonliving cell-wall continuum of a plant; the xylem is a component of the apoplast and long distance apoplastic translocation is via the xylem.

ABSORPTION

A herbicide must enter the plant to be effective. Some plant surfaces absorb a herbicide readily, but others absorb a herbicide slowly, if at all. The chemical nature of the herbicide is also involved. Therefore, differential absorption or selective absorption may account for differences in plant responses.

The two most common sites of entry in plants are the leaves and the roots. In addition, some chemicals are effectively absorbed through stems, including coleoptiles or young shoots as they grow through treated soil. Seeds also absorb herbicides.

Leaf Absorption

Initial leaf penetration may take place either through the leaf surface or through the stomates. The volatile fumes of some herbicides and some solutions enter through the stomates. However, the direct penetration through the leaf surface is of greater importance. For this route, the herbicide must first penetrate the cuticle, which is a nonliving layer that covers plant leaves and stems. Greater cuticular penetration occurs at preferential absorption sites (e.g., thin cuticle) at trichome bases and anticlinal walls. The cuticle is not homogeneous. Externally it is primarily composed of wax and internally it is primarily composed of cutin. There is a continuous gradation of the wax into the cutin. At the cuticle–cell-wall interface, the cutin is in contact with the pectin, and the pectin is in contact with the cellulose of the cell wall.

There is a gradual transition in the polar nature of the cuticle–cell-wall complex from the surface wax to the cellulose. The cuticular wax is the most nonpolar or hydrophobic (water hating), followed by cutin, pectin, and cellulose, respectively. Cellulose is the most polar or hydrophilic (water loving). Therefore, polar herbicides have considerable difficulty entering the cuticular wax, but once they pass this barrier they enter each succeeding phase more readily. In contrast, nonpolar herbicides readily enter the cuticular wax but have increasing difficulty passing into each succeeding phase. Thus the polar nature of the herbicide may have considerable influence in its rate of absorption. The challenge is to design herbicides that are neither too polar nor nonpolar to achieve optimum penetration.

Any substance that will bring the herbicide into more intimate contact with the leaf surface should aid absorption. Surfactants increase leaf absorption by (1) reducing interfacial tension to give better "wetting," (2) modifying the wax-like and oil-like substances of the cuticle, and/or (3) reducing drying of the spray droplets (Hodgson, 1982). The amount of amitrole absorbed by been leaves within 24 hours after application was 13% without surfactant but 78% with surfactant (Freed and Montgomery, 1958). Leaf surfaces differ in their wetting ability as the result of differences in the composition of cuticular waxes and specialized structures such as trichomes of leaf hairs (Figure 5-1).

The addition of surfactant tends to enhance foliar herbicide absorption in all

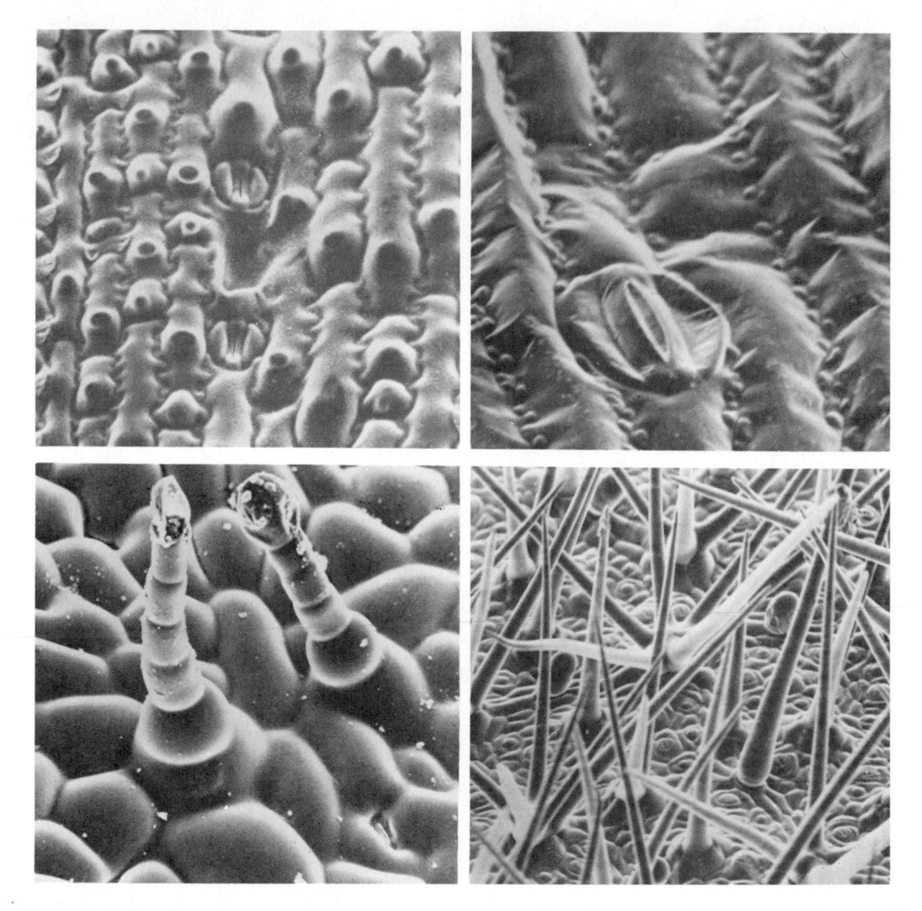

Figure 5-1. Leaf surfaces as they appear on a scanning electron microscope. *Upper left*: Bermudagrass (450 ×). *Upper right*: Nutsedge (1050 ×). *Lower left*: Redroot pigweed (350 ×). *Lower right*: Velvetleaf (170 ×). (D. E. Bayer and F. D. Hess, University of California, Davis.)

types of plants. Therefore, a surfactant may reduce the selectivity of the herbicide if that selectivity is dependent on selective foliar absorption.

Increases in temperature and humidity, within the physiological range, usually increase foliar absorption. High humidity is probably more important than temperature per se. The increased humidity response is mainly related to increased hydration of the cuticle, but decreased drying time of the spray droplets also has a role. The basis for temperature effects alone has been proposed to lie in increased fluidity of the cuticle components. Drought stress due to high temperature may actually reduce absorption. In many areas, these two environmental factors increase concurrently.

Root Absorption

Roots absorb many herbicides from the soil. Comparative studies have shown that roots absorb some herbicides (monuron, simazine) very rapidly and others (dalapon, amitrole) more slowly (Crafts and Yamaguchi, 1964). Some herbicides are absorbed passively (monuron), whereas others (2,4-D) appear to be absorbed actively, requiring on expenditure of energy (Donaldson et al., 1973). However, it is generally considered that the vast majority of herbicide absorption by roots is passive.

The primary region of herbicide absorption by roots is 5 to 50 mm behind the root tip. In this region the xylem is sufficiently differentiated to be functional but the casparian strip is not sufficiently lignified to pose a serious barrier.

Herbicides appear to enter roots by three routes: apoplast, symplast, and apoplast–symplast. The apoplast route involves movement exclusively in cell walls to the xylem. In fully differentiated tissue, this route appears to require that the herbicide pass through the casparian strip. The symplast route involves initial entry into cell walls and then into the protoplasm of the cells of the epidermis, cortex, or both. The herbicide remains in the protoplasm and sequentially passes into the endodermis, stele, and phloem by means of cellular interconnecting protoplasmic strands (plasmodesmata). The apoplast–symplast route is identical to the symplast route except for the fact that the herbicide may reenter the cell walls after bypassing the casparian strip and may then enter the xylem. Although certain herbicides may be restricted to one route of entry, others may enter by more than one route. The chemical and physical properties of each herbicide primarily determine which route is followed.

Under most conditions, there is rapid translocation of herbicides upward from roots in the xylem (transpiration stream), but only limited upward transport in the phloem. Therefore, the entry of foliar active herbicides (e.g., PSII inhibitors) into the xylem is more important than entry into the phloem for root-absorbed herbicides.

Shoot Absorption

Some soil-applied herbicides may be absorbed by young shoots as they develop and grow upward through the soil following germination of seeds. Emerging shoots (coleoptiles) are very important sites of uptake for certain herbicides by grass weeds. For example, the young shoots of barnyardgrass are the main site of EPTC uptake and also the prime site of injury (Dawson, 1963). Soil applied herbicides that are effective in this manner include carbamothiolates, dinitroanilines, and chloroacetamides.

Stem Absorption

Direct application of herbicides to stems of plants is not a common practice except for control of woody plants. However, a foliar-applied herbicide may come in contact with the plant stem. Generally, foliar absorption is much more important than stem absorption because of the greater surface area and more permeable cuticle of leaves.

In stem treatment of woody plants the herbicide may be applied to (1) the bark, usually for small trees or brush, (2) cuts through the bark (cut-surface method) for the control of large trees, and (3) stumps of cut trees or brush to prevent sprouting. Additional information is given in Chapter 22, Woody Plant Control.

TRANSLOCATION

Translocation of herbicides is of major importance in control of weeds. It is particularly important for perennial weeds with belowground reproductive organs. Herbicides are translocated within the plant through the symplastic system and/or the apoplastic system. Some herbicides are primarily translocated in the symplastic system, others in the apoplastic system, and still other in both systems (Table 5-1).

Herbicides move within the plant much the same as endogenous solutes. Figure 5-2 illustrates the movement of herbicides through the symplastic and apoplastic systems.

Symplastic Translocation

When applied to leaves, symplastic mobile herbicides follow the same pathway as sugar formed there by photosynthesis. Such herbicides move from cell to cell in the leaf via the plasmodesmata and/or via the apoplast until they enter the phloem. Then they move out of the leaf and move both up and down via the phloem, accumulating in those areas where sugar is being used for growth. Growth is most rapid in the apical growing point, expanding young leaves, rapidly elongating stems, developing fruits and seeds, and root tips. The direction of flow is governed by the location of leaves and growing points. Upward flow (acropetal) is greatest from mature leaves near the shoot apex and downward flow (basipetal) is predominant from basal leaves. Bidirectional flow occurs from intermediately located leaves.

Translocation through the phloem appears to involve a mass flow of solution. One explanation of this driving force is the difference in tugor pressure between the photosynthetic cells and the cells utilizing the products of photosynthesis, primarily sugar. Photosynthetic cells (high pressure) are often called the *source* and using cells (low pressure) the *sink*. This movement is through the living plasmodesmata and phloem. Herbicides with great acute toxic properties kill these tissues and block transport of the herbicide and photosynthates in the symplastic system. Crafts and Crisp (1971), Hess (1985), and Giaquinta (1985) provide additional information on phloem or symplastic translocation.

Effective use of foliar-applied herbicides to kill the underground parts of perennial weeds largely depends on the maintenance of active photosynthesis and well functioning phloem cells. Translocation of the chemical to belowground

TABLE 5-1. Relative Mobility and Primary Translocation Pathway(s) of Herbicides[1,2]

Good Mobility			Limited Mobility			Little or No Mobility
Apoplast	Symplast	Both	Apoplast	Symplast	Both	
Bentazon	Glyphosate	AMA	Bromoxynil	Oxadiazon	AMS	Bensulide
Carbamothloates		Amitrole	Chloroxuron	Phenoxys	Aryloxyphenoxys	DCPA
Carbamates[3]		Arsenicals[4]	Difenzoquet		Chloramben	Dinitroanilines
Chloroacetamides		Asulam	Diquat		Endothall	Diphenylethers
Diclobenil		Clopyralid	Ethofumesate		Fenac	
Diphenamid		Dalapon	Fluridone		Naphthalam	
Methazole		Dicamba	Paraquat		Propanil	
Napropamide		Fosamine				
Norflurazon		Imidazolinones				
Pyrazon		Pyridines				
TCA		Sethoxydim				
Triazines		Sulfonylureas				
Uracils						
Urea[5]						

[1] Translation rate may vary considerably in different species. Some herbicides may also move from the symplast to the apoplast and vice versa.
[2] When a class of herbicides is given, see class section in herbicide chapters for individual herbicides.
[3] Except for asulam, which has good mobility in both the symplast and apoplast.
[4] Organic arsenicals.
[5] Except for chloroxuron, which has limited mobility in the apoplast.

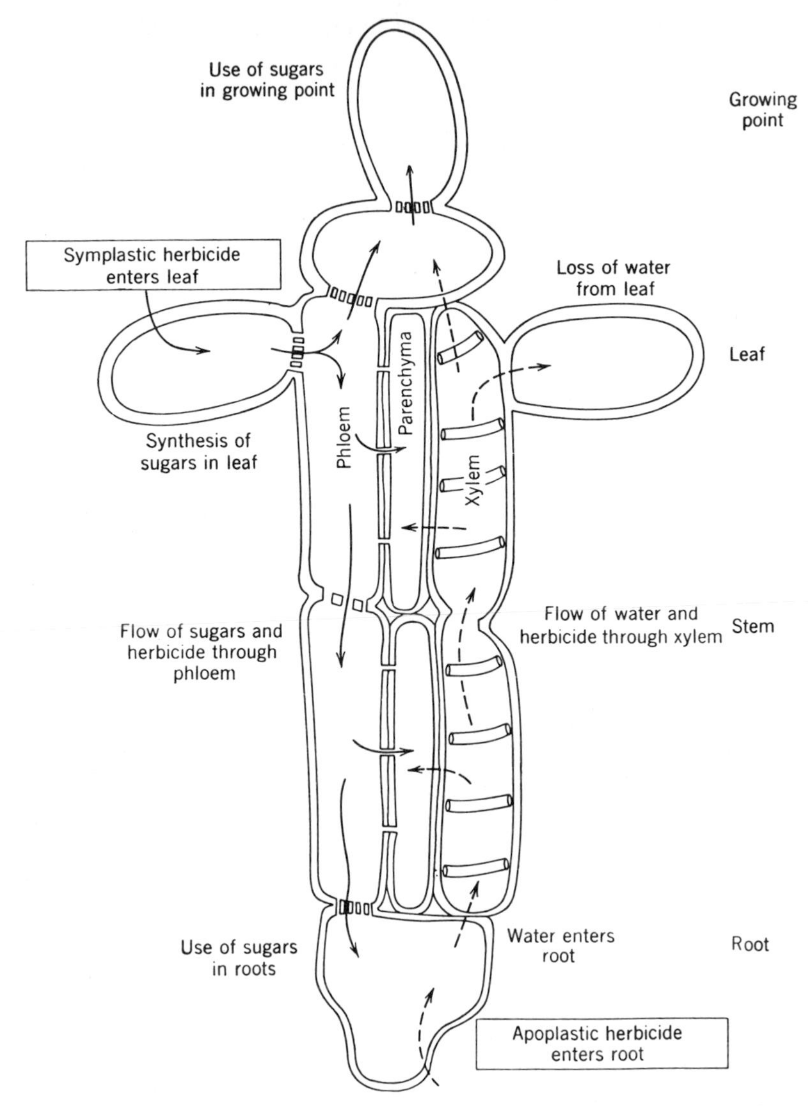

Figure 5-2. Diagram representing routes of translocation of herbicides in plants. (Adapted from Bonner and Galston, 1952.)

parts is most rapid and most effective when large amounts of photosynthates are being moved into these organs. This usually occurs after full leaf development, after sexual reproduction, and/or in the fall during storage of overwintering reserves. An excessive rate of the herbicide may inhibit photosynthesis and/or injury the phloem. Either effect will probably reduce or block translocation of the herbicide into the underground parts resulting in poor weed control. Even though the tops may die, the plant will usually reestablish itself by resprouting from underground buds. Such excessive rates are almost always greater than the recommended rates.

Apoplastic Translocation

Apoplasticly mobile herbicides that are absorbed by roots follow the same pathway as water. They enter the xylem and are swept upward with the transpiration stream of water and soil nutrients. The driving force for this movement is the removal of water from leaves by transpiration.

The xylem, cell walls, and intercellular space are the principal components of the apoplastic system. They are considered to be nonliving when fully differentiated and functioning in the system. Therefore, even high concentrations of very phytotoxic herbicides can be absorbed from the soil and quickly translocated to all parts of the plant. Absorption and translocation may continue for some time even though the herbicide has killed the root.

Interaction of Symplastic and Apoplastic Translocation

Translocation of most herbicides is not restricted to either the symplast or apoplast, but may involve both systems. However, many herbicides appear to be limited primarily to one or the other system. For example, 2,4-D is translocated mainly in the symplast whereas diuron is translocated mainly in the apoplast. Amitrole is readily translocated in both systems and actually appears to circulate in the plant.

As the herbicide moves through the long-distance transport channels, xylem or phloem, some may move into adjacent cells. This may occur by simple diffusion or active uptake. The herbicide may then move from these adjacent cells into the other long-distance transport channel, xylem or phloem, and be translocated in it. Pesticides that are mobile in both the xylem and phloem have been referred to as *ambimoble.*

MOLECULAR FATE

The molecular fate of a herbicide concerns the changes in the chemical structure of the herbicide within the plant. This process has also been referred to as degradation. Most of these changes reduce the phytotoxicity of the molecule and are referred to as *inactivation.* The conversion of simazine to hydroxysimazine is an example of inactivation. However, some changes increase the phytotoxicity of the molecule and are referred to as *activation.* The conversion of 2,4-DB to 2,4-D is an example of activation. Since different species of plants have varying degrees of ability to modify the chemical structure of a herbicide, this difference often determines their tolerance to a given herbicide. Some plants can inactivate a herbicide so rapidly that they are not injured (e.g., atrazine in corn). Conversely, some plants activate a herbicide so slowly that they are not injured (e.g., 2,4-DB in alfalfa).

Many herbicides are not only degraded but may form conjugates before or after an initial modification. These conjugates are formed by a reaction between the herbicide or slightly modified herbicide and a normal plant constituent (e.g., sugars, amino acids).

Higher plants have been shown to alter the molecular configuration of herbicides by a wide variety of chemical reactions. Most of these reactions are probably catalyzed by enzymes; however, some of these appear to be nonenzymatic. In most cases the specific enzyme or enzymes have not been isolated and characterized.

Metabolism of herbicides by plants can occur via a three-phase process (Shimabukuro, 1985). Phase I reactions (oxidation, reduction, or hydrolysis) are initial reactions that generally detoxify herbicides and predispose the resultant metabolite to conjugation (Phase II reactions) with sugars, amino acids, or other natural plant constituents. Phase III reactions are unique to plants and consist of secondary conjugation reactions or the formation of insoluble bound residues. In general, plants "immobilize" these metabolites whereas animals excrete them.

An example of the molecular fate of a herbicide in higher plants is given in Figure 5-3. This figure shows that 2,4-D undergoes a number of Phase I and Phase II reactions. The Phase I reactions include dechlorination, chlorine shifts, hydroxylations, decarboxylation, and dealkylation. The Phase II reactions include conjugation of 2,4-D and/or its metabolites with sugar (glucose) and amino acids (aspartic and glutamic acids). Although Phase III reactions are not illustrated in this figure, insoluble 2,4-D residues in higher plants have been reported by a number of investigators. Additional information on the molecular fate of herbicides in higher plants is available in Ashton and Crafts (1981), Hatzios and Penner (1985), and Kearney and Kaufman (1975, 1976, 1988).

PLANT METABOLISM

Plant metabolism includes the numerous biochemical reactions that occur in the protoplasm of living plant cells. Although most of these take place in all cells (e.g., respiration), some occur only in specific cells such as photosynthesis in cells containing chlorophyll. A given herbicide may initially interfere with a single biochemical event or be relatively nonspecific and interfere with several reactions simultaneously. For example, diuron inhibits a specific step in the electron transport chain of photosynthesis, whereas dalpon is nonspecific and has multifold effects. Biochemical reactions are closely coupled and often when one reaction is altered by a herbicide others are soon affected.

Biochemical processes that may be affected by herbicides are photosynthesis, respiration, carbohydrate metabolism, lipid metabolism, protein metabolism, and nucleic acid metabolism (Ashton and Crafts, 1981). Certain aspects of the

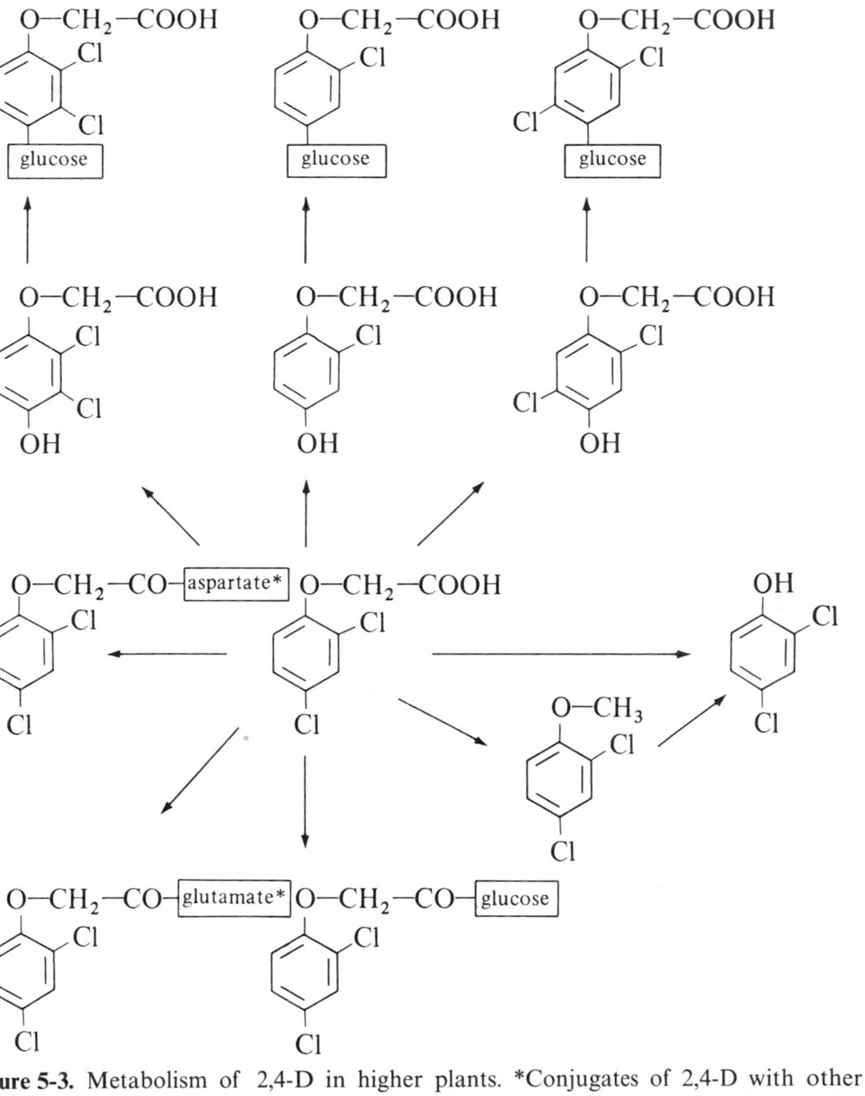

Figure 5-3. Metabolism of 2,4-D in higher plants. *Conjugates of 2,4-D with other amino acids (namely, alanine, valine, leucine, phenylalanine, and tryptophan) have also been reported in soybean callus tissue. (Ashton and Crafts, 1981.)

above processes and others that are associated with cell division and/or cell-wall biosynthesis may also be affected by herbicides. By disrupting any of these biochemical reactions, herbicides may upset plant metabolism and thus injure or kill the plant.

Determining the mechanism of action of herbicides is difficult, at least in part, because even normal plant metabolism is not thoroughly understood. However, considerable progress in understanding how herbicides kill plants has been made in the last two decades due to increased interest and research on this topic. Such studies have usually shown that the herbicide interferes with a number of biochemical events rather than a single reaction. However, the PSII inhibitors have been known to act at a specific biochemical site for some time. More recently, the sulfonylurea and imidazolione herbicides' interference with plant growth has been shown to be triggered by inhibition of a single biochemical reaction (Ray, 1985; Shaner et al., 1985). Additional information on the mechanism of action of herbicides can be found in books by Ashton and Crafts (1981), Audus (1976), Duke (1985), Fedtke (1982), Kearney and Kaufman (1975, 1976, 1988), and Moreland et al. (1982).

GROWTH AND PLANT STRUCTURE

Normal growth and plant structure are the result of previous normal biochemical or biophysical processes. Therefore, abnormal growth and plant structure caused

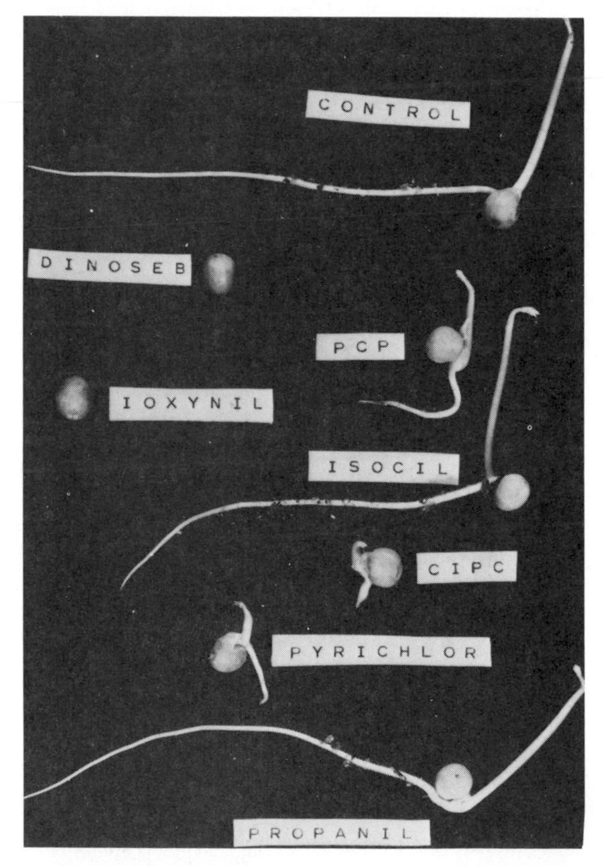

Figure 5-4. Effect of several herbicides on seed germination and root elongation in peas. (van Hoogstraten, 1972.)

Figure 5-5. Grape leaves exposed to 2,4-D. *Left*: Low level. *Right*: High level.

by herbicides must be preceded by altered biochemical or biophysical processes induced by the herbicide.

Herbicides may induce abnormal plant growth through morphological, anatomical, and cytological effects. However, these effects are generally specific for a given herbicide on a particular plant species. Thus a given abnormal growth symptom in the field often suggests which herbicide, class of herbicide, or no herbicide induced the injury. This has become increasingly important with time as it is related to herbicide injury litigation.

Some of the abnormal responses induced by herbicides include (1) seed germination failure (Figure 5-4), (2) leaf chlorosis, (3) abnormal leaf form (Figure 5-5), (4) stem swelling, (5) root swelling or inhibited elongation (Figure 5-4), (6) cell division inhibition, and (7) chloroplast destruction (Figure 5-6).

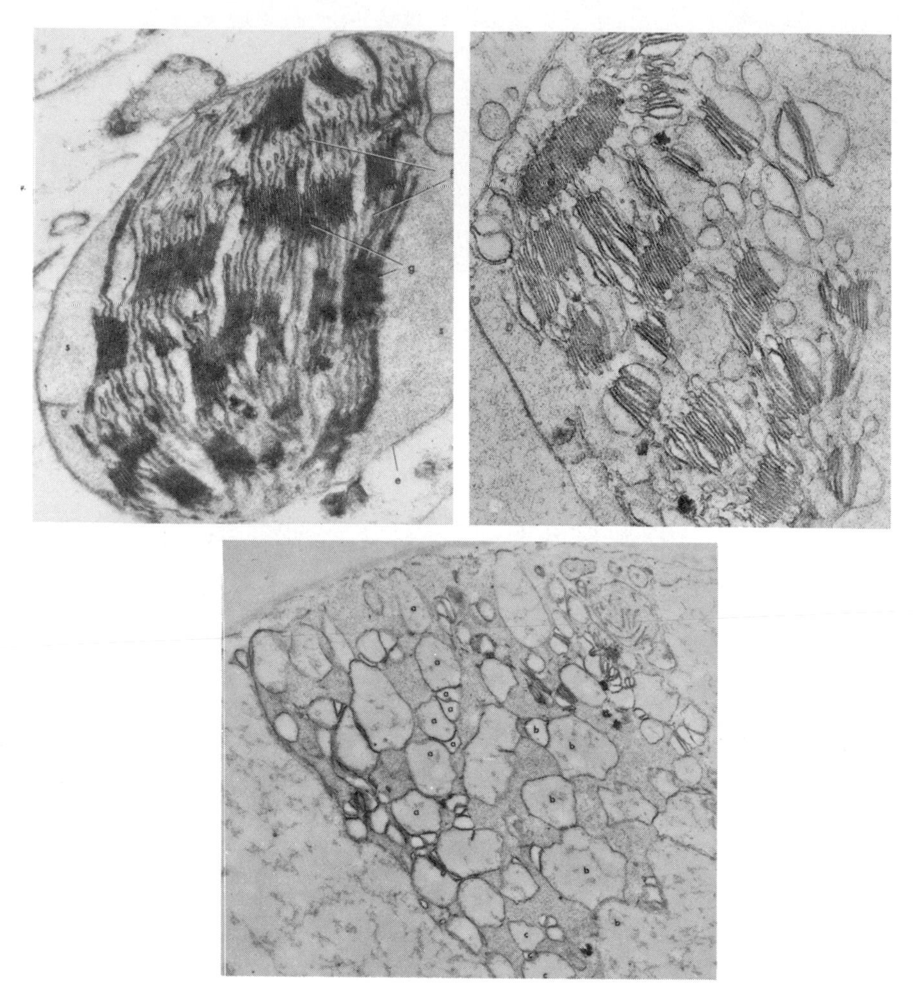

Figure 5-6. Effects of atrazine on the ultrastructure of bean chloroplasts. *Upper left*: Control. *Upper right*: Moderate injury. *Lower*: Severe injury.

SELECTIVITY

Selectivity is perhaps the most important concept of modern weed science. Due to this phenomenon weeds can be controlled in crops with herbicides. This is the primary basis for the extensive use of herbicides today.

A selective herbicide is one that kills or significantly retards growth of an unwanted plant species (weed) without significantly damaging a desired plant species (crop). Retarding weed growth long enough for the crop to become dominant is almost as effective as killing the weed. However, weeds that are only retarded may have a slight competitive affect and may produce seeds that could cause problems in subsequent crops.

A herbicide is selective to a particular crop only within certain limits. These limits are determined by various factors involving a complex interaction between *plants*, the *herbicide*, and the *environment* (Ashton and Harvey, 1987). Because of the numerous variables involved, selectivity is relative not absolute. Crop injury or lack of weed control may occur if the herbicide is used improperly. Selectivity is dependent on using the proper rate with the prescribed method of application and under the appropriate environmental conditions.

Role of the Plant

Seven factors affect plant response (both weeds and crop) to a chemical: age, growth rate, morphology, physiology, biophysical processes, biochemical processes, and inheritance.

Age Younger plants have more penetrable cuticle and a greater relative amount of metabolically active tissue than older plants. Metabolically active tissue is more susceptible to herbicide injury than metabolically inactive tissue. Thus the age of a plant often determines it response to a particular herbicide; young plants are less tolerant to herbicides than older plants. Preemergence treatments that kill germinating weed seeds or seedlings often have little or no effect on established weeds.

Growth Rate The growth rate of plants can have a pronounced effect on their reaction to some herbicides. In general, fast-growing plants are more susceptible to injury from a herbicide treatment than slow-growing plants. However, there are exceptions to this general rule. Even normally tolerant plants (crop) whose growth has been limited by stress conditions such as low temperature or high soil moisture can be injured by an otherwise noninjurous herbicide treatment. Conversely, even normally susceptible plants (weed) whose growth has been limited by stress conditions such as drought are more tolerant of herbicide injury.

Morphology The type of morphology of a plant can determine whether it is killed by a specific herbicide. These morphological difference are found in root systems, location of growing points, and leaf properties.

Root Systems. Annual weeds in some perennial crops that regenerate from adventitious roots (e.g., alfalfa) can be controlled with nonselective contact herbicides. The crop will recover from moderate injury to the above ground parts, whereas annual weeds will be killed. Differential depth of rooting of a crop and a weed in conjunction with the position of the herbicide in the soil can also be the basis of selectivity between two species.

Leaf Properties. Certain leaf properties protect crops treated with herbicides. Liquid spray droplets can adhere to only a small portion of the surface of a narrow, upright leaf or a waxy leaf surface, or a leaf surface with small ridges. The sprays bounce off when they hit such leaves reducing the effect of the herbicide. Broadleaf plants usually have wide, smooth leaf surfaces, extending horizontally from the plant stem. Such leaves intercept and hold more of the spray than upright leaves and are controlled. The same sprays on cereals or onions tend to bounce off and the crop plants are uninjured.

Location of Growing Points. The initial growing point of a cereal is located at the base of the plant and is protected by the surrounding leaves. Any contact spray that remains on cereals may injure the leaves, but will not injure the growing point. In contrast, most broadleaf plants have exposed growing points at the tips of the shoots and in the leaf axils. These growing points are directly exposed to the chemical spray and killed. Annual broadleaf weed and perennial broadleaf weeds can be killed or controlled, respectively, in cereals by this means.

Physiology The physiology of the plant determines how much herbicide is absorbed by the plant and the extent its translocation within the plant. The absorption and translocation of a herbicide vary from species to species. Generally, plants that absorb and translocate the greatest amount of herbicide will be killed. Detailed information on absorption and translocation was previously presented in this chapter.

Biophysical Processes Three biophysical differences may determine whether a plant is killed: adsorption (nonspecific), binding (specific), and membrane stability.

Adsorption. Nonspecific adsorption of herbicides by plant–cell constituents inactivates some herbicides. Studies have shown that movements of some herbicides are slowed down by surrounding plant tissues. In extreme cases, a herbicide may be bound so tightly to some plant constituents that it is not translocated from the point of application to the site of action. It may even be bound so tightly that it is unavailable for herbicidal action.

Binding. Lack of specific binding of a herbicide at its biochemical site of action results in resistance of certain weed biotypes (LeBaron and Gressell, 1982). For example, different biotypes of groundsel, pigweed, and other weeds are susceptible or resistant to control by atrazine depending on whether atrazine is bound to a specific biochemical site in the choroplast.

Membrane stability. Tolerance of crops of the carrot family to oils is one of the oldest examples of biophysical selectivity. Selective oils used for weed control in these crops kill weeds by damaging the cell membranes and allowing the cell sap to flow into the intercellular spaces. Initially the leaves have a water-soaked appearance; later the cells die and tissues become dehydrated. The cellular membranes of members of the carrot family are resistant to this type of damage.

Biochemical Processes Certain differential biochemical reaction among plant species may protect some of them from injury by certain herbicides. These reaction are referred to as inactivation and activation.

Herbicide Inactivation. Metabolism of a herbicide into a nonphytotoxic form is herbicide inactivation. When this metabolism occurs rapidly toxic levels of the herbicide do not accumulate and the plant is not injured. However, when it does not take place or occurs slowly toxic levels of the herbicide accumulate and the plant is killed. The tolerance of corn and the susceptibility of many weeds to atrazine are examples of selectivity based on differential herbicide inactivation. The basis of herbicide antidotes or safeners usually involves accelerated herbicide inactivation by the crop.

Herbicide Activation. Metabolism of a relatively nonphytotoxic chemical into a phytotoxic form is herbicide activation. For example, the relatively nonphytotoxic compound 2,4-DB is rapidly converted into 2,4-D by susceptible species such as cocklebur. However, this reaction occurs only very slowly in the tolerant crops such as alfalfa.

Inheritance The genetic complement of a plant determines it capacity to respond to its environment. These responses vary from genus to genus, but, within a genus, plant reactions to a given herbicide tend to be similar. Notable exceptions are biotypes of certain species. For example, the common biotype is susceptible to a specific herbicide whereas within the population there is a resistant biotype that is not controlled by the same herbicide (e.g., common groundsel-atrazine). The percentage of the resistant biotype in the native population is initially very low, but after several years' use of the same herbicide on a field the resistant biotype becomes dominant. This is because the susceptible biotype has been controlled and the resistant biotype has increased to fill the space. The development of a dominant resistant biotype population can be prevented by not using the same herbicide on the same field year after year.

However, there are also positive aspects of the resistant biotype phenomenon. Through application of this knowledge, it is potentially possible to develop crops that are tolerant to a specific herbicide. This can be accomplished by transferring the gene responsible for resistance to a crop. The approach is to use conventional breeding methods in closely related species or genetic engineering. The latter is a very active area of research by chemical companics and independent molecular biology companies. The book by LeBaron and Gressel (1982) presents comprehensive coverage of herbicide resistance.

Three broad forces interact to determine selectivity of any herbicide. One is the plant, just discussed. The other two are the herbicide and the environment.

Role of the Herbicide

The various aspects of herbicides relative to selectivity include molecular configuration, concentration, formulation, chemical combinations, and how a herbicide is used.

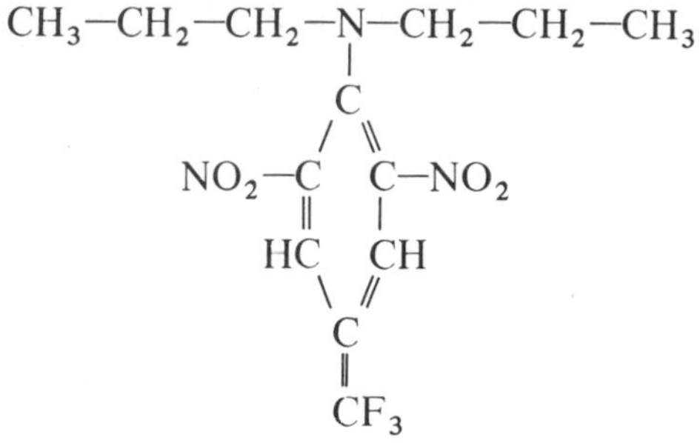

$CH_3-CH_2-CH_2-N-CH_2-CH_2-CH_3$
|
C
/ \\
NO_2-C $C-NO_2$
‖ |
HC CH
\ //
C
‖
CF_3

trifluralin

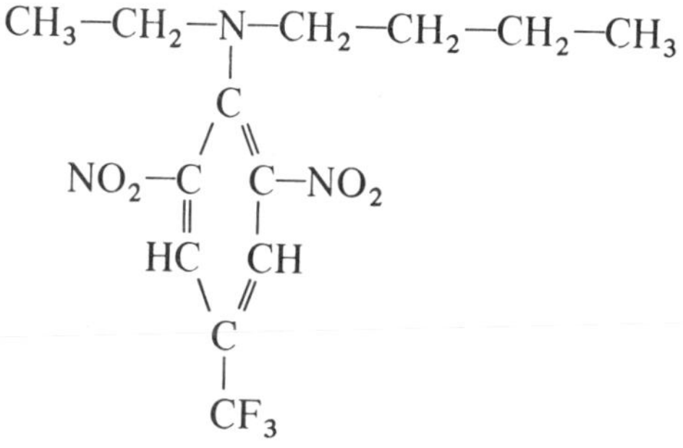

$CH_3-CH_2-N-CH_2-CH_2-CH_2-CH_3$
|
C
/ \\
NO_2-C $C-NO_2$
‖ |
HC CH
\ //
C
|
CF_3

benefin

Figure 5-7. Chemical structures of trifluralin and benefin are quite similar; however, there are major differences in plant selectivity.

Molecular Configuration Variations in molecular configuration of a herbicide change it properties, which in turn modify its effect on plants. This is illustrated by Figure 5-7, which shows the herbicides trifluralin and benefin. The only difference is that a methyl ($-CH_2-$) group is moved from one side of the molecule to the other side. Trifluralin kills lettuce, but benefin will not injure lettuce at the recommended rate. Chemical structures of herbicides are modified during discovery and development by chemical companies to alter phytotoxicity and selectivity.

Concentration Concentration may determine whether a herbicide inhibits or stimulates metabolism and growth of a plant. The endogenous plant growth regulator indole 3-acetic acid (IAA) inhibits respiration and growth at high concentrations but stimulates them at low concentrations. In many ways, including these processes, 2,4-D acts like IAA.

Formulation The formulation of a herbicide is vital in determining whether it is selective or not with regard to a given species. Perhaps the most striking example is the granular form that permits the herbicide to "bounce off" the crop and fall to the soil. Substances known as adjuvants and surfactants are often added to

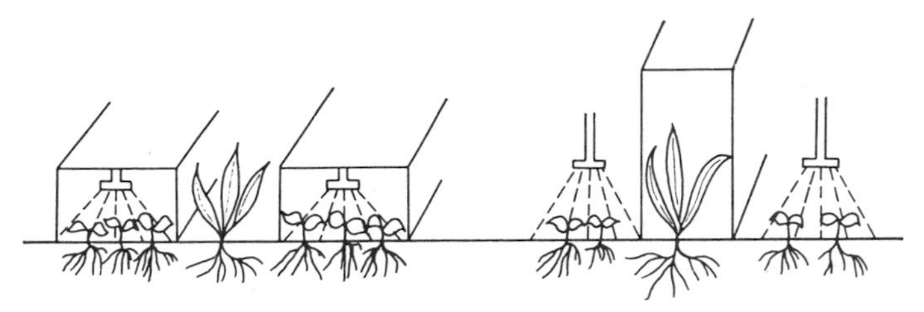

Figure 5-8. Shielded sprays protect crops from being sprayed with herbicides. *Left*: Confine the spray within shields. *Right*: Cover the crop with shields.

improve the application properties of a liquid formulation; these additives may increase or decrease phytotoxicity. The addition of nonphytotoxic oils or surfactants to liquid atrazine or diuron formulations induces foliar contact activity in these normally soil-active herbicides.

The addition of herbicide antidotes, safeners, or protectants (synonymous terms) to formulations is used to increase the tolerance of a crop to a specific herbicide (e.g., dichlormid in carbamothiolates for corn). Antidotes are also used as seed treatments to protect milo from acetanilide injury (e.g., flurazole, oxabetrinil). The book by Pallos and Casida (1978) provides information on antidotes.

Chemical Combinations Herbicides are often mixed with fertilizers, fungicides, insecticides, nematicides, or other herbicides to facilitate application. Occasionally, tank mixing of herbicides with other herbicides, pesticides, or fertilizers can alter selectivity. Therefore, it is advisable to only use such combinations when they are specifically recommended on the product label.

How the Herbicide Is Used A herbicide can be applied so that most of it covers the weed but little of it contacts the crop. This can be done by using shielded or directed sprays or wick-wiper applicators.

Shielded sprays prevent the herbicidal spray from touching the crop while the weeds are covered with the spray. The spray nozzles are simply placed under a hood or the crop is covered with a shield (Figure 5-8). *Directed sprays* are less

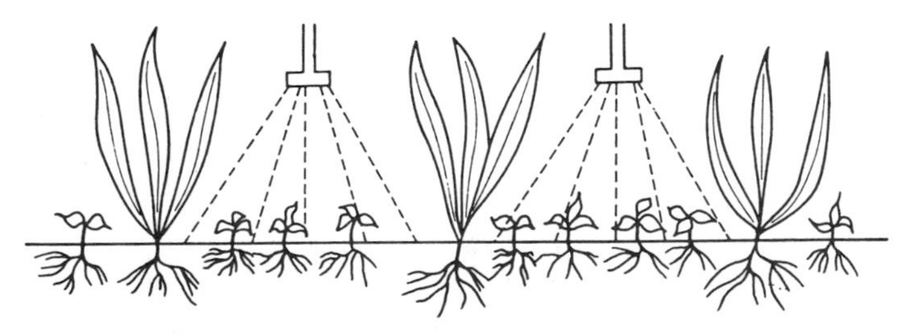

Figure 5-9. Directed sprays are aimed toward the base of the crop plant, favoring minimum coverage of crop and maximum coverage of weeds.

cumbersome than shielded sprays and can be used when the crop can tolerate a small amount of the herbicide. This is accomplished in row crops by using drop nozzles and nozzles that minimize spray drift, carefully controlling the nozzle height and direction of the nozzles, and when the crop is taller than the weeds, e.g., 2,4-D in corn (Figure 5-9). *Wick-Wiper* applicators are used where the weeds are higher than the crop. The herbicide solution is wiped on the weeds; very little, if any, herbicide contacts the crop (Figure 7-11).

Role of the Environment

Dominant environmental factors that affect selectivity include soil type, rainfall or overhead irrigation, and soil–herbicide interactions. Details of soil–herbicide interactions are presented in Chapter 6.

In general, herbicide characteristics, soil type, and the amount of water received after herbicide application from rainfall or overhead irrigation determine the vertical position of a specific herbicide in the soil. Adsorption, the tenacity with which a herbicide molecule is bound to soil particles, will strongly affect its movement in soil. Low adsorption, high water solubility, high amounts of overhead water, and course soil types favor leaching of the herbicide in the soil profile. Some herbicides are extremely resistant to leaching, whereas others readily move with water. This movement is normally downward, but the herbicide may move upward as water evaporates from the soil surface.

Some herbicides not inherently selective may become selective when they are in specific vertical positions in the soil profile (Figure 5-10). Such selectivity depends on different rooting habits of crop and weed. A herbicide that readily leaches below the rooting zone of a shallow-rooted crop can be used to control deep-rooted weeds without injuring the crop. Conversely, a herbicide that

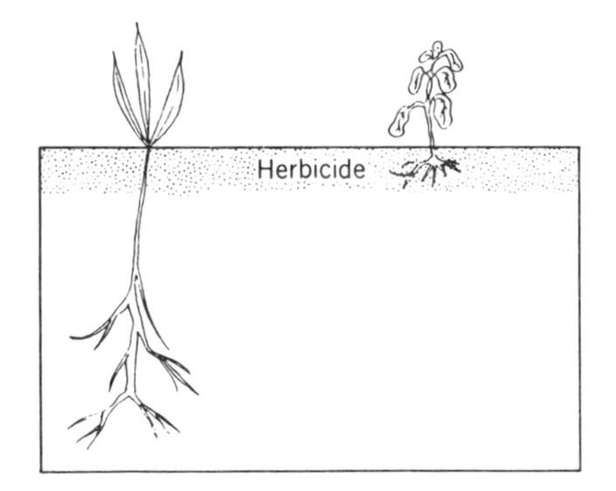

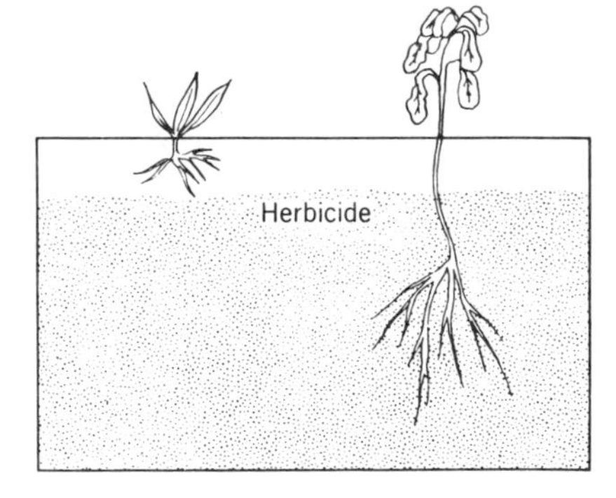

Figure 5-10. Differential leaching by herbicides alters selectivity. *Left*: A herbicide that remains near the soil surface can injure shallow-rooted weeds (*right*) and not injury the deep-rooted crop (*left*). *Right*: A herbicide that leaches from the soil surface into the lower soil profile can injure deep-rooted weeds (*right*) and not injury a shallow-rooted crop (*left*).

remains near the soil surface can control shallow-rooted weeds in a deep-rooted crop.

The temperature of the environment in which a plant is growing has considerable influence on the rate of its physiological and biochemical processes. For example, the optimum temperature for germination of seeds of different species varies greatly, e.g., spinach 41°F and cantaloupe 77°F. The selectivity of various plants to herbicides also varies as the temperature differentially affects their processes.

The effect of temperature on these processes is often expressed the temperature coefficient (Q_{10}). A Q_{10} of 2 means that the rate is doubled with an increase of 10°C. Most chemical reactions have a Q_{10} between 2 and 3, whereas physical reactions have a Q_{10} only slightly greater than 1. Most reactions of herbicides that influence plant growth are chemical in nature. Therefore, a change from 15°C (59°F) to 25°C (77°F) may result in doubling the activity of the herbicide.

SUGGESTED ADDITIONAL READING

Anon., 1968, *Principles of Plant and Animal Pest Control*, Vol. 2, *Weed Control*, Pub. 1975. National Academy of Science, Washington, D.C.

Ashton, F. M., and A. S. Crafts, 1981 *Mode of Action of Herbicides*, Wiley, New York.

Ashton, F. M., and W. A. Harvey, 1987, *Selective Chemical Weed Control*, Bulletin 1919, Univ. of California, Div. of Agric. and Natural Resources, Berkeley.

Audus, L. J., 1976, *Herbicides*, Vols. 1 and 2, Academic Press, New York.

Bonner, J., and A. W. Galston, 1952, *Principles of Plant Physiology*, Freeman, San Francisco.

Crafts, A. S., and C. E. Crisp, 1971, *Phloem Transport in Plants*, Freeman, San Francisco.

Crafts, A. S., and S. Yamaguchi, 1964, *The Autoradiography of Plant Materials*, Division of Agricultural Science, University of California, Berkeley.

Dawson, J. H., 1963, *Weeds* **11**, 60.

Donaldson, T. W., D. F. Bayer, and O. A. Leonard, 1973, *Plant Physiol.* **52**, 638.

Duke, S. O., 1985, *Weed Physiology*, Vols. 1 and 2, CRC Press, Boca Ratan, FL.

Fedtke, C., 1982, *Biochemistry and Physiology of Herbicide Action*, Springer-Verlag, New York.

Freed, V. H., and M. Montgomery, 1958, *Weeds* **6**, 386.

Giaquinta, R. T., 1985, Physiological basis of phloem transport of agrichemicals, in P. A. Hedin, Ed., *Bioregulators for Pest Control*, pp. 7–18, American Chemical Society Symposium Series 276, American Chemical Society, Washington, D.C.

Hatzios, K. K., and D. Penner, 1982, *Metabolism of Herbicides in Higher Plants*, Burgess, Minnesota.

Hatzios, K. K., and D. Penner, 1985, Interactions of herbicides with other agrochemicals in higher plants, in J. S. Bannon, Ed., *Reviews of Weed Science*, Vol. 1, pp. 1–63, Weed Science Society of America, Champaign, IL.

Hess, F. Dan, 1985, Herbicide absorption and translocation and their relationship to plant tolerances and susceptibility, in S. O. Duke, Ed., *Weed Physiology*, Vol. 2, pp. 191–214, CRC Press, Boca Ratan, FL.

Hodgson, R. H., 1982, *Adjuvants for Herbicides*, Weed Science Society of America, Champaign, IL.

Kearney, P. C., and D. D. Kaufman, 1975, 1976, 1988, *Herbicides*, Vols. 1, 2, and 3, Dekker, New York.

LeBaron, H. M., and J. Gressel, 1982, *Herbicide Resistance in Plants*, Wiley, New York.

Meggitt, W. F., R. J. Aldrich, and W. C. Shaw, 1956, *Weeds* **4**, 131.

Moreland, D. E., 1980, *Annu. Rev. Plant Physiol.* **31**, 597.

Moreland, D. E., J. B. St. John, and F. D. Hess, 1982, *Biochemical Responses Induced by Herbicides*, American Chemical Society Symposium Series 181, American Chemical Society, Washington, D.C.

Pallos, F. M., and J. E. Casida, 1978, *Chemistry and Action of Herbicide Antidotes*, Academic Press, New York.

Pfister, K., S. R. Radosevich, and C. J. Arntzen, 1979, *Plant Physiol.* **64**, 995.

Ray, T. B., 1985, *Proc. Br. Crop Prot. Conf. Weeds*, 131.

Shaner, D., M. Stidham, M. Muhitch, M. Reider, and P. Robson, 1985, *Proc. Br. Crop Prot. Conf. Weeds*, 147.

Shimabukuro, R. H., 1985, Detoxication of herbicides, in S. O. Duke, Ed., *Weed Physiology*, Vol. II, pp. 215–244, CRC Press, Boca Ratan, FL.

van Hoogstraten, S. D.,1972, Ph.D. dissertation, University of California, Davis.

6 Herbicides and the Soil

Numerous soil factors, many different kinds of herbicides, and large numbers of plant species and climatic variations make the study of herbicides in soils very complex and diverse. There are at least 10 different soil variables of major importance, 150 different herbicides, and hundreds of different plant species involved. Thus the complexity of herbicide–soil–weather–plant interactions is enormous. Comprehensive reviews of herbicide and pesticide behavior in soils can be found in Hance (1980) and Saltzman and Yaron (1983).

Herbicides are applied directly to the soil as (1) *preplanting* treatments and (2) *preemergence* treatments. The time of application may refer to the crop or to the weed.

Some preplanting treatments are mechanically mixed into the soil, whereas others are applied to the surface. When mechanically incorporated into the soil, the chemical is usually immediately effective on seeds germinating in the area–with no added moisture. When applied to the soil surface, most herbicides must be moved into the soil by water to be effective. This is generally referred to as *activation.*

The success of an incorporated preplanting treatment or a preemergence treatment depends largely on the presence of a high concentration of the herbicide in the upper 2 inches of soil. This is where most annual-weed seeds germinate. Also, there must be a relatively low concentration of the herbicide in the zone where the crop seeds germinate, unless the crop seed is tolerant to the chemical (see Figure 6-1).

The herbicide-treated area may remain weed free long after the chemical has dissipated if all the initially germinating weed seedlings are killed, no further viable weed seeds sprout, and the soil is not disturbed by tillage. This happens because most weed seeds will not germinate if buried deeply in the soil.

For effective soil sterilization, the chemical must remain active in the rooting zone to kill both germinating seeds and growing plants.

PERSISTENCE IN THE SOIL

The length of time that a herbicide remains active or persists in the soil is extremely important because it determines the length of time that weed control can be expected and the length of time that a chemical is present in the environment. A certain amount of persistence is usually desirable for adequate weed control. Residual toxicity is also important because it relates to phytotoxic

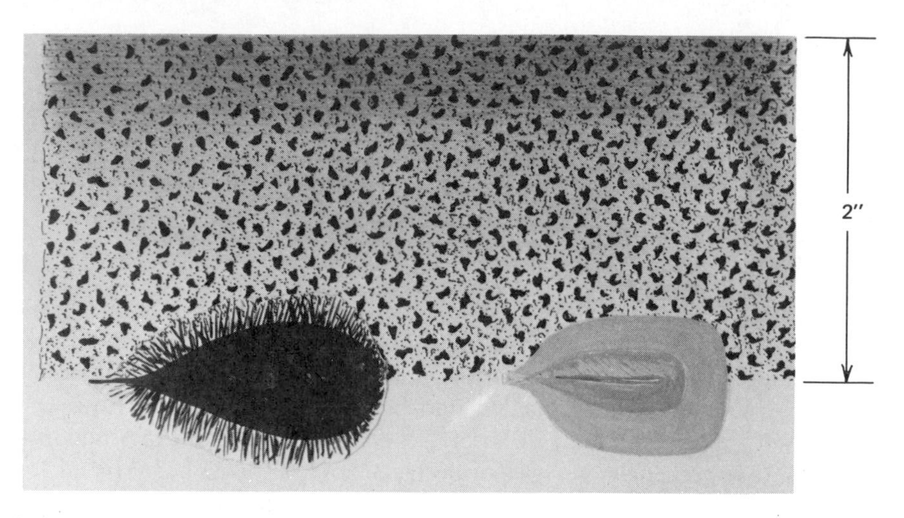

Figure 6-1. Most preemergence type herbicides require soil incorporation to be effective in the absence of adequate rainfall or overhead irrigation. Soil incorporation places the herbicide in the area of the soil profile where most of the weed seeds germinate, upper 2 inches. (North Carolina State University.)

aftereffects (*carryover*) that may prove injurious to subsequent crops or plantings. In such cases, excessive persistence may restrict crop rotation options available to the former and could cause groundwater contamination or other environmental problems.

Factors that affect the persistence of a herbicide in the soil are classified as either degradation processes or transfer processes (Weber et al., 1973). The degradation processes that break down herbicides and change their chemical composition are (1) biological decomposition, (2) chemical decomposition, and (3) photodecomposition. Kearney and Kaufman (1975, 1976, 1988) prepared reviews on the *degradation of herbicides* and they are suggested for additional reading. Transfer processes important in determining what happens to herbicides in the soil are (1) adsorption by soil colloids, (2) leaching or movement through the soil, (3) volatility, (4) surface runoff, (5) removal by higher plants when harvested, and (6) absorption and exudation by plants and animals (see Figure 6-2). Each factor is discussed later in the chapter.

Table 6-1 gives approximate lengths of persistence for various herbicides under a given set of conditions. This table is based on experimental work and general observations. In general, the persistence values are based on conditions favorable for rapid herbicide decomposition. Those herbicides persisting 1 month or less may be used to control weeds present at the time of treatment. Those persisting 1 to 3 months will protect the crop only during a short period early in the growing season. This is generally adequate for many annual row crops that produce a dense canopy and thereby suppress weed growth through shading. Those providing 3 to 12 months of control may provide protection to the crop for

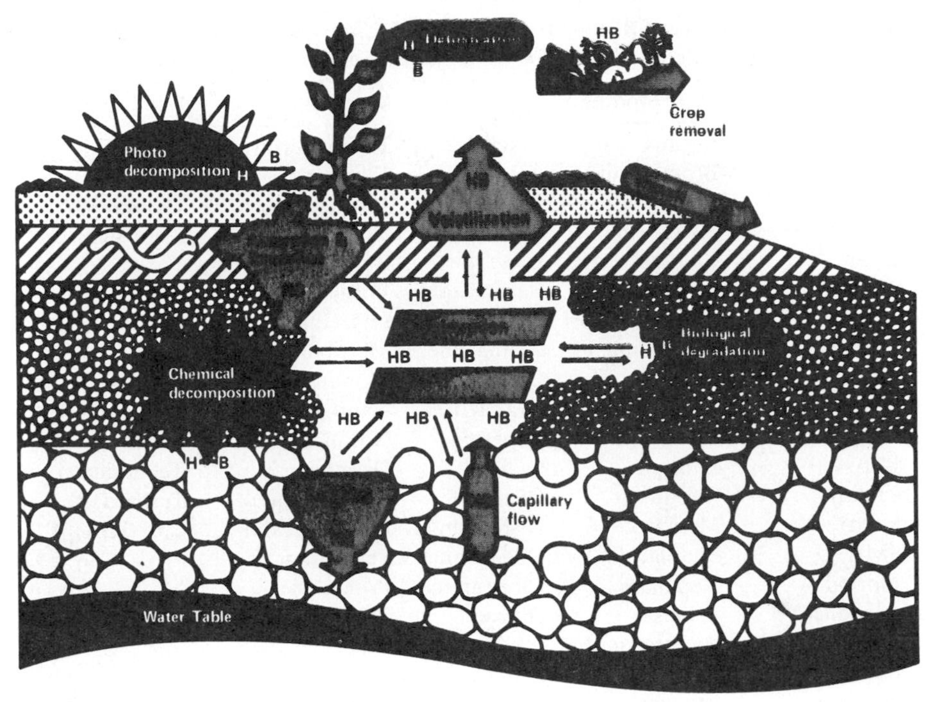

Figure 6-2. Processes influencing the behavior and fate of herbicides in the environment. Transfer processes are characterized by the herbicide molecules (HB) remaining intact. Degradation processes are characterized by splitting of the herbicide molecules [H ("space") B]. (Weber et al., 1973.)

the entire growing season and are useful in perennial crops such as orchards and vineyards. Those providing more than 12 months of control are used primarily for total vegetation control in noncrop situations where persistence is desirable (see Figures 6-3 and 6-4).

Degradation Processes

Biological Decomposition Biological decomposition of herbicides includes detoxification by soil microorganisms and higher plants.

Microbial Decomposition. Microbial decomposition will be discussed first. The principal microorganisms in the soil are algae, fungi, actinomyces, and bacteria. They must have food for energy and growth. Organic compounds of the soil provide this food supply, the exception being a very small group of organisms that feeds on inorganic sources.

Microorganisms use all types of organic matter, including organic herbicides. Some chemicals are easily decomposed (easily utilized by the microorganisms) whereas other resist decomposition. The microorganisms use carbonaceous

TABLE 6-1. Persistence of Biological Activity at the Usual Rate of Herbicide Application in a Temperate Climate with Moist-Fertile Soils and Summer Temperatures[1]

1 Month or Less	1–3 Months	3–12 Months[2]		Over 12 Months[3]
Acifluorfen	Bifenox	Alachlor	Fluometuron	Borates
Acrolein	Bromoxynil	Ametryn	Fluridone[4]	Bromacil
Amitrole	Butachlor	Atrazine	Hexazinone	Chlorates
AMS	Butylate	Benefin	Isopropalin	Chlorsulfuron
Barban	Chloramben	Bensulide	Imazamethabenz	Fenac
Bentazon	Chlorpropham	Buthidazole	Imazaquin	Fluridone[5]
Benzadox	Cycloate	Chlorimuron	Imazethapyr	Hexaflurate
Cacodylic acid	Desmedipham	Clomazone	Metribuzin	Imazapyr
Chloroxuron	Diallate	Clopyralid	Monuron	Karbutilate
Dalapon	Diphenamid	Cyanazine	Napropamide	Picloram
2,4-D	EPTC	Cyprazine	Norflurazon	Prometon
2,4-DB	Linuron	DCPA	Oryzalin	Tebuthiuron
Diclofop	Mecoprop	Dicamba	Oxyfluorfen	Terbacil
Diquat[6]	Methazole	Dichlobenil	Pendimethalin	2,3,6-TBA
DSMA	Metolachlor	Difenzoquat	Perfluidone	
Endothall	Naptalam	Dinitramine	Pronamide	
Fluorodifen	Pebulate	Diuron	Propazine	
Glyphosate	Prometryn	Ethalfluralin	Simazine	
Fluazifop	Propachlor	Fenuron	Sulfometuron	
Fenoxaprop	Proham	Fluchloralin	Trifluralin	
Metham	Pyrazon			
Methyl bromide	Siduron			
MCPA	TCA			
MCPB	Terbutryn			
Molinate	Thiobencarb			
MSMA	Triallate			
Nitrofen	Vernolate			
Paraquat[6]				
Phenmedipham				
Propanil				
Sethoxydim				

[1] These are approximate values and will vary as discussed in the text.

[2] At higher rates of application, some of these chemicals may persist at biologically active levels more than 12 months.

[3] At lower rates of application, some of these chemicals may persist at biologically active levels for less than 12 months.

[4] In water.

[5] In soil.

[6] Although diquat and paraquat molecules may remain unchanged in soils, they are adsorbed so tightly that they become biologically inactive.

Figure 6-3. Total-vegetation control herbicides may provide annual-weed control for up to 2 years. (E. I. du Pont de Nemours and Company.)

Figure 6-4. Annual-weed control for up to 2 years may be provided with 4 to 20 lb/acre of simazine. (Ciba-Geigy Chemical Corporation.)

organic matter in their bodies primarily through aerobic respiration. In the process, O_2 is absorbed and CO_2 is released.

If an organic substance such as a herbicide is applied to the soil, microorganisms immediately attack it. Those that can utilize the new food supply will likely flourish and increase in number. In effect, this hastens decomposition of that organic substance. When the substance has decomposed, the organisms generally decrease in number because their new food supply is gone. However, recent observations of accelerated degradation of thiocarbamate herbicides in soils previously treated with these herbicides suggest that increased capacity for degradation of the herbicides persists even after the herbicides are removed.

Other factors beside food supply may quickly affect the growth and rate of multiplication of microorganisms. These factors are temperature, water, oxygen, and mineral nutrient supply. Most soil microorganisms are nearly dormant at 40°F, 75–90°F being the most favorable for microbial growth. Without water, most microorganisms become dormant or die. Aerobic organisms are very sensitive to an adequate oxygen supply, and deficiency of nutrients, such as nitrogen, phosphorus, or potash, may reduce microorganism growth.

Thus a herbicide may remain toxic in the soil for considerable time if the soil is cold, dry, poorly aerated, or if other conditions are unfavorable to the microorganisms. If the organisms are destroyed by soil sterilization (steam or chemical methods), decomposition of the herbicide may temporarily stop.

Soil pH also influences growth of microorganisms. In general, the bacteria and actinomyces are favored by soils having a medium to high pH, and their activity is seriously reduced below pH 5.5. Fungi tolerate all normal soil pH values. In normal soils, therefore, fungi predominate at pH 5.5 and below. Above pH 5.5, the fungi are reduced through competition with bacteria and actinomyces.

Thus a warm, moist, well-aerated, fertile soil with optimal pH is most favorable to microorganisms. Under these ideal conditions, these organisms can quickly decompose organic herbicides. Microbial decomposition of many herbicides follows typical growth curves for bacterial populations (see Figure 6-5).

At the usual rate of herbicide application on farmlands, the total number of organisms is seldom changed to any great extent because the herbicide may benefit one group of organisms and injure another group. When the herbicides are decomposed, the microorganism population returns to normal. The biological activity of most herbicides applied at rates recommended for cultivated crops disappears in less than 12 months (Table 6-1). Therefore, no long-term effect on the microorganism population of the soil is expected.

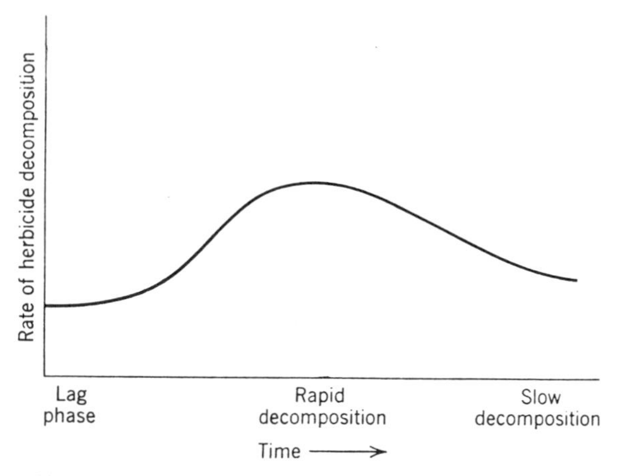

Figure 6-5. Rate of herbicide decomposition by microorganisms. (A. D. Worsham, North Carolina State University.)

Higher Plant Decomposition. Herbicides absorbed from the soil by higher plants are generally changed or metabolized (see Chapter 5). A small amount can remain in the original form and can be stored or exuded. These topics will be dealt with later in this chapter.

Chemical Decomposition Chemical decomposition is the breakdown of a herbicide by a chemical process or reaction in the absence of a living organism. This may involve reactions such as oxidation, reduction, and hydrolysis. For example dalapon will slowly hydrolyze in the presence of water, rendering it ineffective as a herbicide. Another example is the hydrolysis of triazine and sulfonylurea herbicides in low pH or acidic soils.

In soil saturated with water, oxygen will likely become a limiting factor. Under such conditions, anaerobic degradation of organic compounds can be expected. It has not been established whether this is chemical or microbial, but it is probable that both are involved. Trifluralin, under standing water, was completely degraded in 7 days at 76°F in nonautoclaved soils, whereas only 20% had degraded at 38°F.

Photodecomposition Photodecomposition, or decomposition by light, has been reported for many herbicides. This process begins when the herbicide molecule absorbs light energy; this causes excitation of the electrons and may result in breakage or formation of chemical bonds.

Most herbicides absorb radiation in the ultraviolet region (4–400 m). Most herbicides are white, or nearly so, and have peak light absorption in the ultraviolet range (220–324 nm), whereas yellow compounds such as the dinitroanilines and dinoseb have absorption peaks around 376 nm. Solar energy below 295 nm reaching the surface of the earth is considered to be negligible. A "sensitization" process may also be involved; light energy is absorbed by an intermediate molecule and transferred to the herbicide molecule by collision. Thus the effective wavelength of light could be outside the absorption spectra of the herbicide.

Some products of photodecomposition are similar to those produced by chemical or biological means.

Chemicals applied to soil surfaces are frequently lost, especially if an extended period without rain follows herbicide application. It is entirely possible that photodecomposition is responsible for the losses. However, other factors that may account for the loss should not be overlooked. Volatilization accentuated by high soil-surface temperatures, biological and chemical degradation, and adsorption are a few of the factors that should be considered in explaining the disappearance of herbicides from soils.

Transfer Processes

Adsorption on Soil Colloids Colloids (from the Greek work *kolla* meaning "glue") have very high adsorptive capacities. As a soil term, colloid refers to the microscopic (1 μm or less in diameter) inorganic and organic particles in the soil. These particles have an extremely large surface area in proportion to a given volume. It has been calculated that 1 cubic inch of colloidal clay may have 200–500 square feet of particle surface area.

Many clay particles chemically react like the negative radical of a weak acid, such as COO^- in acetic acid. Thus, the negative clay particle attracts to its surface positive ions (cations) such as hydrogen, calcium, magnesium, sodium, and ammonium. These cations are rather easily displaced, or exchanged (from the clay particle) for other cations. They are known as *exchangeable ions*. This replacement is called *ionic exchange* or *base exchange*.

The base-exchange capacity of a soil is expressed as milliequivalents (meq) of hydrogen per 100 g of dry soil. A soil with a base-exchange capacity of 1 meq can adsorb and hold 1 mg of hydrogen (or its equivalent) for every 100 g of soil. This is equivalent to 10 ppm or 0.001% hydrogen.

Adsorptive capacity, and thus exchange capacity, is closely associated with inorganic and organic colloids of the soil. Inorganic colloids are principally clay. There are two principal groups of clay: *kaolinite* and *montmorillonite* (see Table 6-2). Kaolinite is a *nonexpanding* clay; adsorption occurs on the external surface of the clay particle. Montmorillonite is an *expanding* clay; adsorption can occur on external and internal surfaces. Because of this difference, montmorillonite clays have an adsorptive capacity of three to seven times that of kaolinite clays.

Kaolinite clays have a tendency to predominate in areas of high rainfall and warm-to-hot temperatures. Thus, clay soils of the tropics and the Southeastern United States are principally kaolinite. Aluminum and iron oxides are important constituents of kaolinic soils.

The montmorillonite clays predominate in areas of moderate rainfall and moderate temperatures. They are typical of corn belt soils.

Organic colloids involve the humus (organic matter) of the soil. Organic colloids have a very high adsorptive capacity—about 4 times the base-exchange capacity of a montmorillonite clay and perhaps 20 times that of kaolinite on a weight basis.

The interaction of herbicides with negatively charged soil colloids (clays,

TABLE 6-2. Characteristics of Some Common Clay Minerals

Characteristics	Montmorillonite	Vermiculite	Illite	Kaolinite
Type of layering	2:1	2:1	2:1	1:1
Type of swelling	Expanding	Limited expanding	Non expanding	Non expanding
CEC (meq/100 g)	80–120	120–200	15–40	2–10
Specific surface (m^2/g)	700–750	500–700	75–125	25–50

From Weber (1972).

TABLE 6-3. Classification of Herbicides According to Their Ionic Properties

Weak Acids[1]	Cationic[2]	Weak Bases[3]	Nonionic[4]
Aliphatics	Bipyrdiliums	Triazines	Amides
Benzoic		Triazoles	Anilides
Phenoxy		Thiadiazoles	Diphenylethers
Phenylacetic		Pyridinones	Dinitroanilines
Phenoxypropionates			Thiocarbamates
Cyclohexanones			Dithiocarbamates
Phenolics			
Benzonitriles			
Pyridine			Ureas
Amino			Phenylcarbamates
Sulfonylureas			
Imidazolinones			
Organic arsenicals			
Phthalamic			
Phthalic			
Uracils			

[1] At neutral pH, anions or negatively charged forms predominate.
[2] Exists in a cationic or positively charged form.
[3] At acid pH, cations or positively charged forms exist.
[4] At all pH values, uncharged forms exist.

organic matter) is dependent on the chemical nature of the herbicides (see Table 6-3). Positively charged (*cationic*) herbicides such as diquat and paraquat (bipyrdilums) are strongly held to soil by ionic bonds, much like potassium and calcium are held. However, unlike potassium and calcium, paraquat and diquat bound to clays are not readily displaced through ion exchange. Basic herbicides such as the triazines can become cations in low-pH (acid) soils and adsorb to soil particles by ionic bonds. Thus the activity of *s*-triazine herbicides such as atrazine is greater in high-pH (basic) soils than acid soils (Best et al., 1975). This happens because a higher percentage of the herbicide molecules, which are cations in the acid soil, are rendered unavailable to plants through adsorption to soil colloids.

Negatively charged (*anionic*) herbicides such as 2,4-D (phenoxy), dicamba (benzoic), picloram (pyridine), and dalapon (aliphatic) are not readily adsorbed because they have the same negative charges as the soil particles. However, small amounts may be bound to organic matter and positively charged soil colloids such as iron and aluminum hydrous oxides.

Small amounts of neutral or nonionic (*molecular form*) herbicides can be adsorbed by soil particulate matter through relatively weak physical forces. Adsorption of nonionic herbicides generally increases as their water solubility decreases. For example, highly water-insoluble herbicides such as the dinitroanilines are adsorbed in large quantities by the organic matter fraction of soils.

Much evidence supports the fact that organic matter and clay (especially

TABLE 6-4. Monuron Adsorption Correlated with Certain Soil Properties

Soil Property	Correlation Coefficient
Organic matter	0.991
Clay	0.209
Silt	0.358

montmorillonite) play important roles in determining phytotoxicity and residual persistence through adsorption, leaching, volatilization, and biodegradation. Observations in research work as well as in the field have shown the following:

1. Soils high in organic matter require relatively large amounts of most soil-applied herbicides for weed control.
2. Soils high in clay content require more soil-applied herbicide than sandy soils for weed control.
3. Soils high in organic matter and clay content have a tendency to retain herbicides for a longer time than sand. The adsorbed herbicide may be released so slowly that the chemical is not effective as a herbicide.

The adsorption of monuron was studied in various soils; some of the results are shown in Table 6-4. These findings may help to clarify the principles of monuron adsorption. Organic matter was the soil property most highly correlated (99%) with monuron adsorption. Clay and silt were less significant in the adsorption of monuron.

The importance of soil organic matter on the activity of soil applied herbicides was identified in research conducted by Weber et al. (1987). They reported that weed control attained with alachlor, metolachlor, metribuzin, and trifluralin was highly correlated with humic matter or organic matter (see Table 6-5). Scientists have attempted to develop equations for predicting safe, effective rates of herbicides for various soil types. Of the several chemical and physical

TABLE 6-5. Correlation Coefficients (*r*) of Herbicide Rates (kg active ingredient/ha) Required for 80% Weed Control at 4 Weeks after Application and Soil Humic Matter or Organic Matter Content

	Correlation Coefficient (*r*)	
Herbicide	Humic Matter	Organic Matter
Alachlor	0.90	0.87
Metolachlor	0.89	0.91
Metribuzin	0.95	0.88
Trifluralin	0.93	0.88

From Weber et al. (1987).

properties of soils that were measured, organic matter gave the best prediction of performance (Weber et al., 1987). The use of other properties with organic matter in the equations did not greatly improve predictability. Such prediction equations are useful for many soils; however, the theoretical value of a given soil may be considerably different than that observed in practice. Therefore, such values should be used as guidelines but not as absolute recommendations.

"Activated" carbon is one of the most effective adsorptive materials known. It has been used to protect plants from herbicides. Roots of strawberry plants have been coated with "activated" carbon prior to setting the plants in herbicide-treated soil (Poling and Monaco, 1985). Also, bands of "activated" carbon have been placed over previously seeded rows soon after planting and before a preemergence herbicide treatment. Well-decomposed organic matter could presumably have properties similar to those of "activated" carbon.

These facts indicate that a certain amount of herbicide is required to saturate the adsorptive capacity of a soil. Above this "threshold level," greater rates will greatly increase the amount of herbicide in the soil solution, and thus increase the herbicide toxicity to plants.

Therefore, the nature and strength of the "adsorption linkage" or "bonding" are of considerable importance for both cations and anions. Apparently the nature and characteristics of the colloidal organic matter, as well as the clay, may affect the tenacity of this bonding (see Figure 6-6).

In summary, various soils show large differences in their adsorptive capacities. In practice, however, the *range* of herbicidal rates of application is much less than might be predicted from the very wide ranges in adsorptive capacity of the soils.

Leaching Leaching is the downward movement of a substance with water through the soil. Leaching may determine herbicide effectiveness, may explain selectivity or crop injury, or may account for herbicide removal from the soil. Preemergence herbicides are frequently applied to the soil surface. Rain or irrigation leaches the chemical into the upper soil layers. Weed seeds germinating in the presence of the herbicide are killed. Large-seeded crops such as corn, cotton, and peanuts planted below the area of high herbicidal concentration may not be injured (see Figure 6-1). In addition to the protection offered by

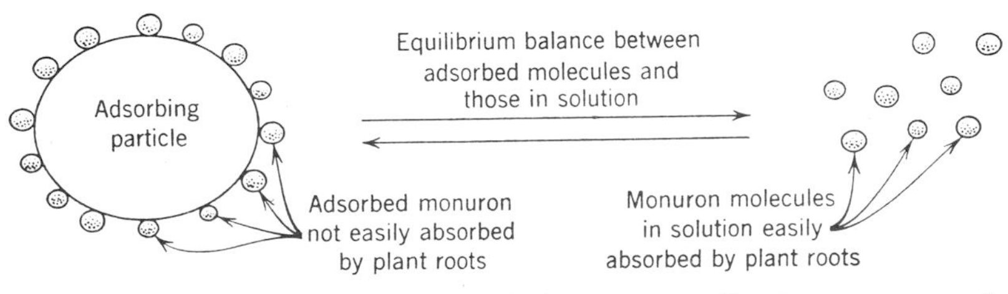

Figure 6-6. Plants absorb monuron in solution more easily than monuron that is adsorbed on soil colloids.

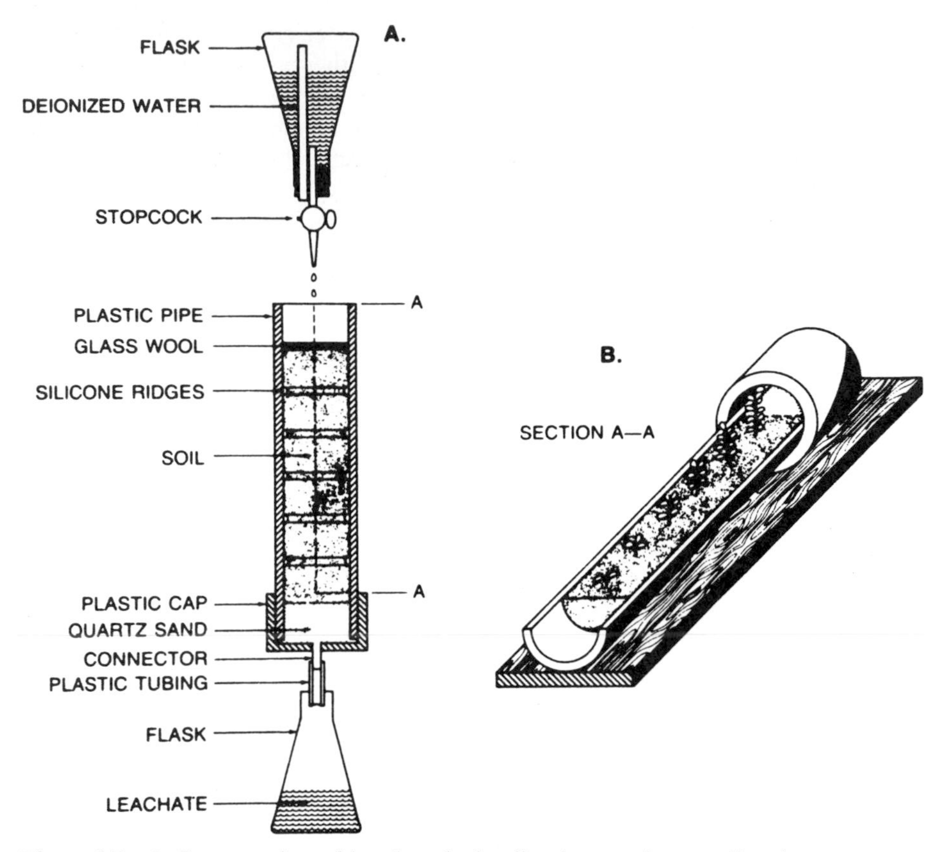

Figure 6-7. *A*: Cross-section of hand-packed soil column using a split-column system. *B*: After leaching, column may be divided longitudinally to provide two continuous half-columns, one of chemical or radiochemical assay and one for plant bioassay. (Weber, 1986.)

depth, crop tolerance to the herbicide through physiological processes is also desirable.

Some herbicides can be removed from the soil by leaching (see Figure 6-7). For example, sodium chlorate can be applied to shallow-rooted turf and the chemical leached to kill deep-rooted plants without injuring the turf (see Figure 5-10).

The extent to which a herbicide is leached is determined principally by

1. Adsorptive relationships between the herbicide and the soil;
2. Solubility of the herbicide in water; and
3. Amount of water passing downward through a soil.

Solubility is sometimes cited as the principal factor affecting the leaching of a herbicide. Simple calculation of the amount of water in a 4-inch rainfall disproves this assumption. A 4-inch rainfall weighs nearly 1,000,000 lb/acre. If you apply 1 pound of herbicide per acre, this equals 1 ppm of the herbicide in water; thus if the herbicide is soluble to the extent of 1 ppm, you might expect a 4-inch rain to remove essentially all of the herbicide from the surface inch of soil.

An example is given to illustrate the point. Monuron at 25°C is soluble in water to a concentration of 230 ppm. Monuron was applied at the rate of 2 lb/acre to a fine sandy soil, and 16 inches of water failed to remove the monuron from the surface inch of soil. This amount of water is capable of dissolving 400 times the amount of herbicide applied (Ogle and Warren, 1954).

The interrelationship between binding of herbicides to the soil, and water solubility can be demonstrated with 2,4-D. Salts of 2,4-D are water soluble and readily leach through porous, sandy soils. Soils with high organic-matter content adsorb the 2,4-D, reducing the tendency to leach. The ester formulations of 2,4-D have a low water solubility. Their tendency to leach is reduced by both low solubility and by adsorption by the soil.

The immobility of paraquat in soil stresses the importance of recognizing the influence of adsorption and water solubility on the leaching process. Paraquat is completely water soluble but does not leach in soils because it is a cation and is held very tightly by soil colloids.

To restate the point, the strength of "adsorption bonds" is considered more important than water solubility is determining the leaching of herbicides. Organic-matter content in the soil is the most important single factor determining the adsorptive capacity of the soil. The second most important is the clay fraction.

Besides downward leaching with water, herbicides are known to move upward in the soil by capillary movement of soil water. If water evaporates from the soil surface, water may move slowly upward and may carry with it soluble herbicides. As the water evaporates, the herbicide is deposited on the soil surface. Volatile herbicides such as EPTC also move upward and laterally in open soil pores in the vapor state.

In arid areas where furrow irrigation is practiced, lateral movement of herbicides in soils with the irrigation water also occurs.

Volatilization All chemicals, both liquids and solids, have a vapor pressure. Water is an example of a liquid that will vaporize and naphthalene (moth balls) is an example of a solid that will vaporize. At a given pressure, vaporization of both liquids and solids increases as the temperature rises.

Herbicides may vaporize and be lost to the atmosphere as either phytotoxic or nonphytotoxic gases. The toxic volatile gases may drift to susceptible plants. The *ester* forms of 2,4-D are volatile, and the vapors or fumes can cause injury to susceptible crops such as cotton or tomatoes (see Figure 5-5).

Certain herbicides may move in a porous soil as a gas. EPTC is thought to move in this way. This was clearly shown in experiments where injecting EPTC into the soil provided a much wider area of weed control than just at the injection point. Adsorbed by the soil, the EPTC may effectively kill germinating seeds.

The importance of volatilization and the loss of a herbicide from the soil surface is often underestimated. In volatility studies, EPTC volatilized from a free-liquid surface at the rate of about 5 lb/acre per hour at 86°F (Ashton and Sheets, 1959). This high rate of vaporization could easily explain the loss of the herbicide. EPTC, trifluralin, and other volatile soil-applied herbicides are usually mechanically mixed into the soil soon after application to reduce loss.

Codistillation with water evaporating from the soil surface (steam distillation) is another means by which a volatile herbicide may be lost. This process has not been extensively studied but may be of considerable importance in view of the immense amount of water lost from the soil surface through evaporation.

Herbicides with very low vapor pressures such as atrazine may also volatize from a surface over an extended period of time, especially if exposed to high temperatures. Soil-surface temperatures have been measured as high as 180°F.

Rain or irrigation water applied to a dry or moderately dry soil will usually leach a surface applied herbicide into the soil, or aid in its adsorption by the soil. Once adsorbed by the soil, the loss by volatility is usually reduced.

Surface Runoff Herbicides applied to the soil surface may dissolve in rainwater and leach into the soil. However, heavy rains may carry the dissolved herbicide away from the treated area. Severe runoff, which causes erosion, can also carry adsorbed herbicides on the eroding soil particles. "*Washoff*" is the term used to describe such losses. Cultural practices which minimize erosion such as conservation tillage (with large amounts of plant residues on the soil surface) or contour plowing will help minimize washoff losses of herbicides.

Removal by Higher Plants Herbicides may be adsorbed by the crop or surviving weeds and stored or given off in their original form. Usually, however, the herbicide molecule is altered in the plants by metabolism, and the herbicide breakdown products are either used by the plant or discharged back into soil solution. In some cases, herbicides are retained within the tissues of the plant, thereby delaying decomposition.

Herbicides may be removed from treated fields if the compounds are present in harvested plant parts, but amounts removed are nearly always insignificant. For example, 1 ppm (part per million) of a herbicide in a 10-ton hay crop amounts to only 0.02 lb/acre removed in the hay. However, sizable amounts can be degraded by plants during a cropping season (Weber et al., 1973).

Removal of herbicides from the soil by plants may not be a major factor in persistence of herbicides under most conditions; however it has been used to help remove persistent herbicides from soils where they were applied as soil sterilants and the planting of ornamentals was desired (e.g., corn for simazine or atrazine removal).

Exudation Herbicides that are absorbed by plants and microorganisms can also be *exuded* or discharged from inside the organism to the surrounding environment. The herbicide can be in an altered form or the original form. Generally, this does not represent a significant percentage of the amount the herbicide absorbed.

Figure 6-2 shows the interrelations of the processes that lead to detoxication, degradation, and disappearance of herbicides in the environment.

SUGGESTED ADDITIONAL READING

Ashton, F. M., and T. J. Sheets, 1959, *Weeds* **7**, 88.

Best, J. A., J. B. Weber, and T. J. Monaco, 1975, *Weed Sci.* **23**(5), 378.

Hance, R. J., 1980, *Interactions between Herbicides and the Soil*, Academic Press, New York.

Kearney, P. C., and D. D. Kaufman, Eds., 1975, 1976, 1988, *Herbicides*, Vols. 1, 2, and 3, Dekker, New York.

Ogle, R. E., and G. F. Warren, 1954, *Weeds* **3**, 257.

Poling, E. B., and T. J. Monaco, 1985, *HortScience* **20**(2), 25.

Saltzman, S., and B. Yaron, 1983, *Pesticides in Soil*, Van Nostrand Reinhold, New York.

Weber, J. B., 1972, *Adv. Chem. Ser.* **111**, 55.

Weber, J. B., 1980, *Weed Sci.* **28**(5), 467.

Weber, J. B., 1986, *Res. Methods Weed Sci.* **3**, 189.

Weber, J. B., T. J. Monaco, and A. D. Worsham, 1973, *Weeds Today* **4**(1), 6.

Weber, J. B., M. R. Tucker, and R. A. Isaac, 1987, *Weed Technol.* **1**(1), 41.

For herbicide use, see the manufacturer's or supplier's label and follow these directions. Also see Preface.

7 Formulations and Application Equipment

A newly synthesized herbicide is not suitable for use in the field. It must be formulated before it can be applied. There are several different types of formulations depending on the characteristics of the herbicide and its uses. Various types of equipment have been developed for application of these several types of formulations. This chapter will cover these formulations and the equipment required for their application.

FORMULATIONS

Herbicides are formulated to facilitate their handling, storage, and application and to improve their effectiveness under field conditions. The formulation chemist can change the formulation of a chemical to affect its solubility, volatility, toxicity to plants, and numerous other characteristics. This is accomplished by changing the chemical form (e.g., acid to ester) or using adjuvants including surfactants. An *adjuvant* is any substance in a herbicide formulation or added to the spray tank to modify herbicidal activity of application characteristics (*Weed Sci.*, 1985). A *surfactant* (surface-active agent) is a material that improves the emulsifying, dispersing, spreading, wetting, or other properties of a liquid by modifying its surface characteristics (*Weed Sci.*, 1985). Adjuvants and surfactants are considered to be *inactive ingredients* even though they can have a pronounced effect on the performance of the product.

The *active ingredient* (*ai*) is the chemical in a herbicide formulation primarily responsible for its phytotoxicity and which is identified as the active ingredient on the product label. The concentration of the active ingredient on the label is commonly given as a percentage for solid formulations and pounds per gallon for liquid formulations. However, the concentrations of certain herbicide derivatives are usually expressed as their *acid equivalent* (*ae*). In this case, the value refers to the concentration of the theoretical mass of the parent acid rather than that of the derivative. Phenoxy-type herbicides are universally handled in this manner, e.g., 2,4-D.

Complimentary information on pesticide formulatons can be found in van Valkanburg (1973), McWhorter et al. (1978), WSSA (1980), and Niessen (1983).

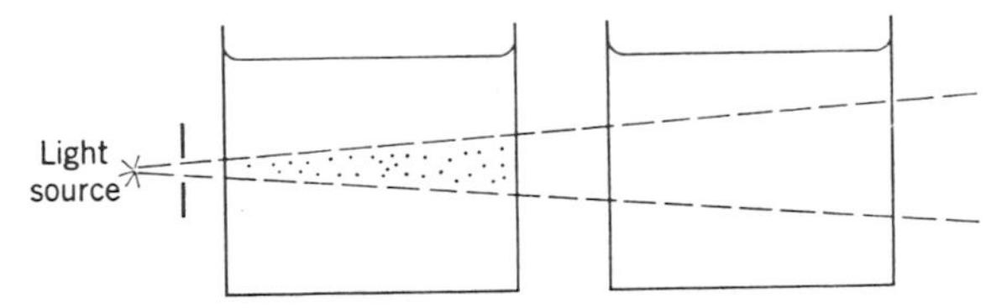

Figure 7-1. Reflective properties of true solutions and colloidal suspensions, the Tyndall effect. *Right*: Light is not reflected when it passes through a true solution. *Left*: Light is reflected from suspended liquid (emulsion) or solid (wettable powder) particles.

Types of Formulations

Herbicides are usually formulated so they can be applied with a liquid or solid carrier. These formulations include (1) water or oil solubles, (2) emulsifiable concentrates, (3) wettable powders, (4) water dispersible liquids and granules, (5) granules, and (6) pellets. The selection of which formulation to use depends on the specific use relative to the species to be controlled, the crop involved, equipment available, and environmental conditions.

Water or Oil Solubles These formulations are liquids or particulate solids that readily dissolve in the carrier, water or oil, to form a solution. A solution is a physically homogeneous mixture of two or more substances and clear in appearance (Figure 7-1). The dissolved constituent is the *solute* and the dissolving substance is the *solvent*.

Liquid water solubles (S or SL) require little tank agitation to dissolve in water while soluble powders (SP or WSP) need more agitation. Once the materials are dissolved, further agitation is not required. These points apply also to oil solubles (OS) except the solvent is oil rather than water.

The salts of most herbicides are soluble water. Conversely, the esters of many herbicides are soluble in oil and it may be used as the carrier. For example, many inorganic and amine salts of 2,4-D are soluble in water whereas their ester formulations are soluble in oil. However, there are also oil-soluble amine formulations of 2,4-D.

Emulsifiable Concentrates Emulsifiable concentrates (EC) are mixtures of the herbicide and emulsifying agents dissolved in an organic solvent. The emulsifying agents enable the emulsifiable concentrate to be dispersed in water forming an emulsion that is used as the spray mixture. An emulsion is one liquid *dispersed* in another liquid, each maintaining its original identity. The droplets (EC) are referred to as the dispersed phase and the liquid (water) they are suspended in is the continuous phase. Emulsions appear milky (Figure 7-1). Emulsifiable concentrates require tank agitation to form the emulsion and maintain it during spraying.

Invert emulsions are mixtures of an emulsifiable concentrate in which oil is the carrier and continuous phase, rather than water. These invert emulsions may be mayonnaise-like and too viscous to spray with conventional equipment. Special equipment has be designed for application of invert emulsions. Although not widely used, they are primarily intended to reduce spray drift.

Wettable Powders Wettable powders (WP) are finely ground herbicide particles that are intended to be suspended in water for spraying. They may also contain clay as a diluent, synthetic silica as an anticaking agent, and various adjuvants. The wettable powder suspension gives a "cloudy" appearance to the liquid and a reflection to a light cone (Tyndall effect) similar to an emulsion (Figure 7-1).

Wettable powders require tank agitation for initial suspension and during application to prevent settling of the solid particles. The smaller the solid particles, the slower the rate of settling. The inert materials in these formulations are often abrasive and can cause excessive wear to nozzles and certain pumps.

Water Dispersible Liquids and Granules Water dispersible liquids (WDL) and water dispersible granules (WDG) are similar to wettable powders. However, WDL is already suspended in a liquid and WDG is an aggregate of granule size made of finely ground particles. Water dispersible liquids are also commonly referred to as liquids (L) or flowables (F) and water dispersible granules are also commonly referred to as dry flowables (DF). These formulations overcome some of the mixing and dust exposure problems of wettable powders. However, like the wettable powders, these formulations require tank agitation for initial suspension and during application to prevent settling of the solid particles.

Granules and Pellets Herbicides can also be applied directly as dry formulations without spraying in a liquid carrier. Some chemicals, such as sodium borates and chlorates, are applied at high enough rates so that crystals of the chemical can be uniformly applied. However, most herbicides are so active that they must be formulated with an inert solid carrier to achieve a uniform application. Many carriers are used including clays, vermiculite, starch, plant residues, and dry fertilizers.

Granules (G) are small than *pellets* (P). Granules are generally spread mechanically and pellets are often spread by hand for spot treatment. The advantages of these formulations over a spray include (1) water is not needed for application, (2) application equipment may be less costly to purchase and maintain, and (3) granules can pass through plant residues to the soil surface in conservation tillage systems. The disadvantages of these formulations relative to a spray include (1) they weigh more and are bulkier, (2) they can be more expensive, (3) application equipment is more difficult to calibrate, and (4) uniform application can be a problem.

Encapsulation Encapsulation is a relatively new approach to herbicide formulation that involves the incorporation of the herbicide into very small capsules, generally 10 μm or less. The capsules are suspended in a liquid system and sprayed. The primary advantage of encapsulation appears to be extending the period of weed control by a herbicide. Differential release times can be accomplished by altering the nature of the encapsulating material. Research on the impact of encapsulation on herbicide behavior in the environment is in progress. The principle of "controlled release," including encapsulation, has been used for some time in a number of fields (Cardarelli, 1976).

Spray Additives

Additional materials are often added to spray mixtures to increase herbicidal activity or aid in the spraying operation. However, they are not always beneficial; some may have no effect or even decrease to desired effect. Therefore, additives should not be placed in the spray mixture unless suggested on the label or by knowledgeable authorities.

Surfactant Chemistry

Surfactants derive their name from the term "surface-active agents." They concentrate and act at the surface of the liquid they are dissolved in because their molecules have both polar and nonpolar segments (Figure 7-2). The polar segment is attracted to water (hydrophilic) and the nonpolar segment is attracted to oil-like compounds (lipophilic). In a mixture of oil and water, the lipophilic portion orients itself into the oil droplets and hydrophilic portion is within the water. This is how emulsifiers, a type of surfactant, facilitate the suspension of an oil-like herbicide in a water carrier. Another type of surfactant, wetting agents, is oriented with the polar segment into water droplets and the lipophilic segment protruding from the water droplets. This allows the herbicide spray to spread over a normally repellant leaf surface.

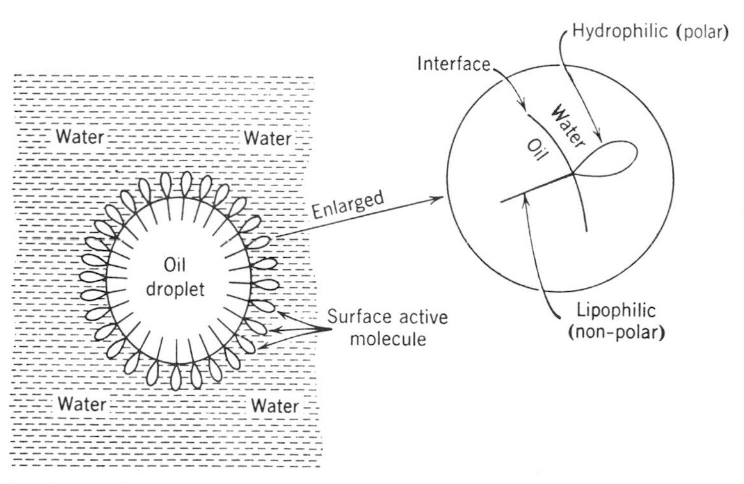

Figure 7-2. The surface-active molecule tends to bind the oil–water surfaces together, reducing interfacial tension.

TABLE 7-1. Approximation of HLB by Water Solubility

Behavior When Added to Water	HLB Range
No dispersibility in water	1–4
Poor dispersion	3–6
Milky dispersion after vigorous agitation	6–8
Stable milky dispersion (upper end almost translucent)	8–10
From translucent to clear	10–13
Clear solution	13 +

From Becher (1973).

Hydrophilic–Lipophilic Balance (HLB) HLB is a quantitative measure of the polarity of surfactant molecules. HLB uses a scale from 0 to 20 with the higher numbers being more hydrophilic than the lower values. The HLB can be determined experimentally or it is possible to estimate the value by calculation. HLB values are very useful for selection of surfactants for formulating emulsions. They also serve to emphasize that all surfactants are not equally suitable for all uses (Table 7-1). It is most important to select a surfactant appropriate for the intended use.

Surfactant Types

Surfactants are generally classified according to the nature of the polar segment of the molecule. Four types of surfactants are anionic, cationic, nonionic, and ampholytic molecules. Nonionic surfactants ionize or disassociate little in water while the others are charged when dissolved in water. Anionic surfactants are negatively (−) charged and the cationic surfactants are positively (+) charged in solution. Ampholytic surfactants can be either positively of negatively charged depending on the water pH.

Nonionic surfactants are the major type used with herbicides. Most emulsifiers are blends of anionic and nonionic types. In general, a blend with a high proportion of the anionic types will improve performance in cold water and soft water, whereas a blend with a predominance of nonionic types will usually perform better in warm water and hard water.

Additives According to Use

McWhorter (1982) classified herbicide additives as activity enhancers, spray modifiers, and utility modifiers. This classification has been generally accepted and will be used in the following discussion.

Activity Enhancers *Surfactants* of the nonionic type are commonly used with contact herbicides such as paraquat to enhance activity at 0.1 to 0.5% by volume of the spray mixture. They are usually available at 50 to 100% active ingredient with an HLB of 10 to 13. Their use can increase wetting, spreading, and other solution characteristics that increase herbicide phytotoxicity. They probably act by increasing herbicide penetration into the leaves. However, surfactants may also possibly modify herbicide behavior in the spray solution, in the cuticle, and inside the leaf.

Crop oils and crop oil concentrates are highly refined paraffinic phytobland oils plus surfactant that are used to increase foliar activity of certain herbicides. Originally the oil–surfactant mixture contained 1 to 2% surfactant and was referred to as crop oil. However, crop oil use has largely been replaced by crop oil concentrate (COC) that contains 17 to 20% surfactant in the oil–surfactant mixture. Crop oil was used at 1 to 4% and COC is used at 1% by volume in the spray mixture. Although the ability of COC to enhance the foliar activity of various herbicides such as strazine, linuron, bentazon, and fluazifop appears to be associated with increased uptake by treated leaves, other factors may also be involved. For example, COC delays crystallization of atrazine on the leaf surface, which keeps the herbicide in an absorbable form for a longer time. COC can also reduce volatile and photodegradative losses of some herbicides.

Spray Modifiers *Wetting agents* increase the spread and cover of a spray mixture on plant surfaces. They act by lowering the surface tension of the liquid. *Surface tension* can be considered the tendency of surface molecules of a liquid to be attracted toward the center of the liquid body. Spray droplet spreading occurs when the surface tension of the droplet is less than the surface tension of the leaf surface. The degree of effectiveness of a wetting agent on droplet spreading can be determined by measuring the contact angle between the droplet and the surface (Figure 7-3).

At low to normal spray volumes (10 to 40 gal/acre), wetting agents usually increase the effectiveness of herbicidal sprays by increasing the coverage of the plant surface; herbicide absorption is thereby also increased. However, at high spray volumes (>100 gal/acre), wetting agents may decrease effectiveness by causing excessive "runoff" from the plant onto the ground.

Herbicidal selectivity may be lost by the addition of wetting agents if the desired selectivity among species depends on differential wetting and thus absorption.

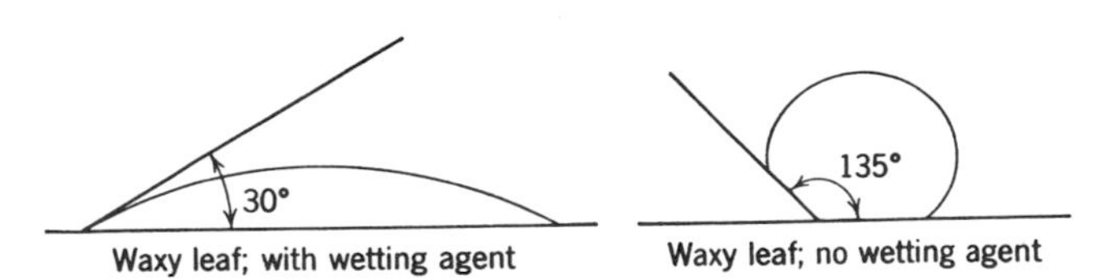

Figure 7-3. *Left*: Water droplets containing a wetting agent will spread in a thin layer over a waxed surface. *Right*: Pure water will stand as a droplet, with only a small area contacting the wax surface.

Stickers increase the adhesion of spray solutions to treated plant surfaces. They are often used in conjunction with wetting agents and referred to as spreader-stickers. Film-forming vegetable gels, emulsible resins, emulsifiable mineral oils, waxes, and water-soluble polymers have been used as stickers.

Drift control agents are materials that thicken the spray solution and thereby increase droplet size and reduce the number of very small satellite droplets. These materials include swellable polymers and hydroxyethyl cellulose or polysaccharide gums and are used at concentrations of 0.1 to 1.0% of the volume. Invert emulsions are also used to reduce spray drift.

Drift control agents are of great value when herbicide applications are make near sensitive nontarget plants, even though they increase application costs. The appropriate spray equipment and operating conditions must be used with these thickened solutions.

Utility Modifiers Utility modifiers are adjuvants that are used to reduce or avoid application problems and/or increase the usefulness of a formulation.

Antifoam agents are used to prevent or reduce excessive foaming in the spray tank. Emulsifiers and other surfactants of herbicide formulations can cause foaming with agitation of the spray mixture. Antifoam agents are typically silicones and used at 0.1% by volume. Kerosene or diesel fuel added to the spray tank at the same rate can often also inhibit foaming.

Compatibility agents are used to help with mixing and/or application problems that may occur when a combination or more than one pesticide is used. They can also be used when herbicides are applied in combination with a suspension, slurry, or true solution of fertilizers. Compatibility agents can counter separation problems that occur with hard or cold water.

HERBICIDE DRIFT

Herbicides may drift through the air from the target site and cause considerable injury if they contact susceptible plants. Movements through the air may result from *spray* or *volatility drift*.

Spray Drift

Spray drift is the movement of airborne spray particles from the target area. The amount of drift depends primarily on (1) droplet size, (2) wind velocity, and (3) height above the ground that the spray is released.

The size of the droplet depends primarily on pressure, nozzle design, and surface tension of the spray solution. In general, low pressures produce large droplets and high pressures produce small droplets. Different nozzle designs produce different droplet sizes, small nozzles produce small droplets. The lower the surface tension of a spray solution, the smaller the droplets. The importance of droplets size on spray drift is illustrated in Table 7-2. The smaller the size of the

TABLE 7-2. Spray Droplet Size and Its Effect on Spray Drift

Droplet Diameter (μm)	Type of Droplet	Number of Droplets/in.^{2} from 1 gal of Spray/Acre	Time Required to Fall 10 ft in Still Air	Distance Droplet Will Travel in Falling 10 ft With a 3-mph Breeze
0.5	Brownian max.	—	6750 min	388 miles
5	Fog	9,000,000	66 min	3 miles
100	Mist	1,164	10 sec	440 ft
200	Drizzle	195	3.8 sec	17 ft
400	Fine rain	28	2.0 sec	9 ft
500 ($\frac{1}{50}$ in.)	Rain	9	1.5 sec	7 ft
1000 ($\frac{1}{25}$ in.)	Heavy rain	1.1	1 sec	4.4 ft

Adapted from Brooks (1947) and Yates and Akesson (1973).

droplet, the longer it takes for the droplet to reach the ground and the greater distance it will travel.

Ideally herbicides should be sprayed when there is no wind. However, since this is not always possible some guidelines should be established to minimize the spray drift hazard. In addition to droplet size and spray release height, these guidelines should consider herbicide characteristics and distance and type of surrounding vegetation. Herbicides should be sprayed only when winds are less than 3 mph and never when winds are greater than 5 mph. Excessive spray drift often occurs at greater wind velocities. Wind velocities are usually lowest just before sunrise and just after sunset and throughout the night.

Airborne herbicide spray damage has also occurred under two unique conditions, an air inversion layer and alternating air flows (e.g., sea–land breeze).

The height above the ground that the spray is released is important for two reasons. First, the greater the height the longer it takes the droplet to reach the ground and the greater the distance of drift. Second, wind velocities are usually lower close to the ground than at higher elevations.

Therefore, spray applications from aircraft present greater drift hazards than ground sprayers. Air currents produced by an aircraft have a major effect on the trajectory of particles released from it. Any aircraft, rotary (helicopter) or fixed wing, produces updrafts at wing-tips and downdrifts under the middle of the aircraft (Figure 7-4).

Volatility Drift

Volatility refers to the tendency of a chemical to vaporize or give off fumes. The amount of fumes or vapors emitted is related the vapor pressure of the chemical. Vapor drift may damage susceptible crops, or may simply reduce, through loss, the effectiveness of the herbicide treatment.

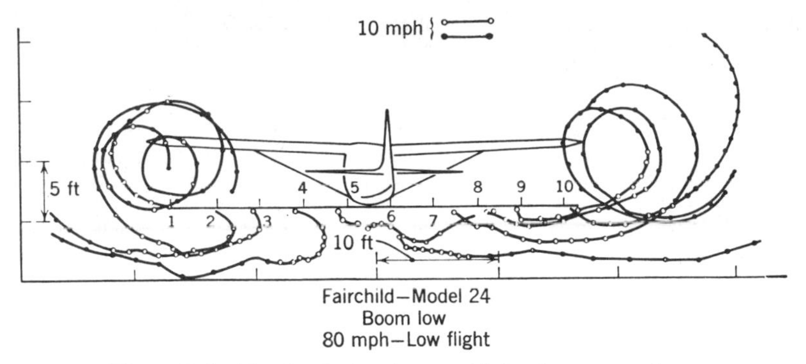

Figure 7-4. Air circulation from a high-wing monoplane.

Figure 7-5. Tomato plants exposed to different 2,4-D formulations. *1.* Sodium salt—slight to no injury. *2.* Diethanolamine salt—no injury. *3.* Triethanolamine salt—no injury. *4.* Butyl ester—serious injury to death. *5.* Ethyl ester—serious injury to death. (Klingman, 1947.)

The voltatility of 2,4-D has perhaps received more attention than that of any other chemical. Figures 7-5 and 7-6 indicate differences found between the salts and esters. The amine and sodium salts of 2,4-D have little or no voltatility hazard. The ester formulations vary from low to high volatility.

The length and structure of the alcohol portion of the 2,4-D ester molecule directly affect its volatility. In general the longer the carbon chain in the part contributed by the alcohol, the lower the volatility. Those esters made from five-carbon alcohols or less are usually considered volatile. Inclusion of an oxygen as an ether linkage in the alcohol portion of the molecule will also reduce the volatility of an ester of 2,4-D.

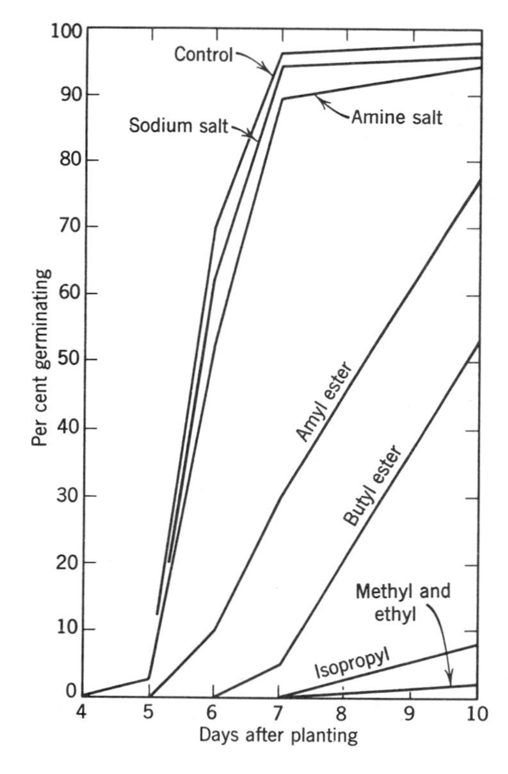

Figure 7-6. Germination of pea seeds after being exposed to different 2,4-D formulations. (Mullison and Hummer, 1949.)

$$
\begin{array}{ccccccccccccccc}
 & & & & H & & H & & H & & & & H & & H & & H & & \\
 & & & & | & & | & & | & & & & | & & | & & | & & \\
H & - & O & - & C & - & C & - & C & - & O & - & C & - & C & - & C & - & H \\
 & & & & | & & | & & | & & & & | & & | & & | & & \\
 & & & & H & & H & & H & & & & H & & H & & H & &
\end{array}
$$

Grover (1976), comparing the relative volatility of various formulations, assigned 2,4-D amine a relative volatility of 1, low volatile esters of 2,4-D (propylene glycol butyl, butoxy-ethanol, and isooctyl) a relative volatility of 33, and high volatile ester of 2,4-D (butyl) a relative volatility of 440. Thus, where 2,4-D-susceptible plants are grown, the 2,4-D amine form should be chosen. Also, application methods that keep spray drift to a minimum should be used.

Soil surfaces exposed to direct sunlight often reach 180°F. At this temperature some chemicals quickly volatilize and may be carried away in the wind, and therefore are hazards to susceptible plants. Also, the herbicidal effect of the treatment may be lost.

APPLICATION EQUIPMENT

Safe and effective use of herbicides requires proper selection, calibration, and operation of the application equipment.

A wide variety of equipment is used to apply herbicides. The selection of the specific type of equipment depends primarily on the weed, crop, herbicide, and formulation. Other considerations include whether the herbicide is to be applied broadcast covering the entire area, in narrow bands, as individual spot treatments, or to a particular part of the plant. Superimposed on these are physiographic, edaphic, and climatic factors.

Herbicides formulated as solutions, emulsions and wettable powders are usually applied to soil or plants as sprays with water as the diluent or carrier. Granular formulations are applied to soil by mechanical spreaders similar to those used of broadcasting seed of fertilizer. Herbicide applications may also involve (1) mechanical incorporation into the top 2 to 6 inches of soil, (2) subsurface layering in a horizontal layer a few inches below the soil surface, (3) injection into the soil or water (lakes, reservoirs, irrigation, drainage), and (4) other specialized techniques discussed in the Practices Chapters in Part III.

Conventional Sprayer

Spraying of the herbicide solution, emulsion, or wettable powder is the most common method of application. Sprays can be applied uniformly to the target with the spray volume varying from 1 to 500 gal/acre. Lower volumes (1 to 5 gal/acre) are usually for aircraft applications and very high volumes (> 100 gal/acre) may be required for through coverage of postemergence applications to dense vegetation. Spray volumes for both soil and foliar applications by ground-rigs are usually in the range of 30 to 60 gal/acre.

The most frequently used equipment to apply herbicide sprays on farms are low pressure sprayers. These sprayers can deliver from close to 0 to 200 psi. These are in contrast to high pressure sprayers that can develop pressures of 600 psi that are used to apply insecticides to tall trees.

Although sprays can be applied by hand-pump type sprayers on limited areas, e.g., home lawns, most spray applications are made using tractor-, jeep-, truck-, trailer-, or aircraft-mounted equipment. This type of equipment has several essential components including tank, pump, nozzles, filters or strainers, pressure gauges, pressure regulators, shut-off valves, and connecting hoses.

Tank An effective tank should hold enough spray mixture to avoid too frequent refillings; be easy to fill, drain, and clean; not corrode; and be equipped with appropriate openings hoses and connections. It also must have markings on transparent tanks or a gauge on opaque tanks to ensure accurate measurement of solution volumes.

A tank for the spray mixture can vary in shape, size, composition, and design depending on nature of the job. It may be a 55 gallon drum or a 1000 gallon trailer-mounted tank. It may be constructed of galvanized steel, stainless steel, aluminum, fiberglass, or polyethylene. Stainless steel is the most durable and corrosion resistant, but most expensive. Glavanized steel is durable and inexpensive, but subject to corrosion. Aluminum is durable and although it is somewhat more expensive than glavanized steel it is much more corrosion resistant. Fiberglass and polyethylene are lightweight and corrosion resistant, but are less durable then the metal tanks and require good support to avoid breakage. Fiberglass costs about the same amount as aluminum and polyethylene is the least expensive of all. However, fiberglass can be repaired whereas this is not possible with polyethylene.

Although there is considerable flexibility in the possible design of the tank, it must have some provision for agitation of the spray mixture. This is necessary because emulsions and wettable powders must have continuous agitation to maintain a homogeneous mixture. Hydrolytic agitation of the spray mixture is most common; however, mechanical agitation with paddles is also used. Hydrolytic agitation is usually accomplished by recycling a portion of the spray mixture from the pump and expelling it into the spray mixture in the tank from either a simple pipe with holes or a special agitator fitting. A separate agitator line should be used, not merely the bypass.

Pumps There are many types of pumps for liquids, each with certain advantages and disadvantages. These include centrifugal or turbine, roller, diaphragm, piston, gear, and flexible impeller types. Since only the first three of these types of pumps are commonly used in the United States for herbicide applications, they are the only ones discussed.

Power supply to drive the pump include (1) tractor power take-off; (2) gasoline engine or electric motors as direct drive, belt drive, or gear box drive; (3) ground wheel traction drive; or (4) on airplanes, a small propeller to drive the pump.

The pump selected should have the necessary capacity in gallons per minute (GPM) to provide the desired volume and agitation requirements plus a 10 to 20% excess to offset lost performance due to wear. Horsepower (Hp) needed to drive the pump, assuming an efficiency of 50 to 60%, can be estimated by the formula

$$\text{Hp} = \frac{\text{GPM} \times \text{psi}}{857}$$

To offset inefficiency in the power source, electric motors and gasoline engines should be rated at 33% and 33 to 67% higher horsepower, respectively.

Centrifugal pumps are the most commonly used pump in low-pressure systems. They develop pressure by centrifugal force created by rapidly rotating blades in a chamber. Since they are nonpositive pumps, they are not self-priming. However, they are easily primed by placing the pump below the tank and using a small vent at the top of the pump to allow trapped air to escape. These pumps wear well even with wettable powders and deliver a high capacity, 70 to 130 GPM at 30 to 40 psi. However, since these pumps need to operate at speeds of 3000 to 4500 rpm, a "step-up" from the power take-off (PTO) is required. *Turbine pumps* are similar to centrifugal pumps but operate at speeds low enough to allow direct drive from the PTO.

Roller pumps are preferred by many operators for low-pressure sprayers because they are inexpensive, operate at PTO rpm, and are easily repaired. They are self-priming positive displacement pumps that move a constant volume of liquid each pump cycle. Therefore, they require a pressure relief or control device to divert unsprayed solution back to the tank. They are capable of delivering 5 to 40 GPM at 40 to 280 psi. Abrasive wear of rollers by wettable powders and other materials in spray solutions is the major disadvantage to roller pumps.

Diaphragm pumps use the movement of a flexible diaphragm to alternatively pull liquid into a chamber through an intake valve and expel it through an outlet valve. They are positive displacement self-priming pumps that need less power to operate than other pumps. They develop moderate pressures and can deliver 15 to 50 GPM. They tend to be more expensive than other type pumps and the diaphragms can be affected by some herbicide solutions. However, they are excellent for the application of wettable powder formulations.

Nozzles Nozzles could be considered to be the most important part of the sprayer and other parts merely facilitate their proper operation. The nozzle converts the spray mixture into spray droplets. Several nozzles are often spaced along the length of a spray boom. The boom should be rigid during the spray operation for an accurate application.

Nozzle design and conditions of its operation, especially pressure, largely determine the uniformity and rate of the application. They also influence the size and uniformity in size of the droplets. *At the ideal pressure*, droplet formation occurs near the nozzle tip with the formation of uniform small droplets across the width of the spray pattern. *At low pressures*, liquid escapes from the nozzle tip as a liquid film. This film ligaments and then forms relatively large droplets at the outer edge as it expands. *At high pressures*, very small droplets are formed immediately at the nozzle tip. These small droplets may be of fog and mist size, subject to drift from the target site creating a potential hazard. Spray pressures for herbicide applications may range from as low as 5 psi (lb/in.^{2}) to as high as 50 psi; however, the usual rate is 20 to 40 psi. Lower pressures are used when the drift hazard is high and higher rates are used when penetration of dense vegetation is desired.

Faulty nozzles or faulty operation of nozzles can cause uneven spray patterns that may result in several-fold variations from the desired application rate. Such unevenness causes crop injury high rates and lack of weed control at low rates. These variations may appear as narrow strips from a few inches to a foot or so wide.

Nozzles are usually constructed of hardened stainless steel, stainless steel, nylon, or brass. Hardened stainless steel and stainless steel are the most and second-most wear resistant and costly, respectively. Nylon and other plastics resist wear somewhat less and may swell with the use of some formulations. Brass spray tips are relatively inexpensive but wear rapidly with abrasive spray solutions and can corrode with some fertilizers.

Although there are many types of nozzles, herbicides are normally applied

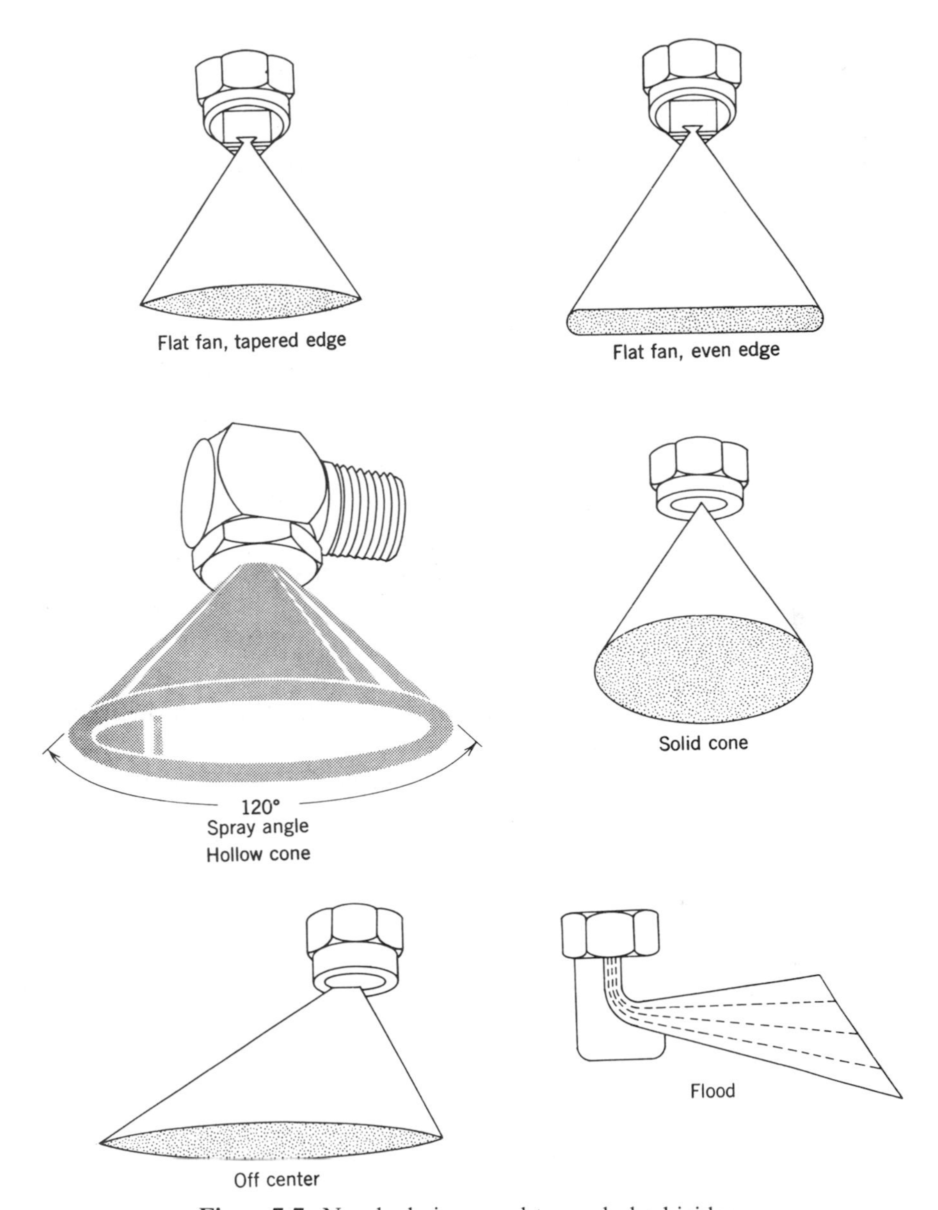

Figure 7-7. Nozzle designs used to apply herbicides.

with regular flat-fan, even flat-fan, flooding, or whirl-chamber hollow cone nozzles (Figure 7-7). Rotary-disc nozzles are also used to some extent when it is essential to minimize spray drift.

Regular flat-fan nozzles are general purpose nozzles suited for broadcast applications of herbicides when penetration of foliage is not needed. These nozzles deliver more spray in the center of the spray pattern than at the edges

Figure 7-8. Raindrop® nozzle spraying at 40 psi. (Photo by Ann Hawthorne, Delavan Corporation.)

(Figure 7-7). The spray patterns must overlap 40 to 50% for uniform applications. Several spray angles are available; selection is determined by nozzle spacing along the boom and height of the nozzle tip above the target. Pressure should be restricted to 15 to 30 psi to minimize drift. "Low-pressure" regular flat-fan nozzles can be operated at 10 to 25 psi and result in larger spray droplets and reduced drift.

Even flat-fan nozzles are similar to the regular flat-fan nozzle except that they deliver an equal amount of spray across the spray pattern (Figure 7-7). These nozzles should be used only to apply a band of herbicide over a crop row. The width of the band is determined by the spray angle and height of the nozzle tip above the target.

Flood nozzles deliver a flat-fan spray pattern similar to flat-fan nozzles but the distribution of droplets is less uniform. They are particularly useful for herbicide and herbicide-liquid fertilizer applications to soil. Large-coarse droplets are produced at their usual operating pressure (8 to 25 psi). Although this reduces drift, 100% overlapping patterns should be used to offset the less uniform spray pattern. These flood nozzles can be operated in any orientation from spraying straight down to straight back, providing the overlap is maintained.

Whirl-chamber hollow cone nozzles are commonly used for spraying directly from a herbicide soil-incorporation implement. These nozzles have a whirl chamber above a cone-shaped outlet and produce a hollow cone pattern with a fan angle. Raindrop® nozzles produce a hollow cone pattern with large droplets and have been designed for both soil and foliar application of herbicides.

Rotary-disc nozzles use centrifugal force to form droplets as the spray solution exits from the nozzle (Figure 7-9). This is in contrast to all previously described nozzles types that use hydraulic energy to create droplets. In the rotary-disc nozzle, the spray solution is directed to the center of a spinning disc and droplets are formed at the disc edge as the solution is thrown off the surface. The disc edge

Figure 7-9. The Micron Herbi® uses a spinning disc to produce uniform-size droplets at low gallonage per acre. Larger units are available for tractor and aircraft application. (Micron Corporation. P.O. Box 19698, Houston, TX 77024.)

may be either smooth or have teeth. A major advantage of a rotary-disc nozzle is a potential reduction of spray drift due to the formation of relatively large, uniform-size droplets. However, lack of a downward force can allow the entire spray pattern to be displaced by the wind.

Other Components *Filters or strainers* are usually used at the tank opening, in the spray lines, and at the nozzles. Strainer size is indicated by mesh size, the larger the mesh number the smaller the size of the strainer openings. Sixteen to 20 mesh screens are used at the tank opening to remove extraneous material and large lumps of herbicide. Strainers of 40 to 50 mesh in the suction line are used to prevent foreign material from the tank from damaging many types of pumps. However, since it is very important that the inlet of a centrifugal pump not be restricted, a strainer no smaller than 20 mesh with a diameter several times larger than the suction line should be used. Strainers are also needed to prevent clogging by particulate material. Nozzle strainers are usually located within the nozzle body and easily removed for cleaning. The desired mesh size is designated by the manufacturer in their catalog, usually 50 to 100 mesh.

A good *pressure gauge* is essential for proper nozzle operation. Its range should be relatively narrow, only slightly exceeding the intended range of intended spraying pressures. A gauge with too broad a pressure range makes accurate pressure determinations difficult.

A *pressure regulator* or *relief valve* is required with positive displacement pumps (e.g., roller pumps) to prevent pump damage that can occur when spraying is interrupted. Although a centrifugal pump does not require a pressure-relief valve, a special throttling valve should be used for accurate control of spray pressure. It can be electric controlled and operated from the tractor cab.

Spray lines or hoses must be of sufficient strength to withstand the pressures or vacuums expected and made of material that will tolerate the chemical to be used. If the sprayer is built so that a vacuum may develop between the spray tank and pump, a heavy-walled hose or metal pipe should be used in place of the hose. Hose is also made with a metal interior to prevent collapse. A vacuum may develop from clogged suction strainers in the hose, the inlet opening held to the wall of the tank, or from collapsed or twisted hose. Insufficient liquid flow will damage some pumps and will reduce the efficiency of all pumps.

Oil-resistant hoses made of neoprene, plastic, or other oil-resistant materials may be most satisfactory with oils and oil-like herbicides. In addition to a longer life span, the oil-resistant hoses will resist absorption of the chemical, making it less difficult to remove the herbicide.

Pressures in a hydraulic system are the same, regardless of hose size, minus any friction loss involved in movement. Therefore, hoses should be chosen that are large enough not to restrict liquid flow (Table A-9).

Calibrating the Sprayer

Several methods can be used to determine the number of gallons of spray applied per acre. However, regardless of the method used, it is imperative that the sprayer is functioning properly and each nozzle delivers the same volume of spray in a uniform spray pattern. Four methods of calibrating a sprayer are listed and briefly discussed.

Spraying a Known Size Area and Measuring the Amount of Spray Applied Perhaps the most accurate method is to actually spray an area of known size and measure the volume of spray mixture used. In practice, one starts with a full tank and measures the gallons required to refill the tank after spraying a specific area. The gallons per acre can be readily calculated from these measurements. Obviously, the size of the area sprayed must be large enough to give an accurate measurement of the area sprayed and spray volume used. In general, aircraft sprayers may require 5 to 10 acres, tractor sprayers 1 to 2 acres, and hand sprayer 1000 square feet.

Measurement of Gallons of Spray Delivered This method uses a measurement of the volume of spray mixture delivered per unit time and the spray width; gallons per acre are calculated by use of the following two formulas.

The acres sprayed per hour can be calculated from the tractor speed, spray width, and two constants as follows:

$$\text{Acres sprayed/hr} = \frac{\text{mph} \times 5280\ (\text{ft/mile}) \times \text{spray width (ft)}}{43{,}560\ (\text{ft}^2/\text{acre})}$$

The gallons per acre can be calculated from the above value (acres sprayed/hr) and the measured spray volume per unit time as follows:

$$\text{Gallons/acre} = \frac{\text{Gallons applied/hr}}{\text{Acres sprayed/hr}}$$

For example, a sprayer travelling at 6 miles per hour with a swath width of $16\frac{1}{2}$ feet will spray 12 acres per hour. If the sprayer applies 60 gallon of spray per hour the sprayer will apply 5 gallon of spray per acre.

Prepared Tables Prepared tables that give nozzle spacing, pressures, speed, and various nozzle sizes giving various gallons of spray per acre are available from some nozzle manufacturers; the proper nozzle size can be selected from these tables. The disadvantage to this method is that there is no assurance that speeds and pressures used in spraying are correct.

Special Measuring Devices Special measuring devices and prepared charts or graphs can be used. The spray is usually collected for a prescribed period of time or distance. The amount of spray collected is then converted, through tables or charts, into gallons per acre. Glass jars with the table printed directly on the jar are available to catch the spray. This has the same disadvantages as those given for prepared tables.

Cleaning the Sprayer

Cleaning the sprayer before storage, even for short periods of time, will usually prove beneficial. Emptying the spray tank and rinsing it with water may be sufficient for short-time storage. Rinsing the pump and tank (both inside and out) with fuel oil will protect most metal parts from corrosion, but oil may injure parts made of natural rubber.

By cleaning the sprayer of leftover chemicals, the possibility of injuring or killing sensitive plants in future sprayings is also reduced. 2,4-D and related products have caused many problems through the neglect of this operation.

Farmer experience has shown that the barrel or tank is by far the most important source of contamination. It is especially difficult to clean barrels lacking a bottom drain.

If the sprayer is to be cleaned, first rinse it with a material that acts as a solvent for the herbicide. Kerosene and fuel oils carry away herbicides known to be oil soluble (chemicals that form emulsions when mixed with water are oil soluble). Following the oil rinse, a rinse with a surfactant in water will help to remove the oil. The oil-soluble herbicides, such as 2,4-D esters, are usually the most difficult herbicides to remove. The 2,4-D salts are water soluble and removed by thorough rinsing with water. Check the spray on susceptible plants, such as tomatoes, to make certain that the 2,4-D has been removed.

For wettable powder herbicides, examine the tank to see that none of the

Figure 7-10. Recirculating sprayer equipped with solid-stream nozzles to apply herbicide to weeds that are taller than the crop. Spray caught in the box is returned to the tank to be sprayed again. (Southern Weed Science Laboratory, USDA-SEA, Stoneville, MS.)

wettable powder remains in the bottom; otherwise, a thorough rinsing with water is usually sufficient.

Recirculating Sprayer, Roller Applicators, and Rope Wicks

These three specialized methods are adapted to treating weeds that are taller than the crop. The recirculating sprayer directs the spray horizontally as a narrow stream above the crop canopy, hitting only the tops of tall weeds. Spray that is not intercepted by tall weeds is caught and recycled through the sprayer (Figure 7-10).

Roller applicators use a rotating, carpeted roller to *wipe* herbicide on tall weeds.

Rope-wick applicators use a straight section of plastic pipe with rows of wicks exposed to wipe tall weeds (Figure 7-11).

Although these applicators do an excellent job on tall weeds, their use must be delayed until the weeds are taller than the crop. Thus crop yields are reduced before this method is used. Glyphosate is the principal herbicide applied this way.

Granular Applicators

Granular herbicides are applied with equipment similar to dry fertilizer spreaders (Figure 7-12). The spreaders can be separate units or mounted on planters or

Figure 7-11. Rope-wick applicator. Herbicide-soaked ropes are attached to a horizontally positioned pipe reservoir. (Southern Weed Science Laboratory, USDA-SEA, Stoneville, MS.)

Figure 7-12. Granular materials being applied in a band over plant rows at planting time. (Gandy Company.)

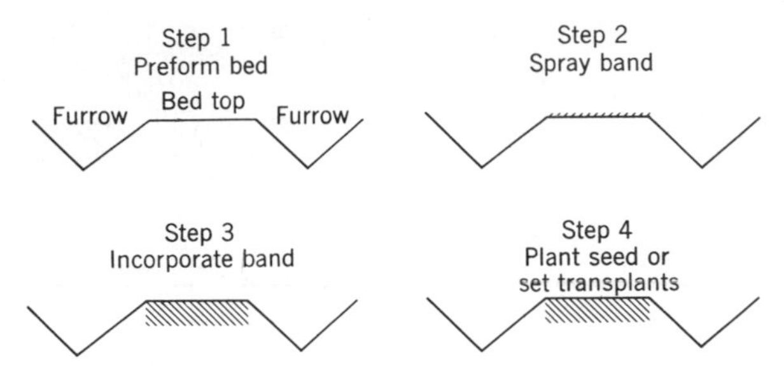

Figure 7-13. Steps in making a band application of a soil-incorporated herbicide to a preformed bed. All of these steps can be performed in a single pass through the field by placing all the equipment required on one tractor.

incorporating equipment. Herbicides impregnated onto fertilizer are spread with a normal fertilizer spreader. It is important that enough bulk is spread and a crossing application is used to ensure uniform application.

Soil Incorporation Equipment

Virtually all types of cultivation equipment have been used to mix herbicides into the soil. However, all implements do not work equally well for every situation or herbicide. For overall broadcast applications, a disc harrow or ground-driven rolling cultivator can be used. For band applications, power-driven rotary tillers are usually used. Figure 7-13 shows the procedure for incorporation a herbicide into a preformed bed.

The disc harrow is perhaps the most common incorporation tool. The second discing is commonly at right angles to the first. Speeds must be fast enough to effectively mix the soil. The small disc harrow (finishing disc) is better adapted to incorporation than the large heavy disc. Disc blades should be spaced no more than 8 inches apart. Large discs recently available may incorporate the chemical too deep in the soil, especially on sandy and sandy loam soils. Depth-gauge wheels set for the disc to cut 4–5 inches deep may help. Following the disc with a heavy spike-tooth harrow may further help to evenly mix the herbicide in the soil. The manufacturer's label should be checked for the depth of incorporation and equipment recommended for each herbicide.

In addition to proper application rate and regardless of the type of soil incorporation equipment to be used, three major factors must be considered: depth of incorporation, soil conditions, and correct ground speed.

Depth of Incorporation The depth of incorporation or mixing of the herbicide into the soil is critical. If herbicides requiring shallow incorporation are placed too deep, they may lose some of their effectiveness by dilution in a greater volume of soil. If volatile herbicides are mixed too shallow some volatility loss may occur. The more volatile herbicides, e.g., EPTC, require deep mixing of 2 to 4 inches.

Figure 7-14. Research plot showing field bindweed control with a subsurface-blade application of trifluralin. (Elanco Products Company, a division of Eli Lilly and Company.)

Less volatile herbicides, e.g., trifluralin, may need to be mixed to only a moderate depth of 1 to 2 inches.

Soil Conditions Weed control may be erratic when many large clods are present. A fine seedbed with clods less than 0.5 inch in diameter is recommended. The herbicide does not readily mix with excessively moist soil, which can also increase volatility losses by decreasing the adsorption of herbicides on soil particles. The development of a plow sole, soil compaction, or loss of soil structure may occur by working soil with excessive moisture.

Ground Speed Ground speeds of 5–6 mph are necessary to obtain adequate mixing of the herbicide into the soil with discs, harrows, and ground-driven rolling cultivators. Lower speeds (1–2 mph) can produce "streaking" and poor weed control. With power-driven rotary tillers, ground speeds of 1.5–3 mph are recommended. Good results have been obtained at a tiller speed of about 500 rpm and 2-mph ground speed. The tillers must be close enough together to provide a "clean sweep" of the soil. This may require L-shaped knives.

For an overall flat preplant soil-incorporation treatment followed by bedding (listing and bed forming), the lister shovels should always be set to run 1–2 inches shallower than the depth of incorporation. This setting prevents placing untreated soil on top of the bed. For a band application of a preplant soil-incorporation treatment to preformed beds, care must be taken not to remove treated soil or place untreated soil on top of the beds by subsequent planting or cultivating operations (Figure 7-13).

Subsurface Layering Equipment

The spraying of a horizontal layer of herbicide a few inches below the soil surface (subsurface layering) has produced effective control of several hard-to-control perennial weeds. One example is field bindweed control with trifluralin or dichlobenil. This method is also called the spray-blade method or simply blading or layering.

The spray-blade method consists of a blade with backward-facing nozzles attached under the leading edge. Modified sweeps and V-shaped blades have been used. The effectiveness of the control also can bc seen in Figure 7-14. The concentrated layer of herbicide acts as a "protective wall," preventing the weed shoots from growing through it; thus the deeper storage roots starve and the plant dies. Any disturbance of this layer by cultivation or natural cracking of the soil often permits the shoots to emerge, and control is therefore greatly reduced.

Injection

Herbicides are injected into soil or water to control terrestrial and aquatic weeds. The most common chemical injected into soil is methyl bromide. This herbicide is usually applied by a commercial applicator because of its toxicity and because the treated area must be covered with a gas-tight plastic tarpaulin for 24–48 hours. Methyl bromide is applied with chisel-type applicators that inject the chemical 6–8 inches below the soil surface. The chisel-injection units are spaced not more than 12 inches apart.

Herbigation

The injection of herbicides into water requires precise metering with adequate mixing. This is particularly important when applying herbicides through a sprinkle or furrow irrigation system. The precision of herbigation application is no better than the accuracy of the water distribution and is usually less accurate than a spray application.

To control aquatic weeds, herbicides are injected into flowing waters (irrigation and drainage canals) and injected from a boat into static waters (lakes and reservoirs). See Chapter 23 for details of aquatic-weed control.

Aircraft

Aircraft sprayers are less commonly used to apply herbicides than other pesticides because of the drift hazards. In addition, the precision of application is somewhat less than that of ground sprayers. Despite these limitations, aircraft are especially adapted for spraying or applying granular formulations to areas not readily accessible to ground equipment, such as utility lines and firebreaks through remote woody areas, flooded rice fields, very large pasture or range areas, and large cereal grainfields.

Both fixed and rotary (helicopter) wing aircraft are used. In general, the components of sprayers for aircraft are similar to those of ground sprayers. Included are a tank, agitators, pump, boom, valves, screens, nozzles, pressure regulator, and pressure gauge. Because the design of aircraft sprayers requires special engineering knowledge, the details of design are not discussed here.

SUGGESTED ADDITIONAL READING

Akesson, N. B., and W. E. Yates, 1976, *Weeds Today* **7**, 8.

Becher, P., 1973, The emulsifier, in W. van Valkenburg, Ed., *Pesticide Formulations*, p. 65, Dekker, New York.

Bennett, H., 1947, *Practical Emulsions*, Chemical Publishing Co., New York.

Bouse, L. F., J. B. Carlson, and M. G. Merkle, 1976, *Weed Sci.* **24**, 361.

Brooks, F. A., 1947, *Agr. Eng.* **28**, 233.

Cardarelli, N., 1976, *Controlled Release Pesticides Formulations*, CRC Press, Cleveland, OH.

Cupples, H. L., 1940, *USDA Bur. Entomology Quarantine Pub.* E504.

Gast, R. and J. Early, 1956, *Agr. Chem.* **11**, 42.

Grover, R., 1976, *Weed Sci.* **24**, 26.

Hess, F. D., D. E. Bayer, and R. H. Falk, 1981, *Weed Sci.* **29**, 224.

Jansen, L. L., 1964, *J. Agr. Food Chem.* **12**, 223.

Klingman, G. C. 1947, *North Carolina Res. Farming* **4**, 3.

McCutcheon, J. W., and R. W. Hummer, 1949, *Bot. Gaz.* **11**, 77.

McWhorter, C. G., 1982, The use of adjuvants, in WSSA, *Adjuvants for Herbicides*, Weed Science Society of America, Champaign, IL.

McWhorter, C. G., T. N. Jordan, A. J. Tafuro, W. W. Abramitis, and J. R. Bishop, 1978, Adjuvant Terminology Subcommittee Report, Adjuvant Terminology Guidelines, *Weed Sci.* **26**, 204.

Mollison, W. R., and R. W. Hummer, 1949, *Bot. Gaz.* **11**, 77.

Niessen, H., 1983, Formulation aids, in K. H. Büchel, Ed., *Chemistry of Pesticides*, Wiley, New York.

Ogg, A. G., 1980, *Weed Sci.* **28**, 201.

Thompson, L., and S. W. Lowder, 1981, *Weed Today* **12**, 19.

van Valkenburg, W., 1973, *Pesticide Formulations*, Wiley, New York.

Welker, W. V., and T. Darlington, 1980, *Weed Sci.* **28**, 705.

Wilson, R. G., and F. N. Anderson, 1981, *Weed Sci.* **29**, 93.

Weed Science, 1985, *Weed Sci. Terminology Supplement* **33**, 1–23.

WSSA, 1982, *Adjuvants for Herbicides*, Weed Science Society of America, Champaign, IL.

Yates, W. E., and N. B. Akesson, 1973, Reducing pesticide chemical drift, in W. van Valkenburg, Ed., *Pesticide Formulations*, p. 275, Dekker, New York.

Yates, W. E., N. B. Akesson, and D. E. Bayer, 1978, *Weed Sci.* **26**, 597.

PART 2
Herbicides

8 Aliphatics, Amides, and Amino Acids

Herbicides may be classified or grouped together in several ways. Some of these include chemical structure, phytotoxic symptoms, translocation patterns, biochemical mechanism of action, persistence in the environment, or how they are used. However, except for a chemical structure system, none of these systems provides an *unequivocal* classification for all herbicides. Therefore, we are presenting the detailed information on the herbicides based on their chemical structure name, in alphabetical order, in this chapter and the next several chapters.

ALIPHATICS

Aliphatic means no rings in the structural formula. The aliphatic herbicides are the chlorinated aliphatic acids (TCA and dalapon), the organic arsenicals (cacodylic acid, MMA, MSMA, DSMA, and MAMA), acrolein, fosamine, and methyl bromide. Except for the aliphatics, most organic herbicides have chemical structures containing one or more rings.

Chlorinated Aliphatic Acids

TCA and dalapon are chlorinated aliphatic acids and are particularly effective on grass weeds. Although closely alike chemically, they show decided differences in phytotoxicity and translocation. Dalapon is about 10 times more phytotoxic than TCA. Dalapon is both phloem and xylem mobile, but TCA moves primarily in the xylem.

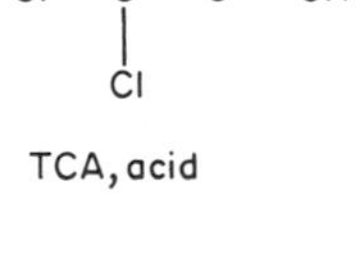

TCA, acid

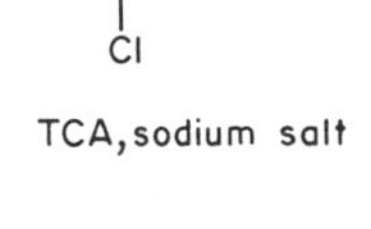

TCA, sodium salt

TCA TCA is the common name for trichloroacetic acid. It is a white deliquescent solid. However, only the sodium salt is used as a herbicide and the following comments refer to this form. Depending on purity, it may be a white solid or a pale yellow to amber liquid with a water solubility of 83.2 g/100 g of water. It is formulated as a water-soluble liquid concentrate, granule, or powder. It is hygroscopic. Exposed to 90 to 95% relative humidity at 70°F, the chemical will absorb its weight in water in 8 to 10 days. Therefore, it must be stored in moisture-proof containers. The acute oral LD_{50} is about 5 g/kg for rats.

Uses. The major use of sodium TCA is as a grass killer. It is primarily active through the soil but has some foliar contact activity at nonselective rates. It has proven useful as a nonselective treatment on perennial weedy grasses such as johnsongrass, common bermudagrass, and quackgrass. Sodium TCA is also used as a selective preemergence treatment in sugar beets to control annual grasses. In sugarcane, one preemergence or early postemergence treatment controls seedling grasses, including johnsongrass seedlings. It has also been used selectively in oilseed rape.

When handling and using sodium TCA, avoid contact with skin and eyes and avoid breathing spray mist. It is irritating to skin, eyes, nose, and throat. Wear protective goggles, respirator, clothing, and impervious gloves.

Soil Influence. TCA is not tightly adsorbed to soil and is subject to leaching, especially by heavy rain. It is slowly degraded by microorganisms. It persists 3 to 10 weeks depending on rate applied and soil moisture and temperature.

Mode of Action. When applied to foliage, TCA often causes rapid necrosis by contact action. It inhibits the growth of both shoots and roots, and causes leaf chlorosis and formative effects, especially to the shoot apex. TCA is readily absorbed by leaves and roots (Blanchard, 1954). It is translocated throughout the plant from the roots, but only small amounts are translocated from leaves. Therefore, it is primarily translocated via the apoplastic system. Perhaps its rapid contact action prevents symplastic movement. TCA is degraded slowly, if at all, by higher plants (Mayer, 1957).

It has been suggested that TCA acts by precipitating proteins since it is commonly used by chemists for this purpose. However, this has not been generally accepted. Perhaps it modifies sulfhydryl or amino groups of enzymes or induces conformational changes in enzymes.

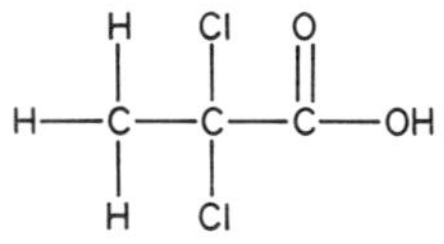

```
   H  Cl  O                 H  Cl  O
   |  |   ||                |  |   ||
H—C—C—C—OH            H—C—C—C—ONa
   |  |                     |  |
   H  Cl                    H  Cl
```

dalapon

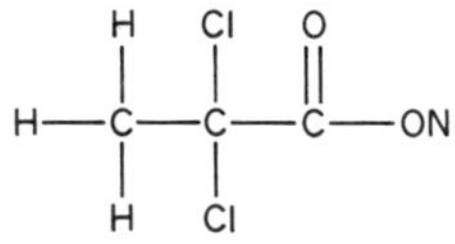

dalapon, sodium salt

Dalapon Dalapon is the common name of 2,2-dichloropropionic acid. Its chemical structure is similar to TCA, except that one chlorine atom in TCA has been replaced by a methyl (—CH_3) group. Only salts of the acid are used as

Figure 8-1. Spot spraying of johnsongrass in cotton with dalapon. (Dow Chemical Company, Midland MI.)

herbicides. Formulations usually contain 85% of the sodium salt (Dalapon 85®) or a mixture of sodium and magnesium salts (Dowpon M®). The latter formulation is less hygroscopic and less subject to caking than the sodium salt of dalapon alone. Other formulations contain other herbicides, e.g., 2,4-D or TCA. Tank mixes with other herbicides are also used, e.g., pyrazon for sugar beets.

The sodium salt of dalapon is a white solid with a very high water solubility of 50.2 g/100 g of water at 25°C. It is hygroscopic and requires moisture-proof containers for storage. When it absorbs water hydrolysis may occur resulting in a loss of herbicidal activity, especially at high temperatures. Spray solutions should be used within 24 hours to avoid excessive hydrolysis. It is irritating to skin and eyes and protective clothing and glasses should be worn by the applicator. The acute oral LD_{50} is 7.6 g/kg for female rats and 9.3 g/kg for male rats.

Uses. Dalapon is an effective grass killer, and considerably more effective than TCA as a foliar application. It is also somewhat less irritating to the skin and eyes, less corrosive to metals, and has a shorter period of residual phytotoxicity in soils than TCA.

Dalapon is used as a foliar application to the target species. It controls annual grasses and many perennial grasses including bermudagrass, quackgrass, and johnsongrass (Figures 8-1 and 8-2). Two or three applications spaced 5 to

Figure 8-2. Quackgrass control in corn with dalapon. *Top*: Dalapon applied at 6 lb/acre 3 weeks before planting corn. *Bottom*: Control, no herbicide applied. (K. P. Buchholtz, University of Wisconsin.)

20 days apart at 5 to 10 lb/acre per application have usually given better control of perennial weeds than one high rate of application. Dalapon is often used on noncrop land, fallow land, and established pastures. Dalapon is also effective for cattail control. An autumn application about 3 to 4 weeks before leaves lose their green color is most effective.

Dalapon is also used in many crops with method, time, and rate of application varying from crop to crop. Preplant applications, especially for perennial grass control, are used in beans (field, kidney, lima, snap), cotton, corn, milo, potatoes, soybeans, and sugar beets. Preemergence applications are used in potatoes and sugarcane. Postemergence applications are used in birdsfoot trefoil, flax, peas, sugar beets and potatoes (after layby), and spot treatments in cotton. Postemergence-directed applications are used in cotton, corn, sugar beets, and sugarcane. Dalapon is also used in citrus (grapefruit, lime, orange, tangerine), deciduous fruit trees (apple, apricot, peach, pear, plum, prune), banana, coffee, grape, macadamia, and pineapple as directed sprays. In asparagus, it is applied before or after cutting; in cranberries, it is applied to shoreline and irrigation and drainage ditches. Limitations for use in various crops may include repeat applications, geographic location, planting date, treatment time, stage of growth, varieties, grazing-forage, or harvest date (see the label). The label also gives specific rates of application for each treatment; they range from 1.7 to 17 lb/acre.

Soil Influence. Dalapon is not tightly adsorbed to soil and is readily leached. It is rapidly and completely degraded by soil microorganisms. Several organisms responsible for its degradation have been isolated and identified (Herbicide Handbook, 1989).

Mode of Action. Dalapon inhibits the growth of both shoots and roots of susceptible species (Ingle and Rogers, 1961; Meyer and Buchholtz, 1963). It interferes with cell division, induces formative in shoots and leaves, and causes leaf chlorosis (Prasad and Blackman, 1964).

Dalapon is absorbed through both the foliage and roots. The rate of absorption by leaves is favored by high relative humidity, moderate temperatures, and surface-active agents. Foliar absorption during the first 6 hours is most important, although absorption usually continues for at least 48 hours. It is translocated in both the phloem and xylem (Foy, 1975). High application rates to leaves can cause acute foliar and phloem injury, thereby greatly diminishing or preventing symplastic translocation. Repeat applications at low to moderate rates gives best control of deep-rooted perennial grasses.

Dalapon probably acts by interfering with the activity of many enzymes by alkylating the sulfhydryl or amino groups (Foy, 1975) and/or inducing conformational changes (Ashton and Crafts, 1981). A rather specific effect is an increase in ammonia. This is most apparent in susceptible species that are not able to form amides at a sufficient rate to prevent the accumulation of toxic levels of ammonia.

Organic Arsenicals

The organic arsenical herbicides include two similar compounds, cacodylic acid and MAA (methylarsonic acid), and their salts.

Cacodylic Acid Cacodylic acid is the common name for dimethylarsinic acid

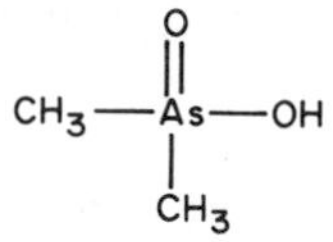

O
‖
CH_3—As—OH
|
CH_3

cacodylic acid

It was one of thc first organic arsenical herbicides introduced. It is a colorless crystalline solid with a very high water solubility of 66.7 g/100 ml. Like other organic arsenical hcrbicides, it has a much lower mammalian toxicity than elemental arsenic. Its acute oral LD_{50} is 830 mg/kg for rats.

Formulations include the acid form, the sodium salt, and combinations of the acid and sodium salt. The trade names are Cotton-Aide®, Monter® and Phytar 560® for the acid form, Bolls-Eye® for the sodium salt, and Kack® and Clean-Boll® for the acid and sodium salt combination.

Uses. Cacodylic acid and its sodium salt are nonselective and used as general contact sprays. They desiccate and defoliate a wide variety of plant species. They are used primarily to control emerged annual weeds in lawn-turf seedbeds, lawn-renovation areas, and noncrop areas. Some control of certain perennial grasses may also be obtained. Do not feed treated forage to live stock or graze treated areas. Certain formulations are also used as a directed spray in nonbearing citrus orchards. Do not treat citrus orchards within 1 year of fruit harvest. Various formulations are also for cotton desiccation and defoliation. Apply 7 to 10 days prior to harvest when over 50% of the bolls are open.

MAA MAA is the common name for methylarsonic acid. It is usually formulated as monosodium methylarsonate (MSMA), disodium methane arsonate (DSMA), or monoammonium methylarsonate (MAMA). Other formulations include the calcium salt (CMA) and the octyl-dodecyl ammonium salt (AMA). There are several suppliers and many trade names for these various formulations of MMA. Some of these are specific for added surfactants or a combination with other herbicides. MAA and its salts are very soluble in water. The acute oral LD_{50} of MAA is 1.8 g/kg for rats.

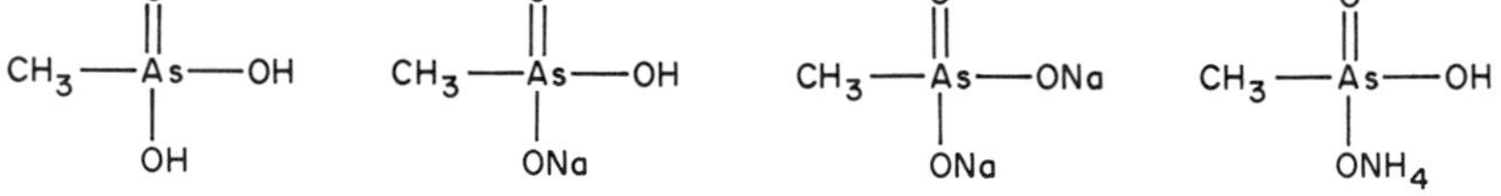

MAA	MSMA	DSMA	MAMA
O ‖ CH_3—As—OH | OH	O ‖ CH_3—As—OH | ONa	O ‖ CH_3—As—ONa | ONa	O ‖ CH_3—As—OH | ONH_4

Uses. The first major use of these compounds was for postemergence control of crabgrass, dallisgrass, and other weedy grasses in lawns and turf. MSMA and DSMA are used to control many annual and perennial grasses and nutsedges on noncrop areas, in lawns and turf, and in citrus (except in Florida). They are also widely used in cotton. MSMA is used preplant in cotton whereas DSMA is used as a directed spray in emerged cotton. These compounds are particularly effective on johnsongrass, dallisgrass, crabgrass, and nutsedges. AMA and MAMA are used in lawns and turf. None of these MMA derivatives is selective to St. Augustinegrass or centipedegrass and should not be used on these species.

Mode of Action. Cacodylic acid desiccates and defoliates many species of plants. MAA and its salts induce foliar chlorosis followed by gradual browning and finally necrosis. They also inhibit growth in general, inhibit the sprouting of rhizome and tuber buds, and cause aberrant cell division.

Cacodylic acid is primarily translocated in the apoplast. Significant symplastic transport is probably prevented by its rapid contact action, which injures the phloem. However, the MAA derivatives are translocated in both the symplast and the apoplast.

These compounds appear to be metabolized in plants by conjugation with sugars, organic acids, and amino acids. However, only the amino acids conjugated have been widely reported. The carbon–arsenic bond of these compounds appears to be very stable in higher plants.

Acrolein

Acrolein is the common name for acrylaldehyde or 2-propenal. The trade name is Magnacide®. It is a colorless liquid and very soluble in water (250,000 ppm). It has an acute oral LD_{50} of 46 mg/kg for rats. It is very toxic to most organisms. It is a flammable liquid subject to explosive reactions under certain conditions. Therefore, it may be applied only by licensed applicators.

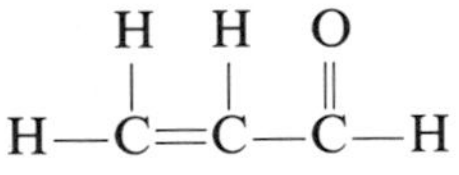

$$H-\overset{\displaystyle H}{\overset{|}{C}}=\overset{\displaystyle H}{\overset{|}{C}}-\overset{\displaystyle O}{\overset{\|}{C}}-H$$

acrolein

Acrolein is injected in water for the control of submersed and floating aquatic weeds. It is a contact herbicide and a general cell toxicant. It acts on enzyme systems through its destructive sulfhydryl reactivity.

Fosamine

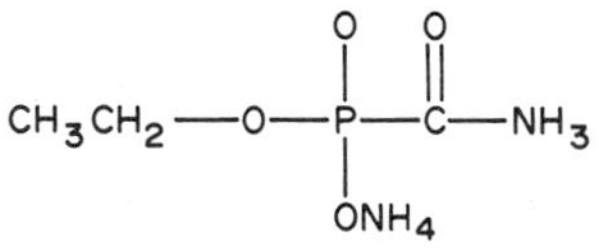

$$CH_3CH_2-O-\underset{\displaystyle ONH_4}{\underset{|}{\overset{\displaystyle O}{\overset{|}{P}}}}-\overset{\displaystyle O}{\overset{\|}{C}}-NH_3$$

fosamine

Fosamine is the common name for ethyl hydrogen (aminocarbonyl)-

phosphonate. The trade name is Krenite®. It is a white crystalline solid, soluble in water, 179 g/100 g water at 25°C, formulated for use as an aqueous spray. Fosamine is relatively nontoxic to mammals; the acute oral LD_{50} of the product is 24.4 g/kg for rats.

Uses Fosamine is relatively nonselective and primarily used for the control of many brush and woody species on noncrop land, it also controls blackberries, brackenfern, and field bindweed. It is applied as a complete-coverage spray to the leaves, stems, and buds for most effective control. Best control of deciduous woody species is obtained when it is applied within 2 months before the foliage changes color in the fall.

Soil Influence Since fosamine is applied to the aboveground parts of the weeds, soil type has little effect on its performance. It is rapidly degraded in soils by microorganisms and has a half-life of about 7 or 10 days under field and greenhouse conditions, respectively. This rapid degradation precludes significant leaching of fosamine under field conditions.

Mode of Action When applied to susceptible deciduous woody plants in the fall, little or no phytotoxic symptoms are observed until the following spring. In the spring, bud development does not occur or only small-malformed leaves develop. In contrast, certain species, e.g., pines and field bindweed, may develop phytotoxic symptoms soon after the herbicide treatment. Some suppression of terminal growth of many species may occur; therefore, spray drift onto desirable plants should be avoided.

Little research on the absorption, translocation, and metabolism of fosamine has been published (Kitchen et al., 1980; Müller, 1981, 1982; Herbicide Handbook, 1989). Fosamine is readily absorbed by leaves, stems, and buds. Retention and penetration are less when the herbicide is applied to rough or hairy leaves than to leaves with smooth surfaces. Translocation from treated areas appears to be relatively slow and limited. This supports the field observations that have shown that complete coverage of the plant is required for control. Translocation studies indicate greater transport in susceptible species than in tolerant species and may explain the differential selectivity among species. The reported translocation patterns suggest that it is transported in both the symplastic and apoplastic systems. Metabolism studies with [^{14}C]fosamine in higher plants indicates that degradation is relatively rapid with a half-life of about 2 to 3 weeks.

Methyl Bromide

Methyl bromide is the common name for bromomethane. It has many trade names. It is a colorless, nearly odorless liquid or gas. At 1 atmosphere of pressure and 38°F, the liquid boils and becomes a gas. The gas is 3.2 times heavier than air

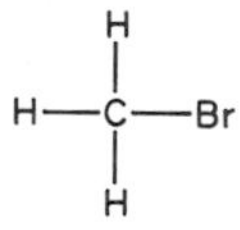

H
|
H—C—Br
|
H

methyl bromide

at 68°F. It is slightly soluble in water and very soluble in alcohol or ether. It is generally considered nonflammable and nonexplosive. However, mixtures containing between 13.5 and 14.5% of the gas in air may be exploded by a spark.

Methyl bromide gas is poisonous to humans and animals and the effects of exposures are cumulative. The acute oral LD_{50} is 100 mg/kg and the inhalation LD_{50} is 3150 mg/liter for rats. For man, the inhalation LD_{50} is 60,000 mg/liter for 2 hours. Because methyl bromide is nearly odorless, 2% chloropicrin (tear gas) is often added as a warning agent. See product label for protective clothing requirements and respiratory protection.

Uses Methyl bromide is used as a temporary soil sterilant. It kills plant tissues and most seeds, insects, and disease organisms. It is an excellent preplant soil treatment for seedbeds or propagating beds of tobacco, flowers, vegetables, turf, and tree seedlings. It is also used preplant for strawberry beds and to sterilize potting soil or compost mixtures. Since methyl bromide treatments are expensive ($1000 to $1500/acre), they are used only in high-value crops for serious pest problems. Formulations of methyl bromide (45%) and chloropicrin (55%) give increased control of certain plant pathogens.

The volatile nature of methyl bromide requires that it be confined by a plastic tarpaulin for about 48 hours to be effective. Planting can usually be safely made 72 hours after removal of the tarpaulin. It is injected into the soil and covered with a plastic tarpaulin in a single operation at a rate of 250 to 300 lb/acre for large areas. For small areas or soil volumes, it is merely released under a plastic tarpaulin at a rate of 1 to 2 lb/100 ft^2 of surface or 100 ft^3 of soil (Figure 8-3).

Although methyl bromide kills most weed seeds, certain seeds are not killed. These include prickly sida, redstem filaree, and certain morningglories, malvas, and clovers. However, most vegetative propagules including bermudagrass and johnsongrass rhizomes and nutsedge tubers are killed by methyl bromide fumigation.

Methyl bromide may kill beneficial microorganisms as well as disease microorganisms. This may result in an inhibition of the normal decomposition of organic matter into ammonia, then nitrite, and finally nitrate. If the microorganisms responsible for one or more of these processes are killed, phytotoxic levels of ammonia or nitrite may accumulate following methyl bromide fumigation. This seldom occurs in low-organic-matter soils; however, ammonium forms of nitrogen fertilizers should not be used. Usually the addition of peat, sawdust, or other organic matter to potting or compost mixtures after

Figure 8-3. Methyl bromide being released under a sealed plastic cover. (Dow Chemical Company, Midland, MI.)

fumigation will provide the microorganism inoculum necessary to prevent this problem.

AMIDES

$$R_1-\overset{\overset{\large O}{\|}}{C}-N\begin{matrix} / R_2 \\ \backslash R_3 \end{matrix}$$

The basic chemical structure of the amide-type herbicides is shown above. However, the substituents in positions R_1, R_2, and R_3 vary greatly, making this a diverse group of chemicals. Likewise, the weeds that the different amide herbicides control and the crops to which they are selective vary widely. Most amides are used as selective herbicides and applied to the soil as a preplant soil-incorporated or preemergence treatment. However, propanil is applied postemergence to the foliage of the weeds to be controlled.

The amides can be classified into three groups: (1) soil-applied chloroacetamides (alachlor, diethatyl, metolachlor, and propachlor), (2) other soil-applied amides (diphenamid, isoxaben, napropamide, naptalam, and pronamide), and (3) foliar-applied amides (propanil).

Chloroacetamides

The chloroacetamides have a monochlorinated methyl group ($Cl-CH_2-$) in the R_1 position of the amide structure. They are usually applied to the soil as a

$$Cl{-}CH_2{-}\overset{O}{\overset{\|}{C}}{-}N\begin{matrix} R_2 \\ R_3 \end{matrix}$$

chloroacetamides

preemergence or preplant soil-incorporated treatment. However, certain ones have postemergence activity on very young weeds.

Alachlor Alachlor is the common name for 2-chloro-*N*-(2,6-diethylphenyl)-*N*-(methoxymethyl)acetamide. The trade name is Lasso®. The chemical is formulated as an emulsifiable concentrate (4 lb/gal), microscopic capsules

$$Cl{-}CH_2{-}\overset{O}{\overset{\|}{C}}{-}N(C_6H_3(C_2H_5)_2){-}CH_2{-}O{-}CH_3$$

alachlor

(4 lb/gal), granules (15%), and combined with atrazine or glyphosate as package products. Alachlor is a cream-coloured solid and moderately soluble in water (242 ppm at 25°C). The acute oral LD_{50} of technical alachlor is 930 mg/kg for rats.

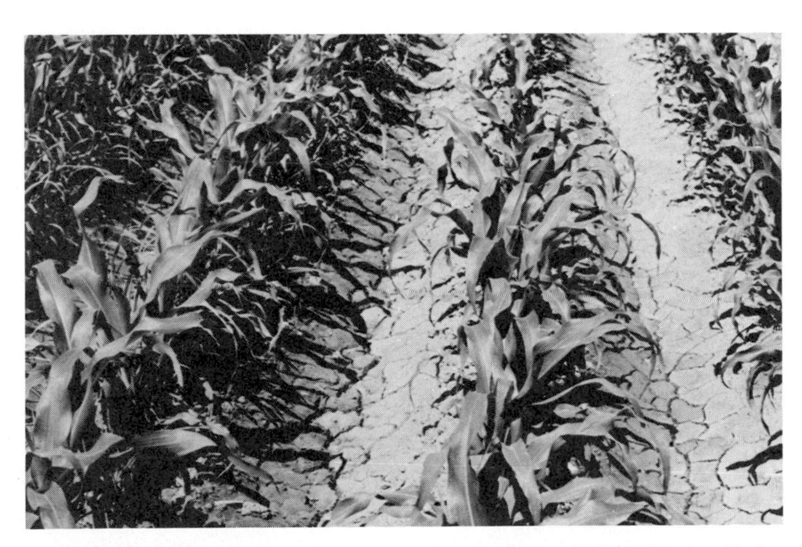

Figure 8-4. Barnyardgrass control in corn with alachlor. (C. L. Elmore, University of California, Davis.)

TABLE 8-1. Alachlor Uses

Crop	Rate (1b/acre)	Combinations	Remarks
Beans	2.5–3.0	—	Dry and green limas
	2.5–3.0	Trifluralin	Dry limas, tank mix
Corn	2.0–4.0	—	Field
	2.0–4.0	Atrazine	Field, package mix
	2.0–4.0	Atrazine, cyanazine, dicamba, paraquat, or atrazine plus cyanazine	Field, tank mix
	2.0–4.0	2.4-D	Field, sequential treatment
	2.0–4.0	—	Sweet and Popcorn
	2.0–4.0	Atrazine or cyanazine	Sweet and popcorn
Cotton	2.0–4.0	—	—
Milo[1]	2.5–4.0	—	—
	2.5–4.0	Atrazine	Package mix
	2.5–4.0	Atrazine, bifenox, or propazine	Tank mix
Peanuts	3.0–4.0	—	—
	3.0–4.0	Naptalam	Tank mix
Peas	2.0–2.5	—	—
Soybeans	2.0–4.0	—	—
	2.0–4.0	Chloramben, linuron, naptalam, chloramben plus metribuzin, or linuron plus paraquat	Tank mix
Sunflower	2.0–4.0	—	—
	2.0–4.0	Chloramben	Tank mix
Ornamentals	4.0	—	Field, junipers and yews, EC formulation Field, junipers and yews, and certain other woody ornamentals, granular formation Container, junipers and holly spp., granular formulation

There are limitations on some of the above uses relative to geographical location, soil type, formulation, and application method; see product labels.

[1] Milo seed must be treated with the protectant Screen® (flurazole) prior to planting.

Uses. Alachlor is applied as a preemergence, early postemergence, or preplant soil-incorporated treatment. It is used to control most annual grasses and many annual broadleaf weeds in beans, corn, cotton, milo, peanuts, peas, soybeans, sunflower, and certain woody ornamentals (Figure 8-4). It is often combined with other herbicides to increase the spectrum of weeds controlled. See Table A-0 for specific weeds controlled and Table 8-1 for detailed information on the crops. Many uses of alachlor are similar to metolachlor.

Soil Influence. Alachlor is adsorbed by clay and organic matter in soils and therefore not subject to excessive leaching in most soils. It is primarily degraded by soil microorganisms. In medium-textured soils with moderate soil moisture and temperature, herbicidal effectiveness usually lasts about 6 to 10 weeks. Persistence is somewhat shorter in sandy soils low in organic matter and somewhat longer in heavy soils.

Diethalyl Diethalyl is the common name for *N*-(chloroacetyl)-*N*-(2,6-diethylphenyl)glycine. It is used in the ethyl ester form, which is referred to as diethalyl ethyl with the trade name of Antor®. It is formulated as an emulsifiable concentrate, 4 lb/gal. Instead of an amide, it could be considered an amino acid derivative. However, its herbicidal characteristics resemble the amides rather than the amino acid derivatives. It is a white crystalline solid with a moderate water solubility of 105 ppm at 25°C. The technical product has an acute oral LD_{50} of 2.3 g/kg for male rats and 3.7 g/kg for female rats.

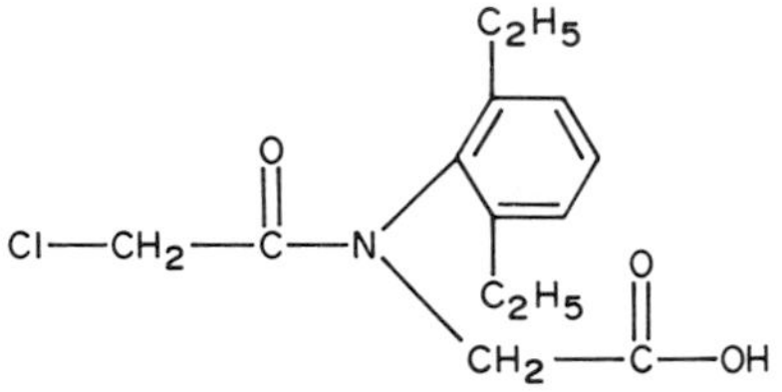

diethalyl

Uses. Diethyl is applied as a preemergence or preplant soil-incorporated treatment. It is used to control most annual grasses and many annual broadleaf weeds in sugar beets, table beets, and spinach. The rate applied varies depending on the soil type and the crop. In sugar beets it is applied at 3 to 6 lb/acre. It is also used as a tank mix with pyrazon or TCA, a tank mix or sequential application with EPTC or ethofumesate, or a sequential application of desmedipham or desmedipham plus phenmedipham in sugar beets. In red table beets for processing, diethalyl is applied at 2 to 6 lb/acre. In spinach, it is applied at 2 to 4 lb/acre.

Soil Influence. Diethalyl is adsorbed by soil colloids and this influences its performance and leaching in soils with various characteristics. Therefore, the higher rates are required on fine textured soils containing significant organic matter; leaching is also less in these soils. Soil microorganisms appear to play a major role in degradation of this compound. The residual herbicidal activity persists about 6 to 10 weeks.

Metolachlor Metolachlor is the common name for 2-chloro-*N*-(2-ethyl-6-methylphenyl)-*N*-(2-methoxy-1-methylethyl)acetamide. The trade names are Dual® and Pennant®. Dual® formulations include an emulsifiable concentrate (8 lb/gal) and granules (25%). Pennant® is a granule formulation (5%) used in ornamental nurseries. Metolachlor is a white to tan liquid with a moderate water

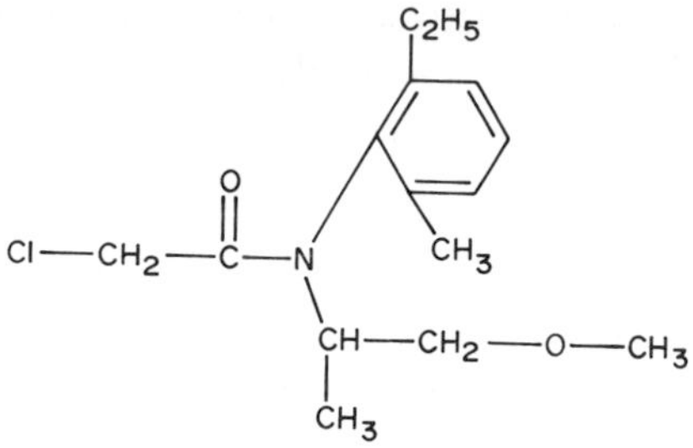

metolachlor

solubility of 530 ppm at 20°C. The acute oral LD_{50} of technical metolachlor is 2.8 g/kg for rats.

Uses. Metolachlor is applied as a preemergence or preplant soil-incorporated treatment for control of most annual grasses, yellow nutsedge, and many broadleaf weeds in beans, chickpeas, cotton, corn, milo, mung beans, okra, peanuts, peas, potatoes, safflower, soybeans, southern peas, and certain ornamentals. It is also used in combination with several other herbicides to increase the spectrum of weeds controlled. See Table A-0 for specific weeds controlled and Table 8-2 for detailed information on the crops. Many uses of metolachlor are similar to those of alachlor.

Soil Influence. Metolachlor appears to behave similar to the other chloroacetamide herbicides in soil. Leaching is restricted by adsorption onto clay and organic matter, with organic matter being the more important restraint. Dissipation of metolachlor in soils is relatively rapid. The half-life (time required to lose one-half its phytotoxicity) is 2 to 3 weeks in the southern United States and 4 to 7 weeks in the north. This is probably related to warmer soil temperatures in the south, which increase microbiological activity. Metolachlor and its degradation products combine primarily with soil humic material; a small amount is converted to carbon dioxide.

Propachlor Propachlor is the common name for 2-chloro-*N*-(1-methylethyl)-

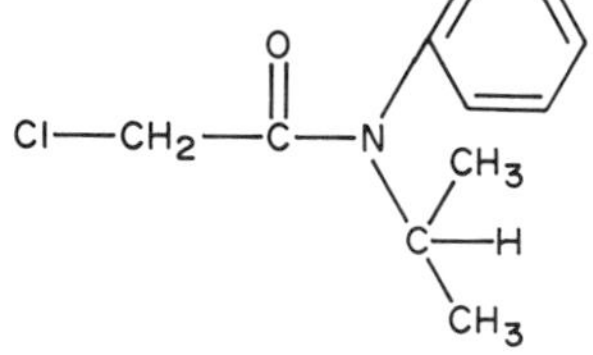

propachlor

N-phenylacetamide. The trade name is Ramrod®. It is formulated as a flowable liquid (4 lb/gal), a granule (20%), and a package mix with atrazine. It is a light-tan solid with a moderate water solubility of 580 ppm at 20°C. The acute oral LD_{50} is 710 mg/kg for rats.

Uses. Propachlor is used as a preemergence treatment to control most annual grasses and many annual broadleaf weeds in corn (field, hybrid seed, silage, sweet), flax, milo, peas (green for processing), pumpkins (processing), and soybeans (seed production). In the corn, it is applied at 4 to 6 lb/acre alone but the rate is reduced to 2.5 to 4 lb/acre when used in combination with atrazine as a tank mix or the package mix. In flax, propachlor is applied at 4 lb/acre. In milo, it is applied at 4 to 5 lb/acre alone but the rate reduced to 2.5 to 4 lb/acre when used in combination with atrazine as a tank mix or the package mix. When used as a tank mix with propazine, the propachlor rate applied is 2.5 to 3 lb/acre. Propachlor is also applied at 3 to 4 lb/acre as a tank mix with bifenox in milo. In green peas for processing propachlor applied at 4 to 5 lb/acre and in pumpkins for processing the use rate is 4 to 6 lb/acre. In soybeans for seed production, propachlor is applied at 4 to 5 lb/acre using the granule formulation. There are some limitations on some of the above uses relative to geographic location, soil type, and formulation (see product label).

Soil Influence. Propachlor is adsorbed by soil clay and organic matter and is therefore not subject to excessive leaching. There are no soil-persistence problems because the chemical is rapidly degraded and not phytotoxic 4 to 6 weeks after application in most soils. However, it persists somewhat longer in organic soils.

Mode of Action of Chloroacetamides The mode of action of all of the chloroacetamide herbicides is probably similar; however, not all phenomenon discussed below have been demonstrated for every one.

These compounds inhibit early seedling growth; this effect is most evident on root growth. These responses appear to be associated with an interference with both cell division and cell enlargement. They do not appear to inhibit seed germination, that is, emergence of the radical through the seed coat. However, they usually kill or affect susceptible weeds before emergence from the soil.

Results differ on whether these compounds are absorbed predominantly by the roots or emerging shoots of the germinating weed seedlings. Evidence suggests that grasses absorb these herbicides mainly through the emerging shoots, whereas broadleaf plants absorb them through both emerging shoots and roots. The different chloroacetamides may vary in this regard. Translocation appears to be mainly in the apoplast; however, limited symplastic transport may also occur.

Most evidence suggests that the chloroacetamides or their degradation products undergo conjugation with glutathione and/or glucose. Hydrolysis of the parent molecule may also occur. Differential rates of detoxification of the chloroacetamide herbicides among plant species by conjugation with glutathione may explain their differential tolerance (Duke, 1985).

TABLE 8-2. Metolachlor Uses

Crop	Rate (lb/acre)	Combinations	Remarks
Beans	1.50–3.0	—	Many types
	1.50–3.0	Chloramben	Tank mix
	1.50–3.0	Triflualin	Tank mix
	1.25–2.5	EPTC	Tank mix or sequential
Chickpeas or southern peas	1.50–3.0	—	—
Cotton	1.50–2.0	—	—
	1.50–2.0	Prometryn	Tank mix
Corn	1.50–4.0	—	—
	1.50–3.0	Atrazine, cyanazine, dicamba, or simazine	Tank mix
	1.50–3.0	Atrazine plus paraquat or glyphosate	Tank mix
	1.50–3.0	Atrazine plus simazine plus paraquat or glyphosate	Tank mix
Milo[1]	1.50–2.5	—	Protectant[1]-treated seed
	1.50–2.5	Atrazine, propazine, or terbutryn	Protectant[1]-treated seed, Tank mix
Mung beans	1.50–3.0	—	—
Peanuts	1.50–3.0	—	—
	1.50–2.0	Benefin	Tank mix

Peas	1.50–3.0	—	—
Potatoes	1.50–3.0	—	—
	1.50–3.0	Linuron	Tank mix
	1.50–3.0	Metribuzin	Tank mix or sequential
Safflower	1.50–3.0	—	—
Soybeans	2.50–3.0	—	—
	1.25–3.0	Linuron or trifluralin	Tank mix
	1.25–3.0	Chloramben or metribuzin	Tank mix or sequential
	1.25–3.0	Chloramben plus metribuzin	Tank mix
	1.25–3.0	Linuron or metribuzin plus paraquat or glyphosate	Tank mix
Ornamentals, certain	2.00–3.0	—	EC formulation: field and liner
	2.00–3.0	—	Granule formulation: field, liner, and container

There are limitations on some of the above uses relative to geographic location, soil type, and application method; see product label.

[1]Concept®

In addition to inhibiting growth, cell division, cell enlargement, and epicuticular wax formation, the chloroacetamide herbicides have been shown to alter membrane structure and function. These changes include increased leakage of solutes, inhibition of ion and amino acid uptake, and altered mitochondria, chloroplast, and other cellular membranes. These changes induced by the chloroacetamides suggest that lipid metabolism modifications may be a major site of action. However, many other metabolic reactions also have been demonstrated to be altered by these herbicides and the primary biochemical site of action has not been unequivocally determined. In fact, there may be several sites of action. Ashton and Crafts (1981), Fedtke (1982), Corbett et al. (1984), and Duke (1985) give additional details on the mode of action of the chloroacetamide herbicides. More recently, alachlor was found to inhibit anthocyanin and lignin accumulation in excised mesocotyl sections of sorghum (Molin *et al.*, 1986).

Other Soil-Applied Amides

In contrast to the chloroacetamides, the other soil-applied amides have at least one ring in the R_1 position rather than a monochloromethyl group. The other soil-applied amides are used as a preemergence or preplant soil-incorporated treatment. They have little, if any, postemergence activity. They tend to control a broader range of annual broadleaf weeds than the chloroacetamide herbicides. However, they are somewhat less effective on annual grasses.

Diphenamid Diphenamid is the common name for *N*,*N*-dimethyl-α-phenyl benzeneacetamide. The trade name is Enide®. It is formulated as a wettable powder (50 and 90%), a liquid dispersion (4 lb/gal), and a granule (5%). It is white to off-white solid with a moderate water solubility of 261 ppm at 27°C. The acute oral LD_{50} is 1.3 g/kg for rats.

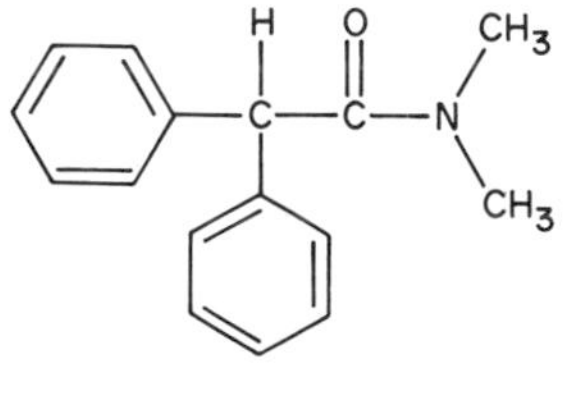

H O CH3
C—C—N
CH3

diphenamid

Uses. Diphenamid is used as a preemergence or preplant treatment to control annual grass and broadleaf weeds in cotton, okra, peanuts, peppers, potatoes, sweet potatoes, tomatoes, tobacco, and certain small fruits, tree fruits, turf, and several ornamental species. See Table A-0 for specific weeds controlled and Table 8-3 for detailed information on the crops.

Soil Influence. Diphenamid is leached fairly rapidly in sandy soils but more

TABLE 8-3. Diphenamid Uses

Crop	Rate (lb/acre)	Combinations	Remarks
Cotton	2.0–4.0	—	—
Okra	3.0–5.0	—	—
Peanuts	4.0–6.0	—	At planting
	2.0–3.0	—	At cracking or 7 days later
	2.0	—	At layby
Peppers	3.0–5.0	—	—
Potatoes	4.0–6.0	—	—
Sweet potatoes	4.0–6.0	—	Field or seedbed
Tobacco	4.0–6.0	—	Field or seedbed
Tomatoes	4.0–6.0	—	1 week before to 1 month after direct seeding or transplanting
	6.0	—	Before winter rains for spring direct-seeded, Calif, only
	6.0	Pebulate	PPI[1], direct seeded, Calif only
	4.0	Trifluralin	PPI[1], direct-seeded, Calif only
Small fruits[2]	4.0–6.0	—	—
Trees, fruit[3]	4.0–6.0	—	—
Turf, dichondria	10.0	—	at seeding or established
Turf, bermuda	4.0	—	Established
Ground covers[4]	8.0	—	Established
Flower crops	6.0–8.0	—	Field, at transplanting
Ornamentals, woody	4.0	—	Field

There are some limitations of some of the above uses relative to geographic location and soil type; see product lables.

[1] Preplant soil-incorporated.
[2] Strawberries and nonbearing blackberries, raspberries.
[3] Apples, peaches, and nonbearing cheries, limes, oranges.
[4] English ivy, ice plant, periwinkle myrtle, strawberries.

slowly in loan or clay soils. It is not tightly adsorbed onto soil colloids. Microorganisms appear to be involved in its degradation in soils. Under warm-moist soil conditions it normally persists 3 to 6 months but lasts longer under low rainfall conditions.

Mode of Action. Diphenamid usually kills susceptible seedling plants before they emerge from the soil but does not appear to inhibit seed germination. At sublethal concentrations, it inhibits root development of many species. In tolerant species such as tomato, marginal leaf chlorosis may occur after normal seed germination and seedling emergence at relatively high rates. Diphenamid is readily absorbed by roots and rapidly translocated to the tops of plants with accumulation in the leaves indicating apoplastic transport. Foliar absorption appears to be limited. Studies with tomatoes and strawberries show that diphenamid is metabolized by higher plants with the major metabolite being *N*-methyl-2,2-diphenylacetamide.

Figure 8-5. Annual weed control in tomatoes with a preemergence application diphenamid, followed by a lay-by treatment with trifluralin. (Elano Products Company, a division of Eli Lilly and Company.)

Glucose conjugates of diphenamid have also been reported. It appears to inhibit RNA synthesis. Diphenamid has been reported to inhibit the uptake of inorganic ions by roots and influence the distribution of calcium within the plant. These effects are discussed in detail by Ashton and Crafts (1981), Fedtke (1982), Corbett et al. (1984), or Duke (1985).

Isoxaben Isoxaben is the common name for *N*-[3-(1-ethyl-1-methylpropyl)-5-

isoxaben

isoxazolyl]-2,6-dimethoxybenzamide. The trade name is Gallery® and Snapshot® as a package mix with oryzalin. Isoxaben is formulated as a dry flowable (75%) and a granule (1%). It is a off-white to tan solid with a very low water solubility of < 1 ppm. The acute oral LD_{50} of the technical material is > 10 g/kg for rats and mice.

Uses. Isoxaben is a relatively new herbicide for the control of annual broadleaf weeds in established turfgrasses, landscape ornamentals, and container grown ornamentals. The growth of certain annual grass weeds is suppressed at the higher use rates. The use rate is in the range of 0.25 to 2 lb/acre. Many orchard and vine crops are relatively tolerant to isoxaben and it may ultimately be used in these crops. Isoxaben alone is for use in turfgrasses and in combination with oryzalin for use in ornamentals.

Soil Influence. Isoxaben is subject to slight leaching. Degradation by microorganisms appears to be the primary means of loss from soils. Dissipation studies indicate that the half-life of isoxaben in soil is 5 to 6 months.

Mode of Action. Seedlings of susceptible species often do not emerge from soils treated with isoxaben but seed germination is not inhibited. It has a marked inhibitory effect on root growth and hypocotyl development. Interference with cell division has been suggested.

Isoxaben is absorbed by roots, transported to the stem and leaves, and has only limited herbicidal activity when applied to leaves. These observations indicate that it is translocated mainly in the apoplastic system. It is primarily degraded in plants by hydroxylation of the alkyl side chain. Unidentified isoxaben conjugates have also been reported. The biochemical site(s) of action of isoxaben has not been determined.

Napropamide Napropamide is the common name for *N,N*-diethyl-2-(1-naphthalenyloxy)propanamide. The trade name is Devrinol®. It is a white crystalline solid with a low water solubility of 73 ppm at 20°C. It is formulated as a wettable powder (50%), emulsifiable concentrate (2 lb/gal), and granule (10%). The acute oral LD_{50} is > 5.0 g/kg for rats.

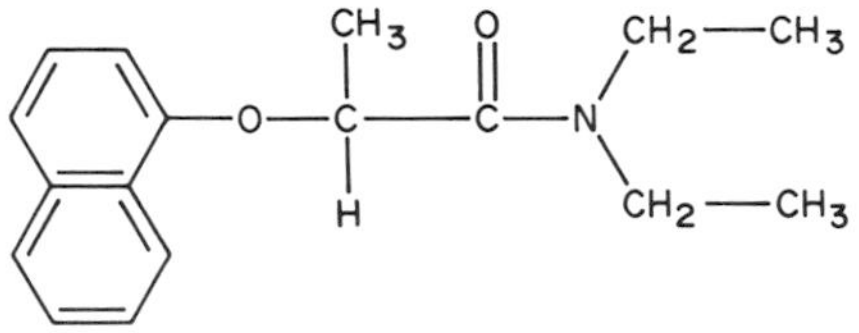

napropamide

Uses. Napropamide is a preemergence or preplant soil-incorporated herbicide

TABLE 8-4. Napropamide Uses

Crop	Rate (lb/acre)	Combinations	Remarks
Orchards	4.0	—	—
(citrus)[1]	4.0	Paraquat	Tank mix
	4.0	Simazine	Tank mix
	4.0	Simazine plus paraquat	Tank mix
Orchards	4.0	—	—
(deciduous)[2]	4.0	Glyphosate	Tank mix
	4.0	Oxyfluorfen	Tank mix, exclude apples
	4.0	Paraquat	Tank mix
	4.0	Simazine	Tank mix; apples, pears, peaches only
	4.0	Simazine plus paraquat	Tank mix; apples, pears, peaches only
Orchards	4.0	—	—
(nuts)[3]	4.0	Glyphosate	Tank mix
	4.0	Oxyfluorfen	Tank mix
	4.0	Paraquat	Tank mix
	4.0	Simazine	Tank mix
	4.0	Simazine plus paraquat	Tank mix
Orchards (subtropical)[4]	4.0	—	—
Small fruits[5]	4.0	—	—
Artichokes	4.0	—	—
Asparagus	4.0	—	Established beds only
Cole crops[6]	1.0–2.0	—	—
Cranberries	4.0–15.0	—	—
Egg plant	1.0–2.0	—	—
Mint	4.0	—	—
Peppers	1.0–2.0	—	—
Tobacco	1.0–2.0	—	Not seed beds
Tomato	1.0–2.0	—	—
	1.0–2.0	Pebulate	Tank mix
Turf	2.0–3.0	—	Established warm season grasses
Ornamentals	4.0–6.0	—	Nursery stock, ground covers, dichondria

There are limitations on some of the above uses relative to geographic location, soil type, formulation, and application method; see product labels.

[1]Grapefruit, lemons, oranges, tangerines.
[2]Apples, apricots, cherries, nectarines, peaches, pears, plums, prunes.
[3]Almonds, filberts, pecans, pistachos, walnuts.
[4]Avocados, kiwi, olives, persimmons.
[5]Blackberries, boysenberries, loganberries, raspberries, strawberries, grapes.
[6]Broccoli, brussels sprouts, cabbage, cauliflower.

used to control most annual grasses and many broadleaf weeds in many crops including orchards, small fruits, vegetable crops, mint, tobacco, turf, and ornamentals. It is also used in combination with other herbicides in some crops. See Table A-0 for specific weeds controlled and Table 8-4 for detailed information on the crops.

Soil Influence. Napropamide is resistant to leaching in most mineral soils. It is slowly decomposed by soil microorganisms. When incorporated into moist loam or sandy-loam soils at 70 to 90°C the half-life was 8 to 12 weeks. However, under conditions limiting microorganism growth, it may persist more than 9 months and cause injury to susceptible crops planted within this period.

Mode of Action. Napropamide inhibits the growth and development of roots. It has been shown to be rapidly absorbed by tomato roots and readily translocated throughout the stem and leaves. However, upward translocation from root absorption in corn was much slower. The distribution pattern suggests that it is primarily translocated in the apoplast. Napropamide is rapidly metabolized to water-soluble metabolites in tomato and several fruit trees. The major metabolites appear to be hexose conjugates of 4-hydroxynapropamide.

It has been reported that napropamide inhibits photosynthesis, RNA synthesis, and protein synthesis but not respiration or lipid synthesis in isolated bean leaf cells. However, these may be secondary responses. In another report using pea roots, cell division and DNA synthesis were more sensitive than protein synthesis and RNA synthesis, oxygen uptake, and ethylene evolution were not affected (DiTomaso, 1987). This study also reported that napropamide increases the levels of the polyamines, putrescine and cadaverine 600 and 100%, respectively, and these high levels may be phytotoxic.

Naptalam Naptalam is the common name for 2-[(1-naphthalenylamino)carbonyl]benzoic acid. The trade name is Alanap®. It is formulated as the sodium salt in water (2 lb/gal) and a granule (10%). It is also formulated

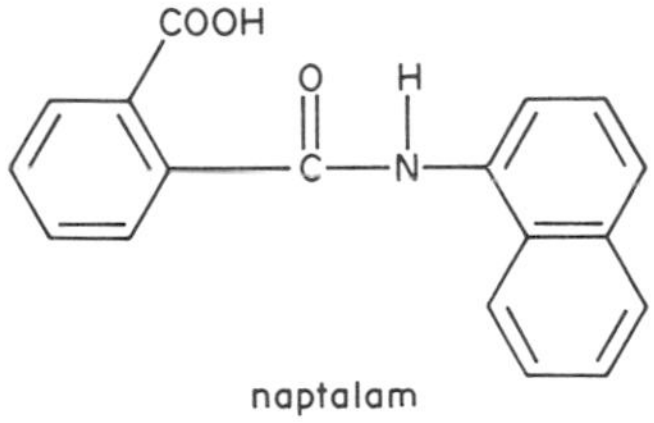

COOH

O H

C—N

naptalam

with 2,4-DB as a liquid package mix with the trade name of Rescue®. It is a purple crystalline solid. The acid form has a water solubility of 200 ppm and the sodium salt a very high water solubility of 230,800 ppm. The acute oral LD_{50} for rats is > 8.2 g/kg for the acid and 1.8 g/kg for the sodium salt.

TABLE 8-5. Naptalam Uses

Crop	Rate (lb/acre)	Combinations	Remarks
Cucurbits[1]	3.0–4.0	—	At seeding and/or 1 month later, after transplanting
	2.0–4.0	Bensulide	Preplant or preemergence
Peanuts	2.0–6.0	—	Preemergence
Soybeans	4.0	—	Preemergence
	1.0–1.5	2,4-DB	Postemergence
Ornamentals, woody	2.0–4.0	—	Established, before or after transplanting

[1]Cantaloupe, muskmelons, cucumbers, watermelons.

Uses. Naptalam is a preemergence herbicide used primarily for the control of broadleaf weeds in cucurbits, peanuts, soybeans, and woody ornamentals. Combinations with other herbicides are also used to broaden the spectrum of weeds controlled in cucurbits and soybeans. See Table A-0 for specific weeds controlled and Table 8-5 for detailed information on the crops.

Soil Influence. Naptalam is subject to extensive leaching in porous soils. If heavy rains occur shortly after seeding and herbicide application, both crop injury and poor weed control may result. It is relatively nonpersistent and presents no soil-residual problem. Weeds are usually controlled from 3 to 8 weeks after application of the herbicide.

Mode of Action. Naptalam has the unique property of acting as an antigeotropic agent; growing shoots and roots have a tendency to lose their ability to grow up or down, respectively. This may be associated with some of its herbicidal action. However, it appears to act primarily as an inhibitor of seed germination and growth responses induced by the normal plant hormones indole-3-acetic acid (IAA) and gibberellic acid (GA) (Ashton and Crafts, 1981). Naptalam is degraded into α-naphthylamine and phthalic acid.

Naptalam is an auxin (IAA) antagonist; it inhibits the polar transport of IAA and at submicromolar concentrations it stimulates IAA absorption (Duke, 1985). The latter results from inhibition of the efflux of IAA from cells and has been proposed to explain the inhibition of polar transport. This may be related to naptalam binding to sites on plant cell membranes.

Pronamide Pronamide is the common name for 3,5-dichloro(*N*-1,1-dimethyl-2-propynl)benzamide. The trade name is Kerb®. It is an off-white solid with a low water solubility of 15 ppm and formulated as a wettable powder (50%). The acute oral LD_{50} is 5.6 g/kg for female rats and 8.4 g/kg for male rats.

pronamide

Uses. As a preemergence, preplant soil-incorporated, or postemergence treatment, pronamide controls many broadleaf and grass weeds in small-seeded legumes, bermudagrass, fallow land, and certain vegetables, tree fruits, small fruits, and woody ornamentals. See Table A-0 for specific weeds controlled and Table 8-6 for detailed information on the crops.

With preemergence applications, pronamide is most effective with an abundance of water from rainfall or sprinkle irrigation. Soil incorporation is required with furrow irrigation. Apparently its relatively low water solubility and considerable affinity for colloidal adsorption sites in soil require considerable water to make it biologically active.

Soil Influence. Pronamide is readily adsorbed on organic matter and other colloidal exchange sites and therefore leaches very little in most soils. It has intermediate persistence in soil, 3 to 8 months.

TABLE 8-6. Pronamide Uses

Crop	Rate (lb/acre)	Combinations	Remarks
Artichoke	0.50–1.00	—	Seed beds
	1.00–2.00	—	New plantings
	2.00–4.00	—	Established plantings
Bermudagrass, established	0.50–1.00	—	PE[1] to weeds, turf/sod
	0.75–1.50	—	Weeds emerged, turf/sod
Fallow land	0.25–0.50	—	Followed by small grains
	0.25–0.38	Chlorsulfuron	Followed by small grains
Legumes[2]	0.50–2.00	—	New or established stands
Lettuce	1.00–2.00	—	PE[1], PPI[1], or post to crop-weeds not emerged
Tree fruit[3]	1.00–4.00	—	Established
Small fruit	1.00–2.00	—	Blueberries
	1.00–3.00	—	Blackberries, raspberries
	1.00–4.00	—	Grapes
Ornamentals, woody	1.00–2.00	—	Established, field grown

There are some limitations of the above uses relative to geographic location, time of application, irrigation, etc.; see product label.

[1] PE, preemergence; PPI, preplant soil-incorporated.

[2] Alfalfa, clover, birdsfoot trefoil, crown vetch, sainfoin.

[3] Apples, cherries, nectarines, peaches, pears, plums, prunes.

Mode of Action. Pronamide inhibits cell division and plant growth. These responses are considered to be related to the loss of cortical and spindle microtubules (Bartels and Hilton, 1973). It is readily absorbed by roots and translocated upward and distributed throughout the plant. Its translocation from leaves is not appreciable (Carlson, 1972). This suggests translocation in the apoplastic system. Pronamide is slowly metabolized in higher plants by means of alterations of the aliphatic side chains (Yih and Swithenbank, 1971).

Foliar-Applied Amides

Propanil Propanil is the common name for *N*-(3,4-dichlorophenyl)-propanamide. The main trade names are Prop-Job® and Stam®; others also exist. Stampede® is the trade name for a propanil-MCPA package mix. It is a light-brown to gray-black solid with a moderate water solubility of about 500 ppm. It is formulated as emulsifiable concentrates. The technical material has an acute oral LD_{50} of 1.4 g/kg for rats.

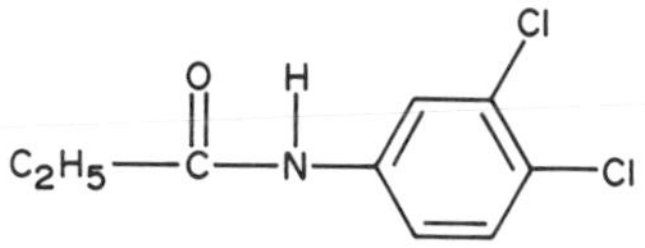

C_2H_5—C(=O)—N(H)—[3,4-dichlorophenyl]

propanil

Uses. Propanil is applied postemergence to both the crop and the weeds. It has little if any herbicidal activity via the soil. It is used alone in rice and in combination with MCPA in spring barley and spring and durum wheat. Propanil controls several annual grasses and certain broadleaf weeds; see Table A-0 for the specific weeds controlled. The time of application is critical for both weed control and crop selectivity. The weeds should be small, grasses in the 1- to 4-leaf stage. Spring barley and durum wheat should not be beyond the 4-leaf stage and spring wheat (hard and red) not beyond the 5-leaf stage. The stage of rice develop is less critical. Uniform coverage of the foliage is essential for good weed control since it is a contact material. Spray drift of the chemical onto sensitive plants is a potential hazard.

Although rice is quite tolerant to propanil, the herbicide may severely injure this crop when certain insecticides (carbaryl or organic phosphates) are applied within 14 days before or after propanil is used. These insecticides inhibit the action of the enzyme in rice that decomposes propanil and thus leads to its selectivity.

Soil Influence. Since propanil is a contact herbicide applied to the foliage, soil type has no effect on its performance. It is rapidly broken down in soils, 1 to 3 days under the warm-moist soil conditions. Therefore, propanil presents no residual problem for subsequent crops.

Mode of Action. Propanil is a contact herbicide and causes chlorosis followed by necrosis when applied to the leaves of susceptible species. A speckled pattern may be observed when it remains as discreet small droplets on the leaves.

It must be absorbed relatively slowly by leaves since a 4- to 8-hour rain-free period after application is suggested. Translocation of propanil within a leaf or from leaves to the rest of the plant is very limited (Yih et al., 1968).

Propanil is rapidly hydrolyzed in rice by an aryl acylamidase enzyme into nonphytotoxic molecules. This reaction is much slower in barnyardgrass and is the basis of differential tolerance between these two species. Lignin and *N*-glucoside conjugates of propanil have also been reported.

Inhibition of photosynthesis is considered to be the main effect of propanil. However, it also inhibits state 3 respiration and uncouples state 4 respiration. Propanil can also alter membrane functions as evidenced by increased betacyanin leakage from red beet tissue, inhibition of ion uptake by roots, and changes in cation permeability of mitochondria. Ashton and Crafts (1981), Fedtke (1982), Corbett et al. (1984), and Duke (1985) give additional details on the mode of action of propanil.

AMINO ACIDS

Herbicides that are derivatives of amino acids are referred to as amino acid herbicides although they are not amino acids per se. They could also be considered to be aliphatic herbicides since they do not contain a ring in their chemical structure. The amino acid herbicides include glyphosate, sulfosate, and glufosinate. Glyphosate is the only one of these compounds commercially available at this time and it will be covered in detail. The other two compounds are in late stages of development and therefore will be discussed only briefly.

Glyphosate

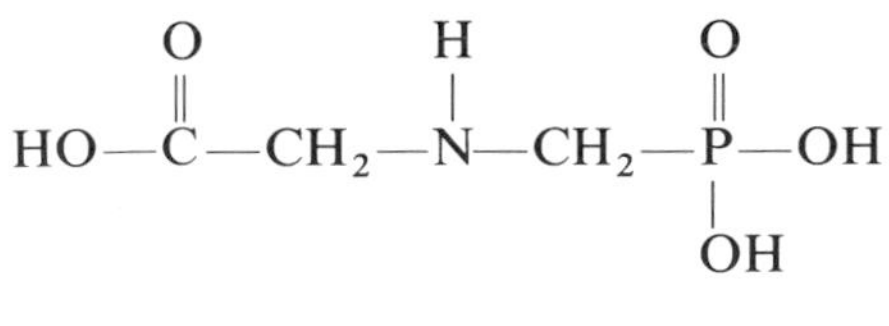

$$HO-\overset{\overset{\displaystyle O}{\|}}{C}-CH_2-\overset{\overset{\displaystyle H}{|}}{N}-CH_2-\underset{\underset{\displaystyle OH}{|}}{\overset{\overset{\displaystyle O}{\|}}{P}}-OH$$

glyphosate

Glyphosate is the common name for *N*-(phosphonomethyl)glycine. It is a white solid with a very high water solubility of 12,000 ppm at 25°C. The herbicide formulation contains 4 lb/gal of the isopropylamine salt of glyphosate, equivalent to 3 lb/gal of the acid, glyphosate. The principal trade name of this product is Roundup®; several other trade names are also used. The isopropylamine salt of glyphosate is even more water soluble than the acid form. The acute oral LD_{50} is 5.6 g/kg for glyphosate and 5.4 g/kg for Roundup®.

Glyphosate was discovered and developed by Monsanto. The herbicidal properties of glyphosate and its salts were first described by Baird et al. (1971). Information on essentially all aspects of glyphosate can be found in the extensive monograph, *The Herbicide Glyphosate*, edited by Grossbard and Atkinson (1985). A more recent review by Duke (1988) emphasizes the mode of action of glyphosate. Additional reviews on glyphosate are cited at the end of this chapter.

Uses Glyphosate is a nonselective broad-spectrum herbicide. It controls annual, biennial, and perennial herbaceous species of grasses, sedges, and broadleaf weeds. It also controls many woody brush and tree species. It is applied to the foliage and has little effect when applied to the soil.

Glyphosate is used in noncrop areas and also in crops when contact with aboveground parts of the crop can be avoided. Directed sprays or the use of recirculating sprayers or wick-wiper applicators are used in crops. Directed sprays are used in orchards, vineyard, and similar crops. Recirculating sprayers and wick-wiper applicators are used when the weeds are taller than the crop plants. Recirculating sprayers direct the herbicide solution horizontally onto the weeds but over the top of the crop (see Figure 7-10). Wick-wiper applicators deposit the herbicide solution on the weeds but on the crop (see Figure 7-11). Spot treatments are also used to control small areas of serious weeds in certain crops. However, severe crop damage usually occurs if the foliage of the crop is sprayed in the treated areas.

Spray drift onto desirable plants must be avoided when glyphosate is applied to either crop or noncrop areas.

Tank mixtures of glyphosate with soil-applied residual herbicides (e.g., ureas and triazines or foliar-applied herbicides (e.g., phenoxys, organic arsenicals, dalapon, and paraquat) may reduce the activity of glyphosate.

Soil Influence Since glyphosate is applied to the aboveground parts of the weeds soil type has little effect on its performance. It is strongly bound to soil particles and therefore has low phytotoxicity via soils and leaching is minimal. Glyphosate is degraded fairly rapidly in most soils by microorganisms and the half-life is usually less than 60 days.

Mode of Action The common symptom of glyphosate injury is foliar chlorosis followed by necrosis. Regrowth of perennial broadleaf and woody plants often shows malformed leaves as well as white spots and striations (Putnam, 1976; Fernandez and Bayer, 1977: Marriage and Kahn, 1978). Multiple shoots ("witches' broom") commonly develop from a bud at a single node (Figure 8-6). At the ultrastructure level, disruption of the chloroplast envelope and swelling of the endoplasmic reticulum with subsequent formation of vesicles have been reported (Campbell et al., 1976).

Glyphosate is readily absorbed by leaves and translocated throughout the plant. It moves via the symplastic system and probably later in the apoplastic system, as evidenced by its high mobility. Glyphosate appears to be degraded

Figure 8-6. Development of multiple shoots ("witches-broom") in bermudagrass 15 days after treatment with 0.5% glyphosate. (C. Fernandez.)

slowly in most higher plants (Gottrup et al., 1976; Duke, 1988); however, Coupland (1985) suggests that it may be extensively metabolized in some species.

Most research on the mechanism of action of glyphosate supports the early findings of Jaworski (1972), which showed that glyphosate interferes with aromatic amino acid biosynthesis. Subsequent research had led to the conclusion that the specific site of action of glyphosate is an enzyme in the shikimic acid pathway, 3-phosphoshikimate-1-carboxyvinyltransferase (PSCV transferase) formally known as 5-enolpyruvoyl shikimate phosphate synthase (EPSP synthase). The strong inhibition of this enzyme results in a decrease in the level of aromatic amino acids. This ultimately leads to a slow cessation of growth and other symptoms. Although other biochemical effects of glyphosate have been reported, none appears to be as consistent and compelling as the PSCV transferase inhibition theory. Much of this research has been reviewed by Corbett et al. (1984) and Duke (1985; 1988).

Sulfosate

Sulfosate (Touchdown®) and glyphosate (Roundup®) are similar in chemical structure and herbicidal activity. They are both salts of *N*-phosphonomethylglycine, sulfosate is the trimethylsulfonium salt and glyphosate the isopropylamine salt. They are both postemergence herbicides that are very effective for the control of a wide range of annual and perennial grass and broadleaf weeds.

Glufosinate

Glufosinate is the common name for DL-homoalanin-4-yl(methyl)phosphine. It is formulated as the ammonium salt and the trade name is Ignite®. It is used as a foliar spray to control a broad spectrum of emerged annual and perennial grass and broadleaf weeds. It is primarily a contact herbicide with limited systemic activity and it has little soil or residual activity. Contact with foliage or green tissue of desirable vegetation can cause severe injury or death. It is being evaluated for nonselective postemergence weed control in soybeans, pome fruit, stone fruit, citrus, vine crops, and noncrop areas.

SUGGESTED ADDITIONAL READING

Ashton, F. M., and A. S. Crafts, 1981, *Mode of Action of Herbicides*, Wiley, New York.

Baird, D. D., R. P. Upchurch, W. B. Homesley, and J. E. Franz, 1971, *Proc. NCWCC* **26**, 64.

Bartels, P. G., and J. L. Hilton, 1973, *Pest. Biochem. Physiol.* **3**, 463.

Blanchard, F. A., 1954, *Weeds* **3**, 274.

Campbell, F. W., J. O. Evans, and S. C. Reed, 1976, *Weed Sci.* **24**, 22.

Carlson, W. C., 1972, Ph.D. dissertation, University of Illinois, Champaign.

Corbett, J. R., K. Wright, and A. C. Baillie, 1984, *The Biochemical Mode of Action of Pesticides*, Academic Press, New York.

Coupland, D., 1985, Metabolism of glyphosate in plants, in E. Grossbard and D. Atkinson, Eds., *The Herbicide Glyphosate*, pp. 25–34, Butterworth, London.

Crop Protection Chemical Reference, 1989, Wiley, New York (revised annually).

DiTomaso, J. M., 1987, Ph.D. dissertation, University of California, Davis.

Duke, S. O., 1985, *Weed Physiology*, Vol. II, *Herbicide Physiology*, CRC Press, Boca Raton, FL.

Duke, S. O., 1988, Glyphosate, in P. C. Kearney and D. D. Kaufman, Eds., *Herbicides*, Vol. 3, pp. 1–70, Dekker, New York.

Farm Chemicals Handbook, 1988, Meister, Willough, OH. (revised annually).

Fedtke, C., 1982, *Biochemistry and Physiology of Herbicide Action*, Springer-Verlag, New York.

Fernandez, C. H., and D. E. Bayer, 1977, *Weed Sci.* **25**, 396.

Foy, C. L., 1975, The chlorinated aliphatic acids,—P. C. Kearney and D. D. Kaufman, Eds., *Herbicides*, Vol. 1, pp. 399–452, Dekker, New York.

Gottrup, O., P. A. O'Sulivan, R. J. Schraa, and W. H. Vanden Born, 1976, *Weed Res.* **16**, 197.

Grossbard, E., and D. Atkinson, 1985, *The Herbicide Glyphosate*, Butterworth, London.

Herbicide Handbook, 1989, Weed Science Society of America, Champaign, IL.

Ingle, M., and B. J. Rogers, 1961, *Weeds* **9**, 264.

Jaworski, E. G., 1972, *J. Agric. Food Chem.* **20**, 1195.

Jaworski, E. G., 1975, Chloroacetamides, in P. C. Kearney and D. D. Kaufman, Eds., *Herbicides*, Vol. 1, pp. 349–376, Dekker, New York.

Kitchen, L. M., R. E. Rieck, and W. W. Witt, 1980, *Weed Res.* **20**, 285.

Marriage, P. B., and S. U. Kahn, 1978, *Weed Sci.* **26**, 374.

Mayer, F., 1957, *Biochem. Z.* **358**, 433.

Meyer, R. E., and K. P. Buchholtz, 1963, *Weeds* **11**, 4.

Molin, W. T., E. J. Anderson, and C. A. Porter, 1986, *Pestic. Biochem. Physiol.* **25**, 105.

Muller, F., 1981, *Mitteilung aus der Biologischen Bundesanstalt für Landund Forstwirtschaft* **203**, 264.

Prasad, R., and G. E. Blackman, 1964, *J. Exp. Bot.* **15**, 48.

Putnam, A. R., 1976, *Weed Sci.* **24**, 425.

Still, G. G., and R. A. Herrett, 1976, Methylcarbamates, carbanilates, and acylanilides, in P. C. Kearney and D. D. Kaufman, Eds., *Herbicides*, Vol. 2, pp. 609–644, Dekker, New York.

Weed Control Manual and Herbicide Guide, 1988, Meister, Willough, OH. (revised annually).

Yih, R. Y., D. M. McRae, and H. F. Wilson, 1968, *Plant Physiol.* **43**, 1291.

Yih, R. Y., and C. Swithenbank, 1971, *J. Agric. Food Chem.* **19**, 314.

For herbicide use, see the manufacturer's or supplier's label and follow these directions. Also see the Preface.

9 Aryloxyphenoxys, Benzoics, and Bipyridiliums and Pyrazoliums

ARYLOXYPHENOXYS

The common name, chemical name, general chemical structure, and unique chemical group substitutions of these herbicides are given in Table 9-1. Various class names have been used for this group of herbicides since the first herbicide of this class, diclofop, was introduced. Diclofop has a phenoxy–phenoxy type structure and therefore it and other experimental compounds of this general chemical structure were referred to as phenoxy–phenoxys. However, subsequent compounds contained heterocyclic rings rather than a benzene ring and phenoxy–phenoxys was not an appropriate term for the class. The term phenoxypropanoates has been considered since they are all derivatives of phenoxypropanoic acid but this has not been widely accepted. Duke and Kenyon (1988) used the term polycyclic alkanoic acids for this class of herbicides and Ross and Lembi (1985) referred to them as aryloxyphenoxy herbicides. Although we consider each of these three term appropriate, we believe that the latter is more meaningful to most weed scientists considering its precedence and the historical development of these herbicides.

Aryloxyphenoxy herbicides include diclofop, fenoxaprop, fluazifop, haloxyfop, and quizalofop. However, haloxyfop is only in the late stages of development and registration is pending on a few crops. The common names refer to their acid form. However, all are formulated as a methyl, ethyl, or butyl ester for herbicidal use. Each is marketed as only one of these esters. They are often referred to by the common name followed by the ester designation, e.g., diclofop-methyl.

Aryloxyphenoxy herbicides are selective postemergence herbicides for the control of grasses in broadleaf crops. Their introduction filled a void in the herbicide arsenal since earlier herbicides did not provide this type of selectivity. Diclofop is also used to control annual grasses, particularly wild oats, in some small grain crops. All of these herbicides except diclofop control both annual and perennial grasses, diclofop controls only annual grasses.

Diclofop

Diclofop is the common name for (±)-2-[4-(2,4-dichlorophenoxy)-phenoxy]propanoic acid. It could be considered a diphenyl ether and placed

TABLE 9-1. Common Name, Chemical Name, and Chemical Structure of the Aryloxyphenoxys Herbicides[1]

$$R{-}O{-}C_6H_4{-}O{-}CH(CH_3){-}C(=O){-}OH$$

Common Name	Chemical Name	Chemical Structure R
Diclofop	(±)-2-[4-(2,4-Dichlorophenoxy)-phenoxy]propanoic acid	Cl, Cl
Fenoxaprop	(±)-2-[4-[(6-Chloro-2-benzoxazoly)-oxy]phenoxy]propanoic acid	Cl, N, C, O
Fluazifop	(±)-2-[4-[[5-(Trifluoromethyl)-2-pyridinyl]oxy]phenoxy]propanoic acid	CF_3, N
Haloxyfop	2-[4-[[3-Chloro-5-(trifluoromethyl)-2-pyridinyl]oxy]phenoxy]propanoic acid	Cl, CF_3, N
Quizalofop	(±)-2-[4-[(6-Chloro-2-quinoxalinyl)-oxy]phenoxy]propanoic acid	Cl, N, N

[1] All of these compoundsare formulated as a methyl, ethyl, or butyl ester for herbicidal use; see text for specific ester form of each herbicide.

in Chapter 11, but its herbicidal characteristics and other aspects of its chemical structure are more similar to the phenoxypropanoate herbicides. Its use form is the methyl ester, diclofop-methyl, and the trade names for this form are Hoelon® (United States) and Hoe-Grass® (Canada). It is a colorless solid with a high water solubility of 3000 ppm at 22°C. It is formulated as an emulsifiable concentrate (3 lb/gal) as well as a package mix with bromoxynil, One Shot® NW, or bromoxynil plus MCPA, One Shot®. Tank mixes of these combinations are also used. The acute oral LD_{50} of the technical product is 557–580 mg/kg for rats.

Uses Diclofop-methyl is a postemergence herbicide mainly used to control wild oats and many other annual grasses in wheat and barley. It is also used for this purpose in flax, lentils, peas, soybeans, and fallow land. Best results are obtained when most wild oat and annual grasses are in the one- to three-leaf stage. For volunteer corn in soybeans, it should be applied after essentially all corn has emerged but before it grows large enough to prevent thorough coverage, including the whorl. It also has phytotoxicity to annual grasses via soils and is

TABLE 9-2. Diclofop-methyl Uses

Crop	Rate (lb/acre)	Combinations	Remarks
Barley	0.50–1.0	—	—
	0.50–1.0	Bromoxynil	Tank mix or One Shot® NW
	0.50–1.0	MCPA	Tank mix or One Shot®
Flax	0.75–1.0	—	—
Lentils	0.75–1.0	—	—
Peas	0.75–1.0	—	Dry field
Soybeans	0.75–1.25	—	Before 6th trifoliate leaf
Wheat	0.75–1.0	—	—
	0.75–1.25	Bromoxynil	Tank mix or One Shot® NW
	1.00–1.25	Bromoxynil plus MCPA	Tank mix or One Shot® NW plus MCPA
	1.00–1.25	Chlorosulfuron	Tank mix, winter wheat
Fallow land	0.75–1.25	—	Spring or early summer

There are limitations on some of the above uses relative in stage of crop growth, geographic locations, use of other herbicides, and grazing or use of forage; see label.

used this way in winter wheat. It does not control broadleaf weeds or perennial grasses. See Table A-0 for specific weeds controlled and Table 9-2 for detailed information on the crops.

Soil Influence Studies have shown that diclofop-methyl does not leach downward or move laterally in soils. It dissipates relatively rapidly in soil; its half-life is 10 days in sandy soils and 30 days in sandy clay soils under aerobic conditions. Under anaerobic conditions it disappears even faster; up to 85% of the parent compound is metabolised within 2 days.

Mode of Action Symptoms from postemergence applications develop slowly. These include foliar and apical meristem modifications. Shoot and root growth are suppressed in wild oats and other susceptible grasses. This is followed by wilting and scattered chlorotic mottling; these spots unite later. Symptoms may differ with species and environmental conditions. Ryegrass develops a rusty color before death. Corn undergoes little color change, but develops a weak stem near the soil surface, which may cause lodging later. In the Pacific North west, wild oats may show leaf-tip burn with necrosis developing toward the leaf base. The apical meristem is often killed before visual symptoms appear on the foliage and the youngest leaf can be pulled out of the whorl of leaves due to the necrosis of the leaf base.

With preemergence applications, the first leaf may emerge from the soil, become purple-like in color, and die. All grass species susceptible to postemergence applications are also controlled with preemergence applications. In fact, crabgrass and johnsongrass seedlings are more sensitive to preemergence applications than to postemergence applications.

Diclofop-methyl is rapidly absorbed by plant foliage but translocation is very limited, less than other members of this class. This is probably the reason that it is ineffective for the control of perennial grasses and differs from the other phenoxpropanoate herbicides in this regard.

Diclofop-methyl is rapidly hydrolyzed to produce the phytotoxic free acid form, diclofop, in both tolerant (wheat) and susceptible (wild oats) species. In species, such as wheat, where further metabolism is the basis for tolerance, diclofop undergoes rapid hydroxylation of one of the rings and subsequently the formation of a stable nonphytotoxic conjugate. In the susceptible species, a secondary diclofop ester is formed by a reversible reaction, which can release diclofop into the plant tissues. This differential metabolism between these two species is considered to be the basis of selectivity (Shimabukuro, 1985).

Diclofop has been reported to cause an increase in membrane permeability and alter lipid content and composition in corn seedlings (Hoppe, 1980). Hoppe and Zacher (1986) also observed that diclofop inhibited fatty acid biosynthesis in isolated corn (susceptible) chloroplasts but not in bean (tolerant) chloroplasts. They concluded that this may be closely linked to its mechanism of action and contribute to its selectivity. This may be related to an inhibition of acetyl-CoA carboxylase in susceptible species and a lack of inhibition of this enzyme in tolerant species (see haloxyfop mode of action). Ashton and Crafts (1981), Fedtke (1982), Corbett et al. (1984), and Duke (1985) give additional details on the mode of action of diclofop.

Fenoxaprop

Fenoxaprop is the common name for (±)-2-[4-[(6-chloro-2- benzoxazolyl)-oxy]phenoxy]propanoic acid. Its use form is the ethyl ester, fenoxaprop-ethyl, and the trade names are Acclaim® for turf and Whip® for rice and soybeans. Both of these products are formulated as emulsifiable concentrates (1 lb/gal). It has an extremely low water solubility of 0.9 ppm at 25°C and the acute oral LD_{50} is 2.4 g/kg for male rats.

Uses Fenoxaprop ethyl is a postemergence herbicide used to control annual grasses in rice, soybeans, and certain established grass turfs. It also controls perennial johnsongrass from rhizomes. Within specified limits, the application rate varies with the stage of growth of the grasses to be controlled. In rice, it is applied at 0.15 to 0.20 lb/acre. In soybeans, it is applied at 0.10 to 0.15 1b/acre and the addition of a crop oil concentrate is required for the control of certain annual grass species. In certain established grass turfs, it is applied at 0.10 to 0.30 1b/acre. Rainfall within 1 hour of application may cause a reduction in grass control. Some of these uses have restrictions in regard to geographic location, stage of crop growth, harvest time after application, and grazing or feeding forage (see label).

Soil Influence Since fenoxaprop is a postemergence herbicide, soil type has no effect on its performance. It appears to disappear relatively rapidly from soils because most crops can be planted 30 days after application, 120 days for small grains.

Mode of Action Fenoxaprop inhibited the biosyntheses of fatty acids in isolated corn (susceptible) chloroplasts but not been (tolerant) chloroplasts (Hoppe and Zacher, 1986). They concluded that this may be closely linked to its mechanism of action and contribute to its selectivity. Fenoxaprop was more inhibitory to this process than diclofop. This may be related to an inhibition of acetyl-CoA carboxylase in susceptible species and a lack of inhibition of this enzyme in tolerant species (see haloxyfop mode of action). Fenoxaprop was reported to be metabolized into a number of water-soluble conjugates and bound residues in 15 days by soybean leaves (Wink et al., 1984).

Fluazifop

Fluazifop is the common name for (±)-2-[4-[[5-(trifluoromethyl)-2-pyridinyl]-oxy]phenoxy]propanoic acid. Fluzaifop-P is the common name for (*R*)-2-[4-[[5-(trifluoromethyl)-2-pyridinyl]oxy]phenoxy]propanoic acid. Fluazifop-P contains only one of the two isomers present in fluazifop. The isomer not present in fluazifop-P is the much less phytotoxic of the two isomers. Both types are formulated as the butyl ester and referred to as fluazifop-butyl (4 lb/gal) or fluazifop-P-butyl (1 lb/gal). The trade names are Fusilade® (fluazifop-butyl) and Fusilade® 2000 (fluazifop-P-butyl).

Fluazifop-butyl is a light straw-colored odorless liquid with a very low water solubility of 2 ppm. The acute oral LD_{50} of fluazifop-butyl is 3.3 g/kg for rats and of the formulated product, Fusilade®, it is 4.8 and 4.4 g/kg for male and female rats, respectively.

Uses Fluazifop-butyl and fluazifop-P-butyl are selective postemergence herbicides for the control of annual and perennial grasses in several broadleaf crops and fallow or noncrop land. The crops include asparagus, cotton, soybeans, small fruits, tree fruits and nuts, and ornamentals. See Table A-0 for specific weeds controlled and Table 9-3 for detailed information on the crops and fallow or noncrop land.

Soil Influence Fluazifop-butyl is of low mobility in soil and not subject to leaching but fluazifop is somewhat more mobile. Fluazifop-butyl is rapidly degraded in moist soils with a half-life of less than 1 week. The major degradation product is fluazifop, which has a half-life of about 3 weeks. Residual phytotoxicity depends on soil type and rainfall but normally susceptible crops can be planted 60 days after fluazifop-butyl is applied.

TABLE 9-3. Fluazifop-butyl and Fluazifop-P-butyl Uses

Crop	Rate (lb/acre)	Combinations	Remarks
Asparagus	0.50 (0.25–0.375)[1]	—	Established, nonbearng
Cotton	0.125–0.50 (0.094–0.375)	—	—
Soybeans	0.125–0.25 (0.094–0.188)	—	—
	0.125–0.25 (0.094–0.188)	Acifluorfen	Tank mix or sequential
	(0.094–0.188)	Bentazon	Tank mix or sequential, only fluazifop-P-butyl
Small fruit[2]	0.50	—	Nonbearing
Tree fruits			
Citrus[3]	0.50 (0.25–0.375)	—	Nonbearing
Deciduous[4]	0.50 (0.25–0.375)	—	Nonbearing
Subtropical[5]	0.50 (0.25)	—	Nonbearing
Tree nuts[6]	0.50 (0.25)	—	Nonbearing
Ornamentals	0.50 (0.25–0.375)	—	Nurseries and flowers, field grown, see label for species
Fallow and noncrop land	0.50 (0.25–0.375)	—	—

There are limitations on some of the above uses relative to stage of crop growth, harvest time, and geographic location; some uses include crop oil concentrate or nonionic surfactant in spray solution; see label.

[1] Fluazifop-P-butyl rates in parentheses, fluazifop-butyl rates not in parentheses.
[2] Berries and grapes.
[3] Grapefruit, lemons, limes, tangelos, tangerines.
[4] Apples, apricots, cherries, nectarines, peaches, pears, plums, prunes.
[5] Avocados, dates, figs, guava, mangos, olives, papayas, pomegranates; also coffee and kiwifruit with fluazifop-P-butyl only.
[6] Almonds, filberts, macadamias, pecans, pistachios, walnuts.

Mode of Action The most obvious phytotoxic symptoms of fluazifop-butyl are foliar chlorosis and necrosis, beginning with the youngest leaves in a few days after application and spreading to all leaves within 2 weeks. It is rapidly absorbed by leaves, and rainfall 1 hour after application results in only slight reduction in activity. Absorption is greater at 30 to 35°C than at 18 to 20°C (Wills and McWorter, 1983; Kells et al., 1984). Translocation was greater in plants exposed to full sunlight than in shade and moisture stress significantly reduced quackgrass control (Kells et al., 1984). It is rapidly metabolized in plants to fluazifop, which is presumably the phytotoxic form. This suggests that the primary function of the ester form is to facilitate foliar penetration. Fluazifop is translocated in both the symplastic and apoplastic systems. Translocation in the symplastic system, via the phloem, from the foliage to the rhizomes and stolons enables this material to effectively control perennial grasses. However, translocation from leaves is limited; only a small percentage of that absorbed by leaves is transported to other parts of the plant (Derr et al., 1985; Hendley et al., 1985). This indicates its high herbicidal activity, especially in perennial grasses. Carr et al. (1985) suggested that the mechanism of action of fluazifop involves altered lipid metabolism. This may be related to an inhibition of acetyl-CoA carboxylase in susceptible species and a lack of inhibition of this enzyme in tolerant species (see haloxyfop mode of action).

Haloxyfop

Haloxyfop is the common name for 2-[4-[[3-chloro-5-(trifluoromethyl)-2-pyridinyl]oxy]phenoxy]propanoic acid. Its use form is the methyl ester, haloxyfop-methyl, and the trade name is Vertict®. It is formulated as an emulsifiable concentrate (2 lb/gal). It is a white crystalline solid with a very low water solubility of 9.3 ppm. The acute oral LD_{50} of technical haloxyfop-methyl is 293 and 599 mg/kg for male and female rats, respectively.

Uses Haloxyfop is a new selective postemergence herbicide that is active against annual and perennial grasses. No activity has been observed on broadleaf weeds of sedges at the suggested uses rate. All broadleaf crops appear to be tolerant to haloxyfop. As noted earlier, haloxyfop is not registered for any use in the United States at this time. However, registration is pending for soybeans and apples and additional uses are being investigated.

Soil Influence Since haloxyfop is a postemergence herbicide, soil type has no effect on its performance. It is adsorbed relatively weakly to soil and it has been suggested that it has the potential for moderate leaching. Haloxyfop-methyl is demethylated to haloxyfop very rapidly in soil, with a half-life of about 24 hours. The half-life of haloxyfop ranges from 27 to 100 days with an average of 55 days in a variety of soil types.

Mode of Action The main site of activity is meristematic tissue. Haloxyfop-methyl is rapidly absorbed by leaves and hydrolyzed to haloxyfop. The latter is readily translocated throughout the plant, presumably accumulating in meristems. Haloxyfop is also readily absorbed by roots of germinating seedlings and soil residual activity usually controls later germinating grasses. Two recent papers reported that haloxyfop inhibited acetyl-CoA carboxylase activity in corn chloroplasts (Burton et al., 1987) and corn shoot extracts (Secor and Cséke, 1988). Burton et al. (1987) also showed that this enzyme was not inhibited in pea chloroplasts. These results suggest that acetyl-CoA carboxylase is the site of action of the aryloxyphenoxy herbicides and its differential response in tolerant and susceptible species is the basis of selectivity. Acetyl-CoA carboxylase catalyzes the first committed step in fatty acid biosynthesis.

Quizalofop

Quizalofop is the common name for (±)-2-[4-[(6-chloro-2-quinoxalinyl)-oxy]phenoxy]propanoic acid. Its use form is the ethyl ester, quizalofop-ethyl, and the trade name for this form is Assure®. It is a white crystalline solid with an extremely low water solubility of 0.3 ppm at 20°C. It is formulated as an emulsifiable concentrate (9.5% = 0.79 lb/gal). The acute oral LD_{50} of the active ingredient is 1.7 and 1.5 g/kg for male and female rats, respectively, and the formulation is 5.0 g/kg for male rats.

Uses Quizalofop-ethyl is a selective postemergence herbicide used to control annual and perennial grasses in cotton and soybeans. The use rate for both of these crops is 1.2–2.0 oz/acre (0.075–0.13 1b/acre) for annual grasses and 1.6–3.2 oz/acre (0.10–0.20 1b/acre) for perennial grasses. Higher rates are recommended in arid regions. Sedges and broadleaf weeds are not controlled by quizalofop-ethyl. Spray solutions of quizalofop-ethyl should always include a nonionic surfactant or a petroleum base oil concentrate. Since quizalofop-ethyl is a relatively new herbicide and many other broadleaf crops are tolerant other uses may be forthcoming.

Soil Influence Soil type has no effect on quizalofop-ethyl performance since it is a postemergence herbicide. It has a very low mobility in most soils and therefore not subject to leaching. Rapid breakdown occurs by microbial action under both aerobic and anaerobic soil conditions.

Mode of Action Visual symptoms of quizalofop-ethyl action include an early chlorosis followed by progressive collapse of the foliage and subsequent death within a few weeks of application. It is rapidly absorbed by the foliage and readily translocated throughout the plant via the symplastic system; apoplastic transport probably also occurs. Quizalofop may act by inhibiting acetyl-CoA carboxylase in susceptible species and not inhibiting this enzyme in tolerant species (see haloxyfop mode of action).

BENZOICS

The benzoic herbicides are derivatives of benzoic acid, shown below. There are only two herbicides of this class currently being used in the United States, however they have many applications. These are chloramben and dicamba. Trichlorobenzoic acid (2,3,6-TBA) was also available until recently. The ring substitutions of chloramben and dicamba are two chlorine atoms and in addition an amino (—NH_2) group in chloramben and a methoxy (—OCH_3) group in dicamba. Zimmerman and Hitchcock (1942) first reported that the substituted benzoic acids had growth-regulating properties in plants. Somewhat latter their herbicidal properties were demonstrated both in England and the Unites States.

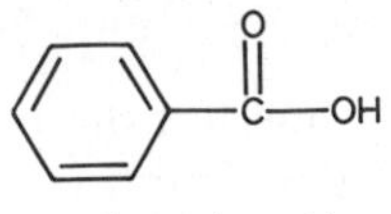

benzoic acid

Chloramben

Chloramben is the common name for 3-amino-2,5-dichlorobenzoic acid. The trade name is Amiben®. It is a white amorphous solid with a moderate water solubility of 700 ppm at 25°C. It is formulated as an ammonium salt and available in this form as a water soluble liquid (2 lb/gal) or as a granule (10%). It is also formulated as a sodium salt available as a water-soluble powder (75%). The ammonium and sodium salts are more soluble in water than the acid. The acute oral LD_{50} of the chloramben free acid is 5.6 g/kg for rats.

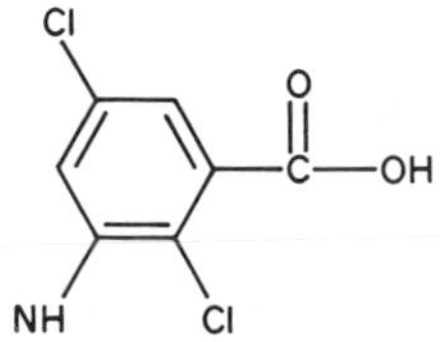

chloramben

Uses Chloramben is used as a preemergence or preplant soil-incorporated herbicide in many vegetable and field crops. It controls many annual broadleaf weeds and grasses. However, it is commonly used in combination with other herbicides as a tank mix or sequential treatment to broaden the spectrum of weeds controlled. See Table A-0 for specific weeds controlled and Table 9-4 for detailed information on the crops.

Soil Influence Chloramben is readily leached in sandy soils and is subject to some leaching in other soils, especially during heavy rains. It is degraded by soil microorganisms and persists from 6 to 8 weeks in most soils.

Mode of Action Chloramben inhibits early seedling growth shortly after seed germination. The emerging seedlings often show epinastic symptoms indicating its auxin-like activity. Root growth is also inhibited. It is absorbed by seeds (Haskell and Rogers, 1960; Swan and Slife, 1965; Rieder et al., 1970) but transport from the soybean seed into the developing seedling was limited (Swan and Slife, 1965). It is readily absorbed by roots but the subsequent amount translocated was species dependent. It is also absorbed by leaves and translocated to roots. It is translocated in both the symplast and the apoplast but transport is limited in some species. Ashton and Crafts (1981) discuss the absorption and translocation of chloramben in detail.

Chloramben is metabolized to conjugates, *N*-glucoside and glucose ester, and

TABLE 9-4. Chloramben Uses

Crop	Rate (lb/acre)	Combinations	Remarks
Asparagus	2.7–3.0	—	PE, direct seeded
Beans[1]	1.8–2.7	—	PE or PPI
	1.8–2.7	Alachlor, EPTC, metolachlor, or trifluralin	PE or PPI, tank mix or sequential
Beans[2]	1.8–2.7	EPTC + trifluralin	PE or PPI, tank mix or sequential
Corn	1.8	—	PE, field
	0.9	Atrazine	PE or PPI, tank mix
Cucurbits[3]	3.0–4.0	—	PE or PPI
Peanuts	1.8–3.0	—	PE
Peppers	3.0–4.0	—	PE, established transplants or layby
Soybeans	1.8–3.0	—	PE
	1.8–2.7	Alachlor, linuron, metolochlor, metribuzin, oryzalin, pendimethalin, trifluralin, or vernolate; or metribuzin + alachlor, metolachlor, pendimethalin, or trifluralin	PE or PPI, tank mix or sequential
Sunflower	1.8–3.0	—	PE
	1.8–3.0	Alachlor, EPTC, pendimethalin, or trifluralin	PE or PPI, tank mix or sequential
Sweet potatoes	3.6–4.0	—	At planting slips or draws
Tomatoes	3.0–4.0	—	Established transplants or layby, granule formulation

There are some limitations of the above uses relative to soil type, formulation, geographic location, and the method of application; see product label.

[1]Adzuki, kidney, lima, navy, pinto, white.
[2]Kidney, lima
[3]Pumpkins, squash.

TABLE 9-5. Dicamba Uses

Crop	Rate (lb/acre)	Combinations	Remarks
Asparagus	0.25–0.50	—	Established
	0.25–0.50	2,4-D	Established, tank mix
Barley, spring	0.094	—	2–3 leaf stage
	0.094	Chlorsulfuron or MCPA	2–3 leaf stage
Barley, fall	0.125	—	Prior to jointing stage
	0.125	Chlorsulfuron, 2,4-D, or MCPA	Prior to jointing stage
Corn[1]	0.25–0.50	—	PP to early post
	0.25	2,4-D	Post, drop nozzles, 36 inches tall or 15 days before tassel
	0.25	2,4-D	Preharvest, after brown-silk stage prior to 7 days before harvest
	0.25–0.50	Alachlor, atrazine, cyanazine, glyphosate, metolachlor, paraquat, phendimethalin, or simazine	PP, PE, or early post
	0.25–0.50	Alachlor, atrazine, butylate +, cyanazine, EPTC, glyphosate, metolachlor, paraquat, phendimethalin, propachlor, or simazine	PE, PPI, overlay, or early post-5 inches tall, some 3-way combinations
	0.25–2.00		Field, cleanup, after harvest–before frost
	0.25–2.00	Atrazine, chlorsulfuron, cyanazine, 2,4-D, glyphosate, metribuzin, or paraquat	Cleanup, residual control, after harvest–before frost

Milo	0.25	—	PP, early post, or preharvest
	0.25	2,4-D	Preharvest, after soft dough stage prior to 30 days before harvest
Fallow land	4.00–6.00	—	Spot treatment
	0.50–2.00	—	Between crop application, > rates from annuals, biennials to perennials
	0.25–0.50	Atrazine, chlorsulfuron, cyanazine, 2,4-D, glyphosate, metribuzin, paraquat, pronamide, or 2,4-D plus glyphosate	Between crop application
Range and pastures	0.25–0.75	—	Grazing restrictions
	8.00–10.0	—	Spot treatment, granules
	0.25–0.75	Picloram	Grazing restriction
Noncrop land			
Herbaceous	2.00–8.00	—	Rate depends on species
Brush	0.50–8.00	—	Rate depends on species and suppression or degree of control
	0.50–8.00	2,4-D	As above
Trees	Dilution[2]	—	Cut surface, frill, girdle, stump
	Dilution	2,4-D	Cut surface, frill, girdle
Water weeds[3]	Dilution	2,4-D	No irrigation for 14 days
	Dilution	2,4-D plus dalapon	As above, plus cattails

There are many restrictions on several of the above uses including rates, crop variety, geographic location, method and time of application, spray volume, grazing, formulation, underseeded legumes in small grains, etc.; see label.

[1]Field, seed, or silage; not registered for sweet or popcorn.
[2]See label.
[3]Still water: ponds, lakes, marshes, shorelines; moving water: irrigation, drainage canals, bayous.

insoluble residues (Hatzios and Penner, 1982). The rate of these reactions is greater in tolerant species than in susceptible species and this appears to be a major factor in its selectivity. Since these metabolites are immobile, this may also explain the differential rates of translocation in various species. Little information is available on chloramben action at the molecular level. However, its auxin-like symptoms suggest that its action may be similar to the phenoxy herbicides that alter nucleic acid metabolism.

Dicamba

Dicamba is the common name for 3,6-dichloro-2-methoxybenzoic acid. The trade name is Banvel®. It is formulated as the dimethylamine salt and available as a water-soluble liquid (1 and 4 lb/gal) and granule (10%). It is also available as a package mix with 2,4-D (Weedmaster®) or as the potassium salt with atrazine (Marksman®). Tank mixes with several other herbicides are also used. The pure acid is a white crystalline solid with a high water solubility of 4500 ppm and the dimethylamine salt has a very high water solubility of 720,000 ppm. The acute oral LD_{50} of technical dicamba is 1.7 g/kg for rats.

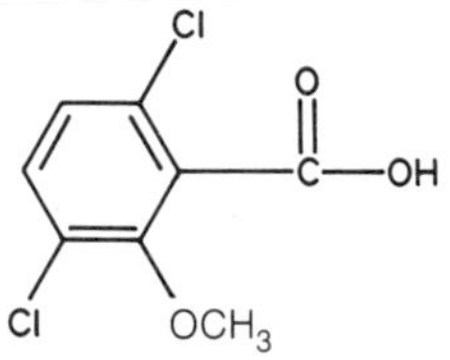

dicamba

Uses Dicamba is used as a preemergence or postemergence herbicide to control many annual and perennial broadleaf weeds, primarily in grass type crops. It is also used for the control of undesirble brush and trees, cut surface treatment, and annual and perennial broadleaf control in fallow land, noncrop land, and water. In general, it is used to control those types of weeds not controlled by 2,4-D; combinations of 2,4-D and dicamba are also used. Combinations of dicamba with several other herbicides are used in some crops to broaden the spectrum of weeds controlled. See Table A-0 for specific weeds controlled and Table 9-5 for detailed information on the crops.

Soil Influence Dicamba is relatively mobile in soil and the degree of leaching is dependent on the amount of rainfall. It is degraded by soil microorganisms. The rate of degradation is most rapid under warm-moist soil conditions and in slightly acid soils, with a half-life of 14 days under ideal conditions. However, under cool-dry soil conditions it persists longer, a few to several months.

Mode of Action Dicamba affects plant growth in much the same way as 2,4-D, with epinasty of young shoots and proliferative growth (see Figure 9-1). It is

Figure 9-1. Modification of leaf morphology of melons induced by dicamba. (A. H. Lange, University of California, Parlier.)

readily absorbed by leaves, stems, and roots and is translocated throughout the plant, accumulating in areas of high metabolic activity. It is translocated in both the apoplast and symplast. It leaks from roots into the surrounding medium and is lost from leaves by volatilization.

Dicamba appears to act like other auxin-type herbicides by altering nucleic acid metabolism (Fedtke, 1982). It has also has been reported to induce nitrite accumulation and cause changes in membranes. The latter involves a decrease in membrane permeability and plasma membrane ATPase activity (Duke, 1985).

BIPYRIDYLIUMS AND PYRAZOLIUMS

These herbicides are heterocyclic (more than one type of atom in the ring) organic compounds belonging to the bipyridylium (diquat, paraquat) or the pyrazolium (difenzoquat) class. The ring structures have a positive charge and are formulated as salts. They are applied as postemergence herbicides. Diquat and paraquat are contact herbicides and relatively nonselective, whereas difenzoquat is selective and used as an overall spray in certain crops.

Several reviews have been published on the bipyridylium herbicides including Calderbank (1968), Akhavein and Linscott (1968), Calderbank and Slade (1976), and Dodge (1982) as well as the monograph, *Bipyridinium Herbicides*, by Summers (1980).

Diquat

Diquat is the common name for 6,7-dihydrodipyrido[1,2-α:2′-1′-*c*]pyrazinediium ion. It is used as the dibromide salt, which is very soluble in water and formulated at 2 lb/gal. The trade names are Ortho® Diquat, Reglone®, and Weedtrine-D®. The pure bromine salt is a yellow solid, but in aqueous solution it turns dark reddish-brown. The acute oral LD_{50} of the diquat ion is 230 mg/kg. It should be handled with care, and the user should avoid breathing spray mist or getting the concentrate on skin. When diquat is used as an aquatic herbicide, there is a wide margin of safety between the recommended dosages and the rates that cause toxic symptoms in fish.

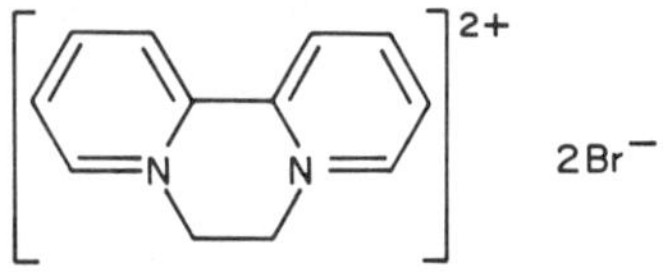

$2Br^-$

diquat

Uses Diquat is a contact herbicide applied to the foliage of plants to control aquatic weeds and weeds in sugarcane and noncrop land. It is also used for preharvest desiccation, thinning alfalfa stands, and flower control in sugarcane. For aquatic weed control, it is used at rates of 1 to 4 lb/acre depending on the weeds to be controlled. There are restrictions on the use of the water within 14 days of the application. For weeds in sugarcane and noncrop land, the use rate is 0.5 to 1.0 lb/acre. For desiccation of potato plants to facilitate harvest, the use rate is 0.5 lb/acre. For preharvest desiccation of alfalfa, carrot, clover, cucurbit, milo, radish, soybean, turnip, and vetch seed crops, the use rate is 0.375 to 0.5 lb/acre. The use of these crops for forage or grazing is prohibited. It is used at 0.25 lb/acre to thin alfalfa stands. For flower control in sugarcane, the use rate is 0.25 lb/acre applied during the flower initiation period. There are geographical restrictions, including only Special Local Need (SLN) registrations, harvest time, and other limitations on some of the above uses (see label).

Paraquat

Paraquat is the common name for 1,1′-dimethyl-4,4′-bipyridinium ion. Trade names are Gramoxone® and Setre® Paraquat. It is used as the dichloride salt, which is very soluble in water and formulated at 2 lb/gal. The pure chloride salt is a white solid, but forms a dark-red aqueous solution. The acute oral LD_{50} of the paraquat ion is 138 mg/kg for male rats. Absorption through intact skin is minimal, but may be facilitated if the skin is damaged. The acute dermal LD_{50} of the paraquat ion is >480 mg/kg for rabbits.

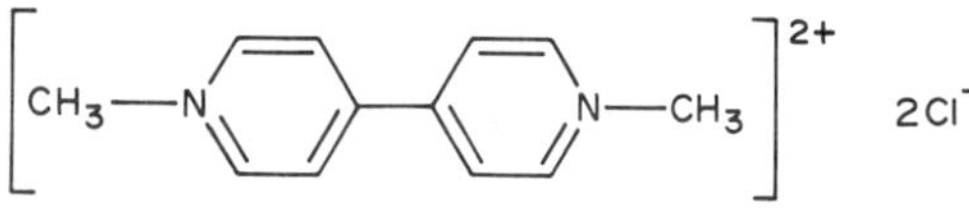

$[CH_3-N(C_5H_4)-(C_5H_4)N-CH_3]^{2+}$ $2Cl^-$

paraquat

Several poisonings and deaths have been reported as a result of ingesting relatively small amounts of the liquid concentrate (Slade and Bell, 1966). Fatalities resulted from progressive pulmonary fibrosis associated with liver and kidney damage (Sinow and Wei, 1973). As a safeguard measure against oral ingestion toxicity, current formulations contain a material that induces vomiting when ingested.

The use of paraquat for the control of marijuana has been the subject of considerable public controversy.

Uses Paraquat is a contact herbicide used to control emerged annual weeds in numerous crops and on noncrop land. The use rate is usually about 0.5 to 1.0 lb/acre. It is registered for use in over 50 crops. [see *Weed Control Manual and Herbicide Guide* (1988) or the label]. In general, it can be used in any crop utilizing techniques that keep sprays off leaves and succulent stems of the crop plant. These techniques include preplant, preemergence, or directed spray treatments. It is also used in certain crops when they are dormant. Paraquat is not selective and crops are usually injured by the spray or spray drift when they come in contact with the crop foliage. Perennial weed growth is suppressed by foliar desiccation but soon recovers. Annual weeds emerging after paraquat application are not controlled.

To obtain control of weeds that emerge after the paraquat application, paraquat is frequently used in combination with one or more soil-active herbicides as a tank mix in certain crops, e.g., field corn, milo, soybeans. It is also used as a preharvest desiccant in certain crops.

Soil Influence, Diquat and Paraquat An important, unique property of both diquat and paraquat is their rapid inactivation in soil (Brian et al., 1958). The inactivation results from a reaction between the positively charged herbicide ion and the negatively charged sites on clay minerals. The presence of the double positive charge on the herbicide molecules causes them to become tightly adsorbed within the clay lattice (Homer et al., 1960). Therefore, these herbicides are essentially nonphytotoxic in most soils. Some phytotoxicity, however, has been demonstrated at high rates in very sandy soils high in organic matter that contain little or no clay. Although nonphytotoxic, the bound diquat or paraquat may persist in soils for some period of time.

Mode of Action, Diquat and Paraquat Bipyridylium-type herbicides cause wilting and rapid desiccation of the foliage to which they are applied, often within a few hours. High light intensities increase the rate of development of the phytotoxic symptoms. Conversely, cloudy conditions can interfere with good activity from these herbicides. Best results in the field have often been obtained by a late-afternoon application, rather than a morning or midday application. This appears to allow some internal transport during the night, before development of acute phytotoxicity induced by light, which could limit movement.

Translocation following a foliar application appears to be almost solely via the apoplastic system (Baldwin, 1963; Slade and Bell, 1966; Wood and Gosnell, 1965).

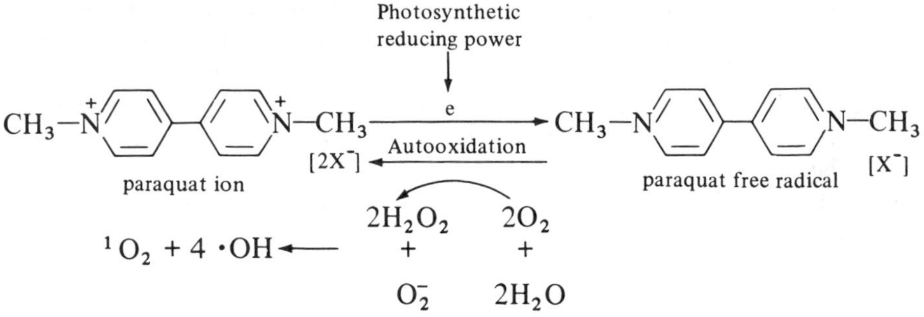

Figure 9-2. Free-radical formation from paraquat ion and autooxidation of free radical yielding H_2O_2 (hydrogen peroxide) and O_2^- (superoxide radical), and subsequently, ·OH (hydroxyl radical), and $^1O_2$ (singlet oxygen). (Ashton and Crafts, 1981.)

However, after loss of membrane integrity induced by the herbicides, they do move into untreated leaves. Presumably they move along with the flow of other cellular contents. They are poorly translocated (Damonakis et al., 1970), presumably because they are tightly bound to cellular components.

These herbicides are not degraded in higher plants in the usual sense. However, they are reversibly converted from the ion form to the free-radical form (Ashton and Crafts, 1981; Dodge, 1982; Duke, 1985). This introversion is cyclic and requires light, molecular oxygen, water, and the photosynthetic apparatus. During autooxidation of the paraquat or diquat ion, four by-products are formed: (1) H_2O_2 (hydrogen peroxide), (2) O_2^- (superoxide radical), (3) ·OH (hydroxyl radical), and (4) $^1O_2$ (singlet oxygen) (Figure 9-2). Although each of these by-products is potentially phytotoxic, the superoxide radical is probably responsible for the phytotoxic symptoms.

Even though paraquat is considered to be nonselective, biotypes of a few weed species have been found resistant to its action. These have appeared in fields that have been treated annually with paraquat for several years (LeBaron and Gressel, 1982). Although several papers have reported on this resistance, including two recent ones (Shaaltiel and Gressel, 1987; Pölös et al., 1988), the basis of the resistance in these species has not been conclusively established.

Difenzoquat

Difenzoquat is the common name for 1,2-dimethyl-3,5-diphenyl-1*H*-pyrazolium. The herbicidal use form is the methyl sulfate salt and has the trade name of

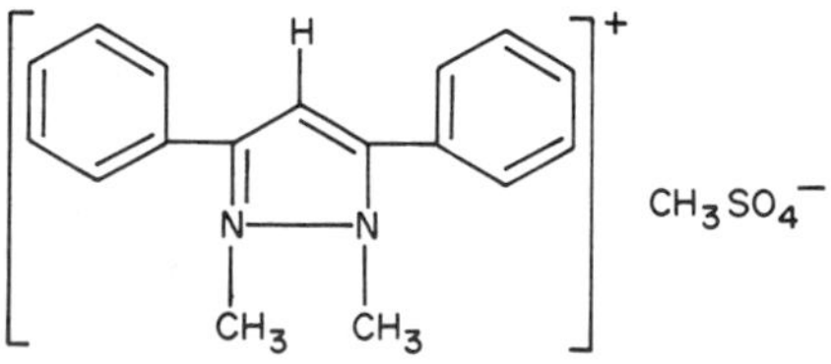

difenzoquat

Avenge®. It is a white to off-white crystalline solid and very soluble in water, 75% (w/w). It is formulated as an aqueous solution, 2 lb/gal (cation). The acute oral LD_{50} is 270 mg/kg in male rats.

Uses Difenzoquat is a selective postemergence herbicide used to control wild oats in barley and wheat. It is applied at 0.625 to 1.0 lb/acre when the wild oats are in the 3- to 5-leaf stage. It can be tank mixed with 2,4-D, MCPA, bromoxynil, chlorsulfuron, or MCPA plus bromoxynil to broaden the spectrum of weeds controlled. See label for crop variety restrictions.

Soil Influence Difenzoquat is strongly adsorbed to soil particles and therefore not subject to leaching. It is not readily metabolized by soil microorganisms. It is readily demethylated photolytically to the relatively volatile monomethyl pyrazole. Difenzoquat residues disappear from soil at a moderate rate and rotation to other crops can be made the following year.

Mode of Action Difenzoquat causes chlorosis and necrosis in leaves. It is readily absorbed by leaves but translocation is limited. It is not significantly metabolized by plants. It has been suggested that difenzoquat may act at the molecular level like diquat and paraquat as described above (Fedtke, 1982; Halling and Behrens, 1983). Cohen and Morrison (1982) indicated that it interferes with active ion transport across the plasma membrane.

SUGGESTED ADDITIONAL READINGS

Akhavein, A. A., and D. L. Linscott, 1968, *Residue Rev.* **23**, 97.

Ashton, F. M., and A. S. Crafts, 1981, *Mode of Action of Herbicides*, Wiley, New York.

Baldwin, B. C., 1963, *Nature (London)* **198**, 872.

Brian, R. C., R. F. Homer, J. Stubbs, and R. L. Jones, 1958, *Nature (London)* **181**, 446.

Burton, J. D., J. W. Gronwald, D. A. Somers, J. A. Connelly, B. G. Gengenbach, and D. L. Wyse, 1987, *Biochem. Biophys. Res. Commun.* **148**, 1039.

Calderbank, A., 1968, *Adv. Pest Control Res.* **8**, 129.

Calderbank, A., and P. Slade, 1976, Diquat and paraquat, in P. C. Kearney and D. D. Kaufman, Eds., *Herbicides*, Vol. 2, pp. 501–540, Dekker, New York.

Carr, J. E., L. G. Davies, A. H. Cobb, and K. E. Pallett, 1985, *Proc. Br. Crop Protection Conf. Weeds* **1**, 155.

Cohen, A. S., and I. N. Morrison, 1982, *Pestic. Biochem. Physiol.* **18**, 174.

Corbett, J. R., K. Wright, and A. C. Baillie, 1984, *The Biochemical Mode of Action of Pesticides*, Academic Press, New York.

Crop Protection Chemicals Reference, 1989, Wiley, New York (revised annually).

Damonakis, M., D. S. H. Drennan, J. D. Fryer, and K. Holly, 1970, *Weed Res.* **10**, 278.

Derr, J. F., T. J. Monaco, and T. J. Sheets, 1985, *Weed Sci.* **33**, 612.

Dodge, A. D., 1982, The role of light and oxygen in the action of photosynthetic inhibitor herbicides, in D. E. Moreland, J. B. St. John, and F. D. Hess, Eds., *Biochemical*

Responses Induced by Herbicides, pp. 57–77, ACS Symposium Series 181. American Chemical Society, Washington, D. C.

Duke, S. O., Ed., 1985, *Weed Physiology*, Vol. II, *Herbicide Physiology*, CRC Press, Boca Raton, FL.

Duke, S. O., and W. H. Kenyon 1988, Polycyclic Alkanoic Acids, in P. C. Kearney and D. D. Kaufman, Eds., *Herbicides*, Vol. 3, pp. 71–116, Dekker, New York.

Farm Chemicals Handbook, 1988, Meister, Willough, OH (revised annually).

Fedtke, C. 1982, *Biochemistry and Physiology of Herbicide Action*, Springer-Verlag, New York.

Halling, B. P., and R. Behrens, 1983, *Weed Sci.* **31**, 693.

Haskell, D. A., and B. J. Rogers, 1960, *Proc. 17th North Central Weed Control Conf.*, p. 39.

Hatzios, K. K., and D. Penner, 1982, *Metabolism of Herbicides in Higher Plants*, Burgess, Minneapolis.

Hendley, P., J. W. Dicks, T. J. Monaco, S. M. Slyfield, O. J. Tummon, and J. C. Barrett, 1985, *Weed Sci.* **33**, 11.

Herbicide Handbook, 1989, Weed Science Society of America, Champaign, IL.

Homer, R. F., G. C. Mees, and T. E. Tomlinson, 1960, *J. Sci. Food. Agric.* **11**, 309.

Hoppe, H. H., 1980, *Pflanzenphysiologie* **100**, 415.

Hoppe, H. H., and H. Zacher, 1986, *Pestic. Biochem. Physiol.* **24**, 298.

Kells, J. J., W. F. Meggett, and D. Penner, 1984, *Weed Sci.* **32**, 143.

LeBaron, H. M., and J. Gressel, 1982, *Herbicide Resistance in Plants*, Wiley, New York.

Nestler, H. J., and H. Bieringer, 1980, *Z. Naturforsch.* **35**, 366.

Pölös, E., J. Mikulás, Z. Szigete, B. Matkovics, D-Q. Hai, A. Párducz, and E. Lehoczki, 1988, *Pestic. Biochem. Physiol.* **30**, 142.

Rieder, G., K. P. Buchholtz, and F. W. Slife, 1970, *Weed Sci.* **18**, 101.

Ross, M. A., and C. A. Lembi, 1985, *Applied Weed Science*, Burgess, Minneapolis, MN.

Secor, J., and C. Cséke, 1988, *Plant Physiol.* **86**, 10.

Shaaltiel, Y., and J. Gressel, 1987. *Plant Physiol.* **85**, 869.

Shimabukuro, R. H., 1985, Detoxication of herbicides, in S. O. Duke, Ed., *Weed Physiology*, Vol. II, *Herbicide Physiology*, pp. 215–240, CRC Press, Boca Raton, FL.

Sinow, J., and E. Wei, 1973, *Bull. Envir. Contam. Toxicol.* **3**, 163.

Slade, P., and E. G. Bell, 1966, *Weed Res.* **6**, 267.

Summers, L. A., 1980, *Bipyridinium Herbicides*, Academic Press, New York.

Swan, D. G., and F. W. Slife, 1965, *Weeds* **13**, 133.

Weed Control Manual and Herbicide Guide, 1988, Meister, Willough, OH (revised annually).

Wink, O., E. Dorn, and K. Beyermann, 1984, *J. Agric. Food Chem.* **32**, 187.

Wills, G. D., and C. G. McWhorter, 1983, *Aspects Appl. Biol.* **4**, 283.

Wood, G. H., and J. M. Gosnell, 1965, *Proc. South African Sugar Tech. Assoc.*, p. 7.

Zimmerman, M. H., and A. E. Hitchcock, 1942, *Contr. Boyce Thompson Inst.* **12**, 312.

Zimmerman, M. H., and A. E. Hitchcock, 1951, *Contr. Boyce Thompson Inst.* **16**, 209.

For herbicide use, see manufacturer's or supplier's label and follow these directions. Also see the Preface.

10 Carbamates, Carbamothioates, and Cyclohexendiones

CARBAMATES

Carbamate herbicides derive their basic chemical structure from carbamic acid (NH_2COOH). Pure carbamic acid is not stable and quickly decomposes to NH_3 and CO_2. Friesen (1929) first observed the effect of esters of carbamic acid on plants. Herbicidal properties of the carbamate herbicide propham were described by Templeman and Sexton (1945). This followed closely the discovery of 2,4-D, the prime forerunner of modern chemical were control. Numerous carbamate compounds have been evaluated for herbicidal properties since 1945 and several have been developed. However, only five are currently available in the United States. Two are primarily applied to the soil (propham, chlorpropham) and three are primarily applied to the foliage (asulam, phenmedipham, desmedipham). The common name, chemical name, and chemical structure of these carbamate type herbicides are given in Table 10-1.

Propham

Propham is the common name for 1-methylethyl phenylcarbamate. The trade name is Chem-Hoe®. In pure form it is a white solid with a moderate water solubility of 250 ppm. The technical product is a light tan solid. It is formulated as a flowable suspension at 3 and 4 lb/gal. The acute oral LD_{50} is 9.0 g/kg for rats.

Uses Propham is used primarily as a preplant and preemergence herbicide or occasionally as a postemergence herbicide to control many annual grasses including volunteer small grain in several broadleaf crops. Most common broadleaf weeds are tolerant but a few broadleaf weed are susceptible. Some annual grasses are tolerant. The type of application may also influence the weed species controlled. See *Herbicide Handbook* (1989) and Table A-0 for specific weeds controlled and Table 10-2 for detailed information on the crops.

Propham is somewhat volatile and thus is more effective at low temperatures due to reduced volatility. At 130°C, 2 to 4 lb/acre will give good weed control whereas at 24°C, 4 to 8 lb/acre will control only the more susceptible species and herbicidal activity will be short in duration (*Herbicide Handbook*, 1989).

TABLE 10-1. Common Name, Chemical Name, and Chemical Structure of the Carbamate Herbicides

$$R_1-\overset{H}{\underset{|}{N}}-\overset{O}{\overset{\|}{C}}-O-R_2$$

Common Name	Chemical Name	R_1	R_2
Asulam	Methyl[(4-aminophenyl) sulfonyl]carbamate	$-S(=O)_2-C_6H_4-NH_2$ (para)	$-CH_3$
Chlorpropham	1-Methylethyl 3-chloro phenylcarbamate	$-C_6H_4Cl$ (meta-Cl)	$-C(H)(CH_3)-CH_3$
Desmedipham	Ethyl [3-[[(phenylamino)-carbonyl]oxy]-phenyl]carbamate	$-C_6H_5$	$-C_6H_4-N(H)-C(=O)-O-C_2H_5$
Phenmedipham	3-[(Methoxycarbonyl)-amino]phenyl(3-methyl-phenyl)carbamate	$-C_6H_4CH_3$ (meta-CH_3)	$-C_6H_4-N(H)-C(=O)-O-CH_3$
Propham	1-Methylethyl phenyl carbamate	$-C_6H_5$	$-C(H)(CH_3)-CH_3$

Soil Influence Although propham is adsorbed by organic matter in soils, the adsorption bond is weak and the compound is subject to leaching. It is readily degraded by soil microorganisms and also lost from soil by volatility. In moist soils and at relatively high temperatures, e.g., 35°C, volatility may be the major means of loss. Volatility losses from dry soil at relatively low temperatures, e.g., 13°C, are negligible. In general, the half-life of propham is about 15 days at 16°C and 5 days at 29°C. Phytotoxic soil residue concentrations usually persist less than 1 month.

Mode of Action The most obvious phytotoxic symptoms of propham are inhibited seedling growth, especially epicotyl growth in broadleaf species, coleoptile growth in grasses, and root growth in both types of plants. These tissues are inclined to be shorter and thicker. The herbicide blocks cell division and induces polyploid nuclei (Ennis, 1948). This mitotic disruption is caused by an abnormal microtubule arrangement into a multipole spindle apparatus (Hess, 1982).

Propham is absorbed through the coleoptiles of emerging grass seedlings and to a lesser degree through the roots of plants. It is also slowly absorbed by leaves.

TABLE 10-2. Propham Uses

Crop	Rate (lb/acre)	Combinations	Remarks
Alfalfa and clovers	3.0–4.0	—	PP, PE, or post
Carrots	3.0–5.0	—	PE or post, seed crop
Lentils	4.0	—	PPI, plant 4 inches deep
Lettuce	4.0–6.0	—	PP or PE
	4.5–6.0	—	Post, 4 or more leaves
Peas	4.0	—	PPI, plant 4 inches deep
Safflower	3.0–4.0	—	PE or PPI, shallow incorporation
Spinach	4.5	—	Post, 2–4 leaf stage
Sugar beets	4.0–6.0	—	PE, PPI, or post, 4–8 true leaf stage
Grasses	3.0–4.0	—	Seed crop, established
	4.0–6.0	2,4-D	Seed crop, apply to fall prepared seedbeds for spring planting
Fallow land	3.0–4.0	—	—
	2.0–4.0	Atrazine	Tank mix
	1.5–4.0	Metribuzin	Tank mix
	3.0–4.0	Paraquat	Tank mix

There are some limitations on some of the above uses relative to geographic location, application method, planting time, irrigation, and grazing; see product labels.

It is primarily translocated in the apoplast and metabolized in plants by aryl hydroxylation and subsequent conjugation with glucose. Other conjugates may also occur. Propham has been reported to alter a number of biochemical reactions, but only at concentrations higher than those interfering with cell division.

Chlorpropham

Chlorpropham is the common name for 1-methylethyl 3-chlorocarbamate. The trade name is Furloe®. Pure chlorpropham is a white solid with a low water solubility of 88 ppm. The technical product has a honey color. It is formulated as an emulsifiable concentrate (4 lb/gal) and as a granule (20%). The oral LD_{50} of technical chlorpropham is 3.8 g/kg for albino rats.

Chlorpropham was developed soon after propham was introduced. It has many of the same herbicidal characteristics as propham, but it differs from propham in that it persists longer in the soil, thus giving longer weed control, and it is less selective than propham to certain crop species, e.g., lettuce.

TABLE 10-3. Chlorpropham Uses

Crop	Rate (lb/acre)	Combinations	Remarks
Garlic	2.0–4.0	—	PE
Legumes	1.0–4.0	—	Hay, established or late summer seeded
	4.0–6.0	—	Established seed crop.[1] PE to dodder, EC formulation
	4.0–10.0	—	Established seed crop,[2] PE to dodder, granule formulation
Onions	4.0–8.0	—	PE or post, directed or before loop stage, not at flag stage
Peas, southern	4.0–6.0	—	PE, plant at least 1 inch deep
Safflower	3.0–6.0	—	PPI
	2.0–3.0	Trifluralin	PPI, tank mix
Small fruit			
Blackberries Rasberries	6.0	—	Established-dormant, avoid wetting canes
Blueberries	8.0–12.0	—	Established-dormant directed
Cranberries	10.0–20.0	—	Dormant, foliage dry, granule formulation
Soybeans	2.0–3.0	Alachlor	PE, tank mix
	2.0–3.0	Trifluralin	PPI, tank mix; PE, sequential
	2.0–3.0	Vernolate	PPI, tank mix; PE, sequential
Spinach	1.0–2.0	—	PE
Sugar beets	4.0	—	PP, after fall bed prep; post or water run, 8 or more true leaves
Tomatoes	4.0	—	PP, after fall bed prep; post, directed spray
Grasses	2.0–3.0	—	Seed crop, established; post
Ornamentals			
flowers	4.0–6.0	—	Field, see label for species
conifers	4.0–8.0	—	Field, directed or granule

There are some limitations of some of the above uses relative to geographic location, soil type, land preparation, time of application, formulation, harvest, forage use, and grazing; see product label.

[1] Alfalfa, clovers (red, ladino, alsike, birdsfoot trefoil, lespedeza).
[2] Alfalfa, clovers (birdsfoot trefoil, ladino).

Uses Chlorpropham is used primarily as a selective preemergence herbicide to control many annual grasses including shallow-seeded volunteer grains and some broadleaf weeds in several broadleaf crops, cane berries, and ornamental trees and shrubs. Its postemergence activity is limited to a relatively few species. Established perennial grasses grown for seed are generally tolerant. Chlorpropham is one of the few herbicides that controls the parasitic weed dodder. See Table A-0 for specific weeds controlled and Table 10-3 for detailed information on the crops.

Soil Influence Chlorpropham is adsorbed readily to organic matter is the soil but little is adsorbed to montmorillonite or kaolinite clay. Therefore, the degree of leaching is mainly influenced by the amount of organic matter in the soil. Soil microorganisms readily degrade chlorpropham and phytotoxic levels persist for only 1 to 2 months.

Mode of Action The mode of action of chlorpropham is similar to that of propham. Chlorpropham inhibits cell division, thus causing multinucleate root cells and inhibits root elongation (Ennis, 1949). This very likely involves an abnormal microtubule arrangement into a multipole spindle apparatus as shown for propham. Chlorpropham is readily absorbed by roots and translocated to the shoots via the apoplastic system (Prendeville et al., 1968). It is also absorbed by emerging shoots as they pass through the treated soil (Knake and Wax, 1968). Its absorption and translocation from foliar applications are much less than from soil treatment. The vapors of chlorpropham have been shown to be absorbed by seeds (Ashton and Helfgott, 1966) and are probably absorbed by other organs as well, e.g., emerged dodder plants (Slater et al., 1969). In the latter case, chlorpropham prevents the parasitic dodder plant from attaching to the host plant.

Chlorpropham is rapidly metabolized in plants by hydroxylation and subsequent conjugation of these compounds with glucose and/or glutathione (Shimabukuro, 1985). Although a number of biochemical reactions have been

Figure 10-1. Shepherdspurse control in garlic with chlorpropham. *Left*: Treated. *Right*: Control, no herbicide treatment.

reported to be altered by chlorpropham, none has been generally accepted to be related to its primary site of action. However, Macherel et al. (1986) recently suggested that the isopropyl carbamilates are multi-site inhibitors, separately inhibiting electron transfer in both mitochondrial respiration and chloroplastic photosynthesis as well as cell division.

Asulam

Asulam is the common name for methyl[(4-aminophenyl)sulfonyl]carbamate. The trade name is Asulox®. It is a white crystalline solid with a high water solubility of 5000 ppm at 20 to 25°C. It is formulated as an aqueous solution of the sodium salt at 3.34 lb/gal. The acute oral LD_{50} is > 8.0 g/kg for rats.

Uses Asulam is a postemergence herbicide used to control several perennial grasses including johnsongrass. However, certain perennial turf grasses are tolerant. It also controls brackenfern, tansy ragwort, and certain annual grass and broadleaf weeds; see Table A-0 for other weeds controlled. It is used in sugarcane at 2.5 to 3.34 lb/acre as a broadcast application and 2.5 to 5.0 lb/acre for spot treatments. At 2.1 lb/acre, it controls several annual grasses in St. Augustinegrass and specific types of bermudagrass lawns. At 3.34 lb/acre, it is used in field grown conifer, juniper, and yew nurseries; see label for geographic and species limitations on conifers. It is also used for christmas tree plantings and site preparation and conifer release in reforestration areas. In noncrop land it is used at 3.34 to 6.68 lb/acre. The addition of surfactants to the spray solution increases the effectiveness of asulam for some of these uses but also reduces its selectivity for some uses; see the label for instructions on surfactant use.

Soil Influence Since asulam is applied to the foliage of the target species, soil type does not affect its performance. It is not persistent in soils since it has a half-life of only 6 to 14 days.

Mode of Action Asulam inhibits meristem growth. This involves an interference with cell division and is related to its effect on microtubule assembly or function (Fedtke, 1982). It is readily absorbed by susceptible species, but absorption is increased by the addition of certain surfactants in some, but no all, species (Catchpole and Hibbitt, 1972; Babiker and Duncan, 1975). It appears to be translocated in both the symplast and the apoplast, but the degree of translocation seems to be species dependent.

The inhibition of folic acid biosynthesis is considered to be the major site of action of asulam (Killmer et al., 1980; Stephen et al., 1980; Veerasekaran et al., 1981a, b; Kidd et al., 1982). It inhibits 7,8-dihydropteroate synthetase, an enzyme involved in folic acid synthesis. Folic acid is required for biosynthesis of purine nucleotides, which are components of both DNA and RNA. This concept supports the previous findings that asulam inhibits nucleic acid and protein synthesis.

Desmedipham and Phenmedipham

Desmedipham and phenmedipham are similar in both their chemical structure and crop use but differ somewhat in their effectiveness for the control of certain weeds. Both compounds have two carbamate groups in their chemical structure (see Table 10-1).

Desmedipham is the common name for ethyl [3-[[(phenylamino)-carbonyl]oxy]phenyl]carbamate. The trade name is Betanex®. It is a colorless to light-yellow crystalline solid with a very low water solubility of about 7 ppm. It is formulated as an emulsifiable concentrate (1.3 lb/gal) and as a one to one ratio combination with phenmedipham, Betamix®. Pure desmedipham has an acute oral LD_{50} of 10.3 g/kg for rats and the formulated product of desmedipham alone has an acute oral LD_{50} of 3.7 g/kg for rats.

Phenmedipham is the common name for 3-[(methoxycarbonyl)-amino]phenyl(3-methylphenyl)carbamate. The trade name is Spin-Aid® and the trade name of the phenmedipham–desmedipham combination is Betamix®. Phenmedipham is a colorless crystalline solid with a very low water solubility of about 5 to 10 ppm. The pure compound has an acute oral LD_{50} greater than 8.0 g/kg for rats. The formulated combination has an acute oral LD_{50} of 4.1 g/kg for rats. The combination is formulated as an emulsifiable concentrate with 8% of each compound giving a total of 1.3 lb/gal.

Uses Desmedipham is used alone or in combination with phenmedipham to control many broadleaf weeds in sugar beets. Phenmedipham also controls also controls many broadleaf weeds plus green and yellow foxtail but it does not control redroot pigweed. However, the redroot pigweed is controlled by desmedipham. See Table A-0 for weed species controlled by either herbicide. They are applied postemergence when the sugar beets are past the two-true leaf stage and the weeds are at the two-true leaf stage. For kochia control, the weed should be in the rosette stage and less than 1 inch in diameter. The rate used is 0.75 to 1.25 lb/acre for desmedipham alone; for the combination, the same total amount is applied but one-half is desmedipham and one-half is phenmedipham. Sensitivity of sugar beets may be increased at temperatures over 85°F combined with moisture stress, or if certain preplant or preemergence herbicides are used before desmedipham or desmedipham–phenmedipham application. Carbamothioate herbicides, e.g., EPTC, increase sugar beet sensitivity to these herbicides due to a decrease in the epicuticular wax on the leaves following carbamothiolate use. The lower amounts of wax increase the absorption of desmedipham and phenmedipham by sugar beets.

Phenmedipham can also be used for weed control in red beets and spinach grown for processing or seed production. The use rate is 0.5 to 1.0 lb/acre when the crop is in the four- six-leaf stage. The 1.0 lb/acre rate should be used only on well-established crops that are not under stress. Use on spinach only when the temperatures are below 75°F. For split applications 4 to 6 days apart, the use rate is 0.5 lb/acre for each application in both crops and spinach should be in thc two-leaf stage.

Soil Influence Since these two herbicides are applied to the foliage, soil type does not influence their performance. They are adsorbed by soils and not subject to excessive leaching, usually remaining in the top 2 inches. They are degraded in soils and usually have a half-life of about 25 days.

Mode of Action Both of these herbicides are readily absorbed through the foliage but rain falling within a few hours after application may reduce their effectiveness. Translocation from the treated leaves is limited since they are primarily transported via apoplast (Kassenbeer, 1970; Hendrick et al., 1974). Kassenbeer (1970) also reported that phenmedipham is rapidly metabolized by sugar beet seedlings but inactivated much slower in many susceptible weeds. Desmedipham is also metabolized by plants (Knowles and Sonawane, 1972; Hendrick et al., 1974). Both of these herbicides are considered to act by inhibiting photosynthesis. See Ashton and Crafts (1981) for additional details on the mode of action of desmedipham and phenmedipham.

CARBAMOTHIOATES

Carbamothioate herbicides include butylate, cycloate, diallate, EPTC, molinate, pebulate, thiobencarb, triallate, and vernolate. This class of herbicides was previously referred to as thiocarbamates. The common names, chemical names, and chemical structure of the carbamothioate herbicides are given in Table 10-4. They are derivatives of carbamic acid with one of the oxygen atoms replaced by a sulfur atom as well as other substitutions. Metham is a carbamodithioate with both oxygen atoms of carbamic acid substituted by sulfur atoms.

Most carbamothioate herbicides are relatively volatile. If not immediately incorporated into the soil by tillage equipment or applied in irrigation water, much of the applied to the soil surface will be lost. They are primarily used as selective herbicides in a wide variety of crops.

Butylate

Butylate is the common name for *S*-ethyl bis(2-methylpropyl)carbamothioate. It is an amber liquid with a low water solubility of 45 ppm at 22°C and relatively volatile. The acute oral LD_{50} of the technical product is 4.7 and 5.4 g/kg for male and female rats, respectively. It is formulated as an emulsifiable concentrate (6.7 lb/gal) and a granule (10%) with an inert safener (e.g., dichlormid; 2,2-dichloro-*N*,*N*-di-2-propenylacetamide, also known as R-25788). This safener or protectant reduces butylate's phytotoxicity to corn but not to most weeds. These formulated products have the trade names of Sutan® + and Genate® Plus. Other trade names are used for certain package mix combinations with other herbicides.

Uses Butylate formulations are selective to corn and primarily used to control annual grasses at 4 lb/acre as a preplant, soil-incorporated treatment. Certain

TABLE 10-4. Common Name, Chemical Name, and Chemical Structure of the Carbamothioates Herbicides

$$R_1R_2N{-}C({=}O){-}S{-}R_3$$

Common Name	Chemical Name	Chemical Structure R_1	R_2	R_3
Butylate	*S*-Ethyl bis(2-methylpropyl)carbamothioate	$CH_3{-}CH(CH_3){-}CH_2{-}$	$CH_3{-}CH(CH_3){-}CH_2{-}$	$C_2H_5{-}$
Cycloate	*S*-Ethyl cyclohexylethylcarbamothioate	$C_2H_5{-}$	cyclohexyl—	$C_2H_5{-}$
Diallate	*S*-(2,3-Dichloro-2-propenyl)bis(1-methylethyl)carbamothioate	$CH_3{-}CH(CH_3){-}$	$CH_3{-}CH(CH_3){-}$	$Cl{-}C(H){=}C(Cl){-}CH_2{-}$
EPTC	*S*-Ethyl dipropylcarbamothioate	$C_3H_7{-}$	$C_3H_7{-}$	$C_2H_5{-}$
Molinate	*S*-Ethyl hexahydro-1*H*-azepine-1-carbothioate	hexahydroazepin-1-yl (N—)[1]		$C_2H_5{-}$
Pebulate	*S*-Propyl butylethylcarbamothioate	$C_2H_5{-}$	$C_4H_9{-}$	$C_3H_7{-}$
Thiobencarb	*S*[(4-Chlorophenyl)methyl]diethylcarbamothioate	$C_2H_5{-}$	$C_2H_5{-}$	$Cl{-}C_6H_4{-}CH_2{-}$
Triallate	*S*-(2,3,3-Trichloro-2-propenyl) bis(methylethyl)carbamothioate	$CH_3{-}CH(CH_3){-}$	$CH_3{-}CH(CH_3){-}$	$Cl{-}C(Cl){=}C(Cl){-}CH_2{-}$
Vernolate	*S*-Propyl dipropylcarbamothioate	$C_3H_7{-}$	$C_3H_7{-}$	$C_3H_7{-}$

[1] The nitrogen atom in the molinate ring structure is the nitrogen atom of the parent thiocarbamic acid molecule; there is only one nitrogen atom in molinate.

perennial grasses and nutsedges may also be controlled or suppressed, often using special techniques and somewhat higher rates. Butylate can be applied when combined with certain liquid fertilizers and butylate-impregnated dry bulk fertilizers are also used. Center pivot sprinkler and subsurface applications (geographic limits) are also used. To broaden the spectrum of weeds controlled, tank or package mixes with atrazine, cyanazine, simazine, or atrazine plus cyanazine, and/or sequential treatments with 2,4-D or dicamba are used. Butylate plus and safener can be used on all types of corn (field, silage, seed, sweet, pop), but there are limitations for some types with some of the combinations.

Cycloate Cycloate is the common name for *S*-ethyl cyclohexylethylcarbamothioate. The trade name is Ro-Neet®. It is a colorless liquid with a low water solubility of 85 ppm at 22°C and is relatively volatile. It is formulated as an emulsifiable concentrate (6 lb/gal) and a granule (10%). The acute oral LD_{50} of the technical material ranges from 2.0 to 3.1 g/kg and 3.2 to 4.1 g/kg for male and female rats, respectively.

Uses Cycloate controls most annual grasses, several broadleaf weeds, and yellow and purple nutsedge in sugar beets, table beets, and spinach. It is applied as a preplant soil-incorporation treatment at 3 to 4 lb/acre in the beet crops and 3 lb/acre in spinach. There are geographic and time-of-year limitations on some of these uses. It may be combined with compatible fluid fertilizers for all of these crops and impregnated on dry bulk fertilizers for sugar beets.

Diallate

Diallate is the common name for *S*-(2,3-dichloro-2-propenyl)bis(1-methylethyl)-carbamothioate. The trade name is Avadex®. It is an oily liquid with a very low water solubility of 14 ppm at 25°C. It is formulated as an emulsifiable concentrate (4 lb/gal) and a granule (10%). The acute oral LD_{50} is 395 mg/kg for rats. Diallate is a Restricted Use Pesticide.

Uses Diallate is used to control wild oats in corn, lentils, peas, and sugar beets. In corn, lentils, and peas it is applied before or after planting with shallow soil incorporation. In sugar beets it is applied as a preplant soil-incorporation treatment. The use rates are 1.5 lb/acre for corn (field, silage) and lentils, 1.25 lb/acre for peas, and 1.5 to 2.0 lb/acre for sugar beets. The emulsifiable concentrate formulation can be used in all of these crops but the granular formulation is used only in sugar beets. There are geographic limitations on its use in corn.

EPTC

ETPC is the common name for *S*-ethyl dipropylcarbamothioate. The trade names are Eptam®7E, Drexel® EPTC-7, Genep®7E, Eradicane®, and Eradicane® Extra. The latter two formulations contain the corn antidote dichlormid and the last formulation also contains a compound that inhibits the degradation of EPTC in soil. EPTC is a light-yellow colored liquid with a moderate water solubility of 370 ppm and is relatively volatile. It is formulated as emulsifiable concentrates (7 lb/gal), an emulsifiable concentrate (6.7 lb/gal) plus the corn antidote, an emulsifiable concentrate (6.0 lb/gal) plus the corn antidote and soil extender, and granules (10%). The acute oral LD_{50} of EPTC is 1.7 g/kg for male rats and 3.2 g/kg for male mice.

Uses EPTC was the first carbamothioate developed. Its volatile nature resulted in highly variable weed control when first used as a surface-applied preemergence herbicide. Soil incorporation corrected this defect and provided the first general use of this technique. Subsequent research established that soil incorporation increased weed control of several other herbicides. The power-driven rotary hoe and ground-driven rotary tiller have been most effective. Soil incorporation by disc, sweep-type cultivator, or drag harrow has also been used. Soil incorporation places the herbicide in the weed-seed-germinating area of the soil where it is most effective. Soil incorporation does not require rainfall or overhead irrigation to leach the herbicide into the weed-seed-germinating area of the soil. However, soil incorporation represents an additional expense and *improper* incorporation can reduce weed control by streaking, excessive dilution, or burying of the herbicide.

EPTC is also applied through sprinkler irrigation systems and by metering into furrow-irrigation water. It may be combined with compatible fluid fertilizers or impregnated on dry bulk fertilizers for application.

EPTC is used against a wide array of weeds including annual grasses and a variety of annual broadleaf weeds. It also controls or suppresses the growth of yellow and purple nutsedges plus perennial grasses such as quackgrass and johnsongrass. See Table A-0 for specific weeds controlled and Table 10-5 for detailed information on the crops.

Molinate

Molinate is the common name for S-ethyl hexahydro-1*H*-azepine-1-carbothioate. The trade name is Ordram®. It is a liquid with a moderate water solubility of 800 ppm at 20°C and is relatively volatile. It is formulated as an emulsifiable concentrate (8 lb/gal) and a granule (10%). The acute oral LD_{50} of technical molinate is 720 mg/kg for rats and 795 mg/kg for mice.

Uses Molinate is used in rice and is particularly effective for the control of barnyardgrass. However, it also controls certain other weeds of rice (see Table A-0). The rates used and the time and methods of application vary considerably depending primarily on local cultural practices. The rates range between 3 and 6 lb/acre. It is used in both water-seeded and dry-seeded rice and applied preplant or postemergence to the rice. It is applied by ground rigs, airplanes, or in the irrigation water. Preplant applications usually use some type of soil

TABLE 10-5. EPTC Uses

Crop	Rate (lb/acre)	Remarks
Beans		
Dry, green	3.0	PPI or post, soil incorporated
Dry	2.2–3.0	PPI, trifluralin combination
Castor	2.0	PPI
Beets		
Sugar	2.0–3.0	PPI or post at layby with soil incorporation, subsurface injection, or in furrow irrigation water
	4.0–4.5	PPI, before ground freezes
Table	2.0	PPI
Cotton	2.0	Post, established-nonirrigated, subsurface injection
Corn	3.0–6.0	PPI[1]
	3.0–6.0	PPI[2]
Flax	4.0–4.5	PPI, before ground freezes
Fruit and nut Crops[3]	3.0	Post, in irrigation water after cultivation
Nursery[4]	6.0	Post, established nursery stock, soil incorporated after lining out
Grapes	3.0	Post, in irrigation water after cultivation
Legumes		
Small seeded[5]	3.0–4.0	PPI
	3.0	PE or established for hay, in irrigation water
Small seeded[6]	3.0	Post, established for seed, in irrigation water
Peas	2.0	PPI, processing type
Potatoes	3.0–6.0	PPI, directed spray at drag-off; last cultivation; may be applied in sprinkle irrigation
Safflower	3.0	PPI
Sunflower	3.0	PPI
Sweet potatoes	3.0	PPI
Tomatoes	3.0	Post; layby, soil incorporation

There are some limitations for some of these treatments relative to geographic location, varieties, soil type and treatment, harvest, or grazing times; see product label.

[1] Eradicane®, formulation contains a crop protectant.

[2] Eradicene® Extra, formulation contains a crop protectant and soil extender.

[3] Grapefruit, lemons, oranges, tangerines; almonds, walnuts.

[4] Grapefruit, lemons, oranges.

[5] Alfalfa, birdsfoot trefoil, clovers (alsike, ladino, red), lespedeza.

[6] Alfalfa, ladino clover.

incorporation. Postemergence applications with propanil are also used in some areas. There are limitations on geographic areas for the various practices and application time prior to harvest (see label).

Pebulate

Pebulate is the common name for S-propyl butylethylcarbamothioate. The trade name is Tillam®. It is yellow liquid with a low water solubility of 60 ppm and is relatively volatile. It is formulated as an emulsifiable concentrate (6 lb/gal). The acute oral LD_{50} is 0.9 to 1.1 g/kg for male rats and 1.5 to 1.8 g/kg for male mice.

Uses Pebulate is a selective soil-incorporated herbicide for the control of many annual grasses and broadleaf weeds in sugar beets, tobacco, and tomatoes. See Table A-0 for the specific weeds controlled. In sugar beets it is used as a preplant soil-incorporated treatment at 4 to 6 lb/acre. In tobacco it is used alone or in combination with isopropalin as a pretransplant soil-incorporated treatment at 4 lb/acre. In direct-seeded or transplanted tomatoes it is used as a preplant or layby soil-incorporated treatment at 3 to 6 lb/acre, but when used in combination with napropamide the use rate is 4 to 6 lb/acre. It is more effective for the control of hairy nightshade than other herbicides applied to tomatoes. However, it does not give adequate control of black nightshade at registered rates. It also controls dodder. There are geographic limitation on some of these uses in tomatoes (see product label).

Thiobencarb

Thiobencarb is the common name for *S*-[(4-chlorophenyl)methyl]-diethylcarbamothioate. The original common name, benthiocarb, was used prior to March 1977. The trade name is Bolero®. It is a light-yellow to brownish-yellow liquid with a low water solubility of 30 ppm at 20°C. It is formulated as an emulsifiable concentrate (8 1b/gal) and granules (10%). Technical thiobencarb has an acute oral LD_{50} of 0.92 to 1.9 g/kg for rats.

Uses Thiobencarb is used in rice and is particularly effective for the control of barnyardgrass and sprangletop. However, it also controls certain other weeds of rice (see Table A-0). It is used in both water-seeded and dry-seeded rice late preemergence, 1 to 5 days before rice emergence, or early postemergence when the weeds are very small. The rate applied is usually 4 lb/acre. However, 3 lb/acre of the granular form is used in some areas and when thiobencarb is used in combination with propanil as a tank mix. Local cultural practices lead to specific use instructions for various geographic areas (see product label). Injury to the rice plant may occur if it is under stress. Do not plant sensitive crops in thiobencarb-treated fields within 6 months of the last application.

Triallate

Triallate is the common name for *S*-(2,3,3-trichloro-2-propenyl)bis(methylethyl)carbamothioate. Far-Go® is its trade name. It is an amber oily liquid with a very low water solubility of only 4 ppm at 25°C. It is formulated as an emulsifiable concentrate (4 lb/gal) and a granule (10%). Mammalian acute studies indicate that the acute oral LD_{50} of technical triallate is 1.1 g/kg.

Uses Triallate is used to control wild oats in barley, lentils, peas, and wheat. It is applied in all of these crops before the wild oat seeds germinate and soil incorporation is required. In barley, postplant applications are used. The 10G form is applied in the fall or spring at 1.25 to 1.50 lb/acre and the EC form is applied in the fall at only 1.25 lb/acre. The EC form is also used at 1.0 lb/acre in combination with trifluralin as a tank mix to increase the spectrum of weeds controlled. In lentils, triallate is applied up to 3 weeks before planting or immediately after planting at 1.25 lb/acre for the EC form or 1.25 to 1.50 lb/acre for the 10G form. In spring wheat (durum), the EC form is applied at 1.0 lb/acre and the 10G form applied at 1.25 to 1.50 lb/acre; both forms can be applied within 3 weeks prior to "freeze-up" in the fall or postplant in the spring. It is also applied at 1.0 lb/acre as a tank mix combination with trifluralin before weed emergence in spring wheat. In winter wheat, either form of triallate can be applied just before seeding to soon after seeding at 1.25 to 1.50 lb/acre. There are grazing restrictions for some of these crops and specific planting and/or soil incorporation instructions on the product label.

Vernolate

Vernolate is the common name for *S*-propyl dipropylcarbamothioate. The trade names are Vernam® and Surpass®; the latter product also contains the corn antidote dichlormid previously discussed in this chapter. Both are formulated as an emulsifiable concentrate; Vernam® contains 7 lb/gal of vernolate and Surpass® contains 6.7 lb/gal of vernolate. Vernolate is a liquid with a low water solubility of 90 ppm at 20°C. The acute oral LD_{50} of technical vernolate is 1.78 g/kg for male albino rats.

Uses Vernolate is a selective soil-applied herbicide that requires soil incorporation. It is used in peanuts and soybeans and when dichlormid is added to the formulation it can be used in corn. It controls most annual grasses, many annual broadleaf weeds, and yellow and purple nutsedge (see Table A-0 for specific weeds controlled). In peanuts, vernolate can be applied preplant, at planting, or postplanting at 2.0 to 2.5 lb/acre. It is also used as a preplant soil-incorporation treatment in combination with benefin or trifluralin as a tank mix. The vernolate use rate is 2.0 to 2.26 lb/acre with benefin and 2.0 lb/acre with trifluralin in peanuts. In soybeans, vernolate is used at 2.0 to 3.0 lb/acre alone or in a tank mix combination with trifluralin as a preplant soil-incorporated treatment. In corn (field, silage, sweet, pop), vernolate plus antidote is used as a preplant soil-incorporation treatment alone or in a tank mix combination with atrazine or cyanazine. The use rate is 3.0 to 4.0 for the control of most annual weeds but may be increased to 6 lb/acre for the control or suppression of certain hard-to-control annual or perennial weeds. There are some geographic limitations for certain of these uses (see product label).

Soil Influence of Carbamothioates

In general, carbamothiolate herbicides are adsorbed by the clay and organic matter in soils and not readily leached (*Herbicide Handbook*, 1989). However, adsorption and leaching do vary somewhat within this class of herbicides. The relative leachability of some of these compounds is molinate > EPTC > pebulate > cycloate > butylate. Quantitative data are generally lacking on the leachability of these compounds; however, it has been reported that cycloate is leached 3 to 6 inches with 8 inches of water in a loamy sand soil (*Herbicide Handbook*, 1989). Vernolate is less subject to leaching than EPTC, but data on its relationship to the other compounds are not available. Likewise, the relative leachability of thiobencarb and triallate to the other carbamothioates is unknown.

The carbamothiolate herbicides compete with moisture for the adsorption sites on soil particles. Therefore, they are readily adsorbed on dry soil but poorly adsorbed on wet soil. Volatilization losses can be substantial when they are applied to the surface of wet soil.

Degradation of the carbamothiolate herbicides by microorganisms is the major cause of their disappearance from soils (*Herbicide Handbook*, 1989). They have a relatively short period of persistent in soils under aerobic conditions, loss of phytotoxicity within 6 weeks or less depending on the specific herbicide. The half-life of six of these compounds range from 1 to 35 weeks under standard conditions (Table 10-6). Diallate has a half-life of 30 days under intermountain irrigated soil conditions. Triallate persists 6 to 8 weeks under northern Great Plains use situations. Thiobencarb has a half-life of 2 to 3 weeks under aerobic conditions but under anaerobic conditions the half-life increases to 6 to 8 months.

TABLE 10-6. Half-Life of Six Carbamothioate Herbicides in Moist Loam Soil at 70 to 80°F

Herbicide	Weeks
EPTC	1.0
Vernolate	1.5
Pebulate	2.0
Butylate	3.0
Molinate	3.0
Cycloate	3.5

Mode of Action of Carbamothioates

The mode of action of the carbamothioate herbicides is similar. However, there are some unique crop selectivities with some of these herbicides. This has resulted in the recommended use of certain of these herbicides only for specific tolerant crops. In general, the differential crop selectivities among the compounds of this class were determined by greenhouse and/or field evaluations and their basis is usually not readily apparent.

The symptoms of carbamothioates herbicides on annual grasses are the lack of seedling emergence or grossly distorted emerging shoots. The edges of the coleoptile of the emerging shoot are often fused and distorted young leaves may emerge through the side of the coleoptile. Figure 10-2 (Dawson, 1963; Parker, 1963). The young leaves may also be dark green, glossy, and brittle (Parker, 1963; Banting, 1967). Broadleaf weeds may develop cupped leaves with necrotic tissue around the edges of the leaf. They also reduce epicuticular wax formation on leaves resulting in the loss of "bloom" of crucifers; this can increase the absorption of other foliar-applied herbicides and growth regulators subsequently applied (Hess, 1985). They also inhibit growth in general by interfering with cell division and/or cell enlargement in meristems.

Carbamothioates are absorbed by seeds, roots, and/or emerging shoots in contact with treated soil (Parker, 1963; Appleby et al., 1965; Gray and Weierich,

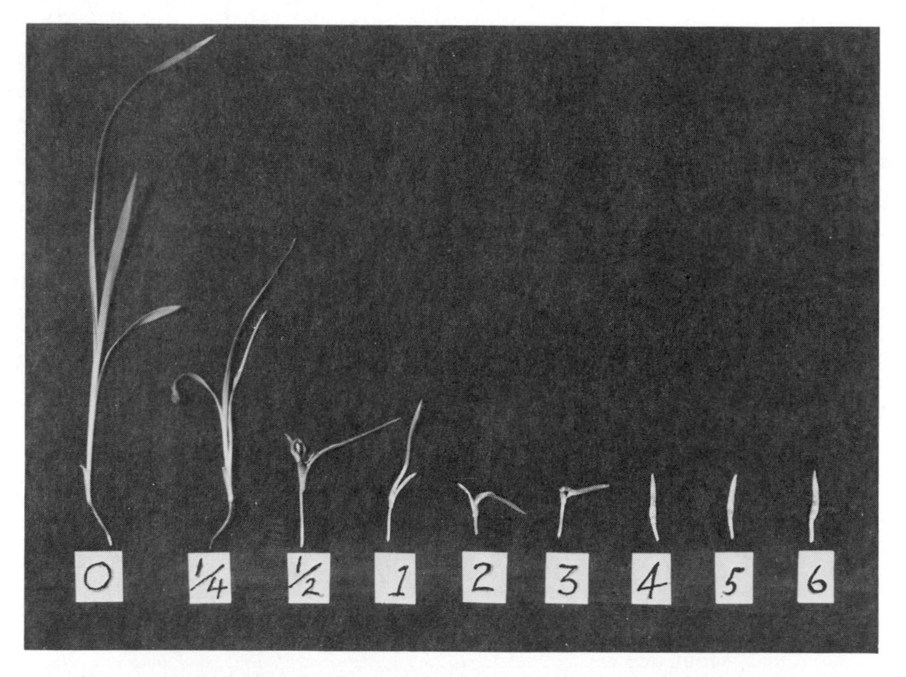

Figure 10-2. Barnyardgrass seedlings grown in sandy loam soil containing the indicated parts per million of EPTC. (J. H. Dawson, USDA, Prosser, WA.)

1969). The relative importance of these three sites of entry varies with the plant species. They are translocated in the apoplastic system.

These compound are metabolized by most plant species to nonphytotoxic metabolites. However, the rate of metabolism varies among species and rapid metabolism appears to be the basis of tolerance for some crops. Hatzios and Penner (1982) reviewed the research on the metabolism of these herbicides. Additional information on the metabolism of these compounds can be found in Duke (1985) and Fedtke (1982). The metabolites include a number of degradation products and conjugates. One of the metabolites, a sulfoxide, is considered to be more phytotoxic than the parent molecule (Lay and Caside, 1976; Pallos and Casida, 1978). The sulfoxide is subsequently conjugated with glutathione to form a nonphytotoxic product.

Extensive research has been conducted on the use of antidotes to protect corn from the phytotoxicity of these compounds. Antidotes are also referred to as safeners or protectants. Two monographs and a review have been written on herbicide antidotes (Pallos and Casida, 1978; Hatzios, 1985; Hatzios and Hoagland, 1988). Dichlormid (2,2-dichloro-*N*,*N*-di-2-propenylacetamide) is an antidote for corn when applied with butylate, cycloate, EPTC, or vernolate (Figure 10-3). It is commonly present in the formulation of these herbicides when they are to be used in corn. It is also effective as a seed treatment for corn and wheat prior to planting. This antidote is species specific; the control of most weeds is not reduced by combining dichlormid with the carbamothiolate herbicides. Crops other than corn or wheat are not protected from carbamothiolate herbicide injury by dichlormid. Other compounds have also been shown to be antidotes for these herbicides. Although the antidotal action of dichlormid and other compounds is not fully understood, they appear to act by competing with

Figure 10-3. Protection of corn from EPTC injury by the antidote dichlormid. *Left to right*: Control, EPTC, EPTC plus dichlormid. (G. R. Stephenson.)

the herbicide for a common site of action or by stimulating the metabolic detoxification of the herbicide in the protected crop (Hatzios, 1985).

Several monographs have cited original research that demonstrated the modification of many normal metabolic processes in plants by the carbamothiolate herbicides (Fang, 1975; Ashton and Crafts, 1981; Fedtke, 1982; Corbett et al., 1984; Duke, 1985). However, the reduction and alteration of lipid synthesis appear to be the major biochemical changes induced by the carbamothioate herbicides (Duke, 1985). Even so, this does not appear to explain the common important symptom induced by these herbicides, the inhibition of cell division and/or cell enlargement.

Metham

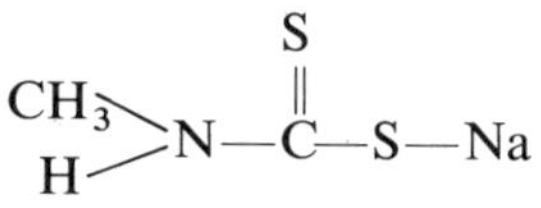

$$CH_3(H)N{-}C({=}S){-}S{-}Na$$

metham

Metham is a carbamodithioate. Both oxygen atoms of the parent carbamic acid molecule are replaced by sulfur atoms. This is in contrast to the carbamothioate-type herbicides previously discussed in this chapter that contain only one sulfur atom.

Metham is the common name for methylcarbamodithioic acid. However, its herbicidal use form is the sodium salt and the following information applies to this salt. The trade name is Vapam®. Pure metham is a white crystalline solid, but since it is unstable, the commercial formulation (Vapam®) is a stable concentrated 32.7% aqueous solution. Its acute oral LD_{50} is 820 mg/kg for male rats.

Uses Metham is a temporary soil fumigant used to control nematodes, garden centipedes, soil-borne disease organisms, and most germinating weed seeds and seedlings. It has also been used to control certain shallow perennial weeds (e.g., nutsedges) and kill roots in sewers. This herbicide is used in the field and for potting soil. When used in the field, the soil should be cultivated before application to allow diffusion of the gaseous toxicant.

Metham may be applied in various ways depending on the size of the area to be treated and the equipment available. For small areas, a sprinkling can or hose proportional diluter may be used. For large areas, soil injection, spray application with immediate soil incorporation, or application through a sprinkler-irrigation system may be used.

Metham is most effective when it is possible to confine the vapors with a plastic tarp; however, the water seal method (saturating the top 2 inches of soil with water) may also be used. When using a tarp, the treated area should be covered for 48 hours or longer. Seven days after treatment the area should be cultivated to a depth of 2 inches. At least 14 to 21 days should pass after application before the treated area is seeded to a crop.

CYCLOHEXENDIONES

Cyclohexendiones are selective postemergence herbicides that control most annual and perennial grasses without injuring most broadleaf plant or sedges. They are relatively new and most of them are still under development. They include alloxydim, cloproxydim, cycloxydim, clethodim, and sethoxydim. Sethoxydim is the only one fully registered in the United States at this time; however, registration for the use of clethodim is expected soon.

Sethoxydim

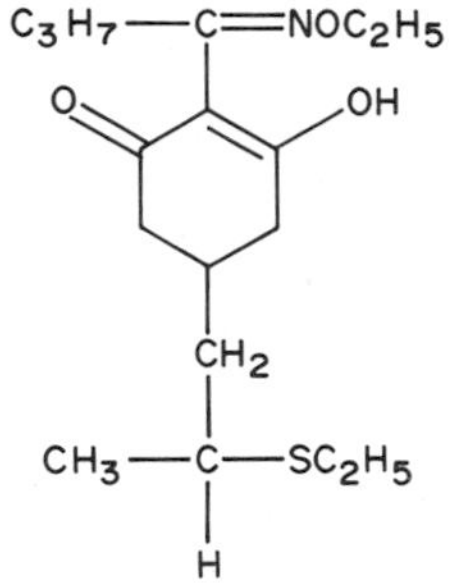

$C_3H_7—C=NOC_2H_5$

O OH

CH_2

$CH_3—C—SC_2H_5$

H

sethoxydim

Sethoxydim is the common name for 2-[1-(ethoxyimino)butyl-5-[2-(ethylthio)-propyl[3-hydroxy-2-cyclohexen-1-one. The trade name is Poast®. It is a light amber liquid with a low water solubility of 48 ppm at 25°C. It is formulated as an emulsifiable concentrate (1.5 lb/gal). The acute oral LD_{50} is 2.7 to 3.1 g/kg for rats.

Uses The addition of a nonphytotoxic oil concentrate at the rate of 2 pints/acre to the spray solution is essential for all the following uses. For best control of certain hard-to-kill species, *in soybeans only* ammonium sulfate can also added to the spray solution at the rate of 2.5 lb/acre.

Sethoxydim registered for use in soybeans, cotton, sugar beats, and many nonbearing food and nonfood crops. Refer to product label for sethoxydin use in tolerant species of nonbearing food crops, ornamentals, nursery, and other nonfood crops.

Since soybeans, cotton, and sugar beets are quite tolerant to sethoxydim injury, use rates vary more relative to the weed species and their size rather than with the specific crop. In general, 0.19 to 0.38 lb/acre (1 to 2 pints of product) are used for annual grass control and 0.28 to 0.47 lb/acre (1.5 to 2.5 pints of product) are used in the first application for perennial grass control. One or more additional applications, at about one-third reduction in rate, can be used for perennial grass control depending on the perennial grass to be controlled and the crop. For postemergence broadleaf and grass weed control *in soybeans*, tank mix or separate applications of sethoxydim, bentazon, and/or acifluorfen may be made. There are a number of limitations on some of the above used relative to time of application and harvest, geographic location, rate and total amount per season, and grazing and feeding forage and ensilage to livestock (see product label).

Soil Influence The efficacy of sethoxydim is not dependent on soil type since it is a postemergence herbicide, but soil activity prevents the germination of grasses immediately after application. Soil persistence is relatively short and pH dependent, 4 to 5 days at pH 6.8 and 11 days at 7.4

Clethodim

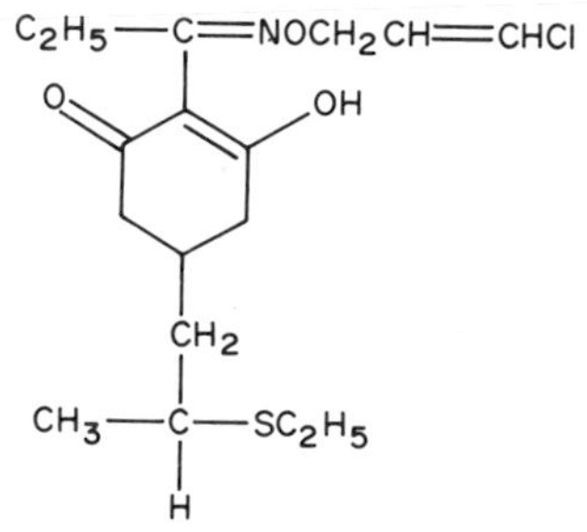

clethodim

Clethodim is the common name for (*E*, *E*)-(±)-2-[1-[[(3-chloro-2-propenyl)oxy]-imino]propyl]-5-[2-(ethylthio)propyl]-3-hydroxy-2-cyclohexen-1-one. The trade name is Select®. Clethodim is an amber liquid and its water solubility is dependent on pH. It is formulated as an emulsifiable concentrate (2 lb/gal). The acute oral LD_{50} is >3.6 g/kg and >2.9 g/kg for male and female rats, respectively.

Uses Clethodim is not registered for any uses at this time but some registrations are expected soon. However, suggested evaluations include many broadleaf field crops, vegetable crops, trees and vines, alfalfa, and conservation tillage programs as an early postemergence treatment for the control of annual grasses. Pre-emergence treatments, preplant or preplant soil incorporation, are also suggested for certain vegetable crops. Refer to the technical information bulletin for specific crops. Oil concentrate should be added to all postemergence applications.

Mode of Action of Cyclohexendiones

Sethoxydim is absorbed fairly rapidly through the foliage and an oil concentrate in the spray solution increases efficacy, possibly through increased uptake (*Herbicide Handbook*, 1989). Following foliar uptake, sethoxydim translocates rapidly both acropetally and basipetally. It acts in the meristematic region of grasses. It is metabolized rapidly is soybeans by oxidation of the parent molecule, structural rearrangement, and conjugation. Presumably the other cyclohexendione herbicides act similarly to sethoxydim in regard to the above processes.

Alloxydim, clethodim, cyloxydim, and sethoxydim appear to act by inhibiting lipid biosynthesis and possibly flavonoid biosynthesis. They have been shown to be a potent inhibitor of acetyl-CoA carboxylase, an enzyme common to both of these processes (Burton et al., 1987; Focke and Lichtenhaler, 1987; Rendina and Felts, 1988).

SUGGESTED ADDITIONAL READING

Appleby, A. P., W. R. Furtick, and S. C. Fang, 1965, *Weed Res.* **5**, 115.

Ashton, F. M., and A. S. Crafts, 1981, *Mode of Action of Herbicides*, Wiley, New York.

Ashton, F. M., and S. Helfgott, 1986, *Proc. 18th Calif. Weed Conf.*, p. 9.

Babiker, A. G. T., and H. J. Duncan, 1975, *Pestic. Sci.* **6**, 655.

Benting, J. D., 1967, *Weed Res.* **7**, 302.

Burton, J. D., J. W. Gronwald, D. A. Somers, J. A. Connelly, B. G. Gengenbach, and D. L. Wyse, 1987, *Biochem. Biophys. Res. Commun.* **148**, 1039.

Catchpole, A. H., and C. J. Hibbitt, 1972, *Proc. 11th British Crop Prot. Council*, p. 77.

Corbett, J. R., K. Wright, and A. C. Baillie, 1984, *The Biochemical Mode of Action of Pesticides*, Academic Press, New York.

Crop Protection Chemical Reference, 1989, Wiley, New York (revised annually).

Dawson, J. H., 1963, *Weeds* **11**, 60.

Duke, S. O., 1985, *Weed Physiology*, Vol. II, *Herbicide Physiology* CRC Press, Boca Raton, FL.

Ennis, W. B., Jr., *Am. J. Bot.* **35**, 15.

Ennis, W. B., Jr., 1949, *Am. J. Bot.* **36**, 823.

Fang, S. C., 1975, Thiocarbamates, in P. C. Kearney and D. D. Kaufman, Eds., *Herbicides*, Vol. 1, pp. 323/348, Dekker, New York.

Farm Chemicals Handbook. 1987, Meister, Willough. OH (revised annually.)

Fedtke, C., 1982, *Biochemistry and Physiology of Herbicide Action*, Springer-Verlag, New York.

Focke, M., and H. K. Lichtenthaler, 1987, *Z. Naturforsch* **42c**, 93.

Friesen, G., 1929, *Planta* **8**, 666.

Gray, R. A., and A. J. Weierich, 1969, *Weed Sci.* **17**, 223.

Hatzios, K. K., 1985, *Adv. Agron.* **36**, 265.

Hatzios, K. K., and D. Penner, 1982, *Metabolism of Herbicides in Higher Plants*, Burgess, Minneapolis.

Hendrich, L. W., W. F. Meggett, and D. Penner, 1974, *Weed Sci.* **22**, 179.

Herbicide Handbook, 1989, Weed Science Society of America, Champaign, IL.

Hess, F. D., 1982, Determining causes and categorizing types of growth inhibition induced by herbicides, in D. E. Moreland, J. B. St. John, and F. D. Hess, Eds., *Biochemical Responses Induced by Herbicides*, pp. 207–230. Symposium Series 181, American Chemical Society, Washington, D.C.

Hess, F. D., 1985, Herbicidee absorption and translocation and their relationship to plant tolerances and susceptibility, in S. O. Duke., *Weed Physiology*, Vol. II, *Herbicide Physiology*, pp. 191–214, CRC Press, Boca Raton, FL.

Kassenbeer, H., 1970, *Z. PflKrankh. PflPath.PflSchutz* **77**, 79.

Kidd, B. R., N. H. Stephen, and H. J. Duncan, 1982, *pl. Sci. Lett.*, **19**, 203.

Killmer, J. L., J. M. Widholm, and F. W. Slife, 1980, *Pl. Sci. Lett.* **19**, 203.

Knake, E. L., and L. M. Wax, 1968, *Weed Sci.* **16**, 393.

Knowles, C. O., and B. R. Sonawane, 1972, *Bull. Environ. Contem. Tox.* **8**, 73.

Lay, M. M., and J. E. Casida, 1976, *Pestic. Biochem. Physiol.* **6**, 442.

Macherel, D., M. Tissut, F. Nurit, P. Ravanel, M. Bergon, and J.-P. Calmon, 1986, *Physiol. Veg.* **24**, 97.

Pellos, F. M., and J. E. Casida, 1978, *Chemistry and Action of Herbicide Antidotes*, Academic Press, New York.

Parker, C., 1963, *Weed Res.* **3**, 259.

Prendeville, G. N., Y. Eshel, C. S. James, G. F. Warren, and M. M. Schreiber, 1968, *Weed Sci.* **16**, 432.

Rendina, A. R., and J. M. Felts, 1988, *Plant Physiol* **86**, 983.

Shimabukuro, R. H., 1985, Detoxication of herbicides, in S O. Duke, Ed., *Weed Physiology*, Vol. II, *Herbicide Physiology*, pp. 215–240., CRC Press, Boca Raton, FL.

Stephen, N. H., G. T. Cook, and H. J. Duncan, 1980, *Ann. Appl. Biol.* **96**, 227.

Slater, C. H., J. H. Dawson, W. R. Furtick, and A. P. Appleby, 1969, *Weed Sci.* **17**, 238.

Still, G. G., and R. A. Herrett, 1976, Methylcarbamates, carbenilates, and acylanilides, in P. C. Kearney and D. D. Kaufman, Eds., *Herbicides*, Vol. 2, pp. 609–664, Dekker, New York.

Templeman, W. G., and W. A. Sexton, 1945, *Nature* (*London*) **156**, 630.

Veerasekaran, P., R. C. Kirkwood, and E. W. Parnell, 1981a, *Pestic. Sci.* **12**, 325.

Verrasekaran, P., R. C. Kirkwood, and E. W. Parnell, 1981b, *Pestic. Sci.* **12**, 330.

Weed Control Manual and Herbicide Guide, 1988, Meister, Willough, OH (revised annually).

For herbicide use, see manufacturer's or supplier's label and follow these directions. Also see the Preface.

11 Dinitroanilines, Diphenyl Ethers, and Imidazolinones

DINITROANILINES

The herbicidal properties of the 2,6-dinitroanilines were first reported by Eli Lilly and Company scientists in 1960 (Alder et al., 1960). They subsequently developed several dinitroaniline herbicides including benefin, ethalfuralin, isopropalin, oryzalin, and trifluralin. Several other companies also attempted to develop herbicides from this class of compounds but only one of these is currently available in the United States, namely pendimethalin. Table 11-1 gives the chemical structure and name of these six dinitroaniline herbicides.

Trifluralin

Trifluralin is the common name for 2,6-dinitro-*N*,*N*-dipropyl-4-(trifluoromethyl)benzenamine. The trade name is Treflan®. Trifluralin is a yellow-orange crystalline solid with an extremely low water solubility of 0.3 ppm at 25°C. The compound is formulated as an emulsifiable concentrate (4 lb/gal), nonfreezable formulation (4 lb/gal), liquid flowable formulation (5 lb/gal), and granules (5 and 10%). The acute oral LD_{50} is 5 g/kg for rats.

Uses Trifluralin was the first dinitroaniline developed and marketed and is one of the most important herbicides used selectively in crops. Although the major use has been in cotton and soybeans, it is used on more than 40 other crops (see Table 11-2).

Trifluralin is usually applied as a preplant or preemergence treatment. Both of these require soil incorporation with the application, or shortly thereafter, to prevent volatility losses of the herbicide from the soil surface. In some crops, such as tomatoes, sugar beets, cantaloupes, cucumbers, and watermelon, the herbicide is too phytotoxic to be applied preplant or preemergence in the direct seeded crop, but can be used on transplants or established plants.

Trifluralin controls most weeds shortly after the seed germinates. This includes nearly annual grasses and many annual broadleaf species; see Table A-0 for specific weed species controlled. It also controls certain perennial weeds (e.g., field bindweed and johnsongrass from rhizomes) when used in accordance with special rate and application techniques.

TABLE 11-1. Common Name, Chemical Name, and Chemical Structure of the Dinitroaniline Herbicides

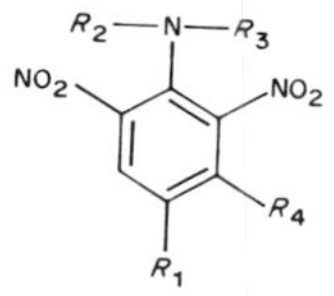

Common Name	Chemical Name	R_1	R_2	R_3	R_4
Benefin	*N*-Butyl-*N*-ethyl-2,6-dinitro-4-(trifluromethyl)benzenamine	$-CF_3$	$-C_2H_5$	$-C_4H_9$	$-H$
Ethalfluralin	*N*-Ethyl-*N*-(2-methyl-2-propenyl)-2,6-dinitro-4-(trifluoromethyl)benzenamine	$-CF_3$	$-C_2H_5$	$-CH_2-C(CH_3)=CH_2$	$-H$
Isopropalin	4-(1-Methylethyl)-2,6-dinitro-*N*,*N*-dipropylbenzenamine	iso-C_3H_7	$-C_3H_7$	$-C_3H_7$	$-H$
Oryzalin	4-(Dipropylamino)-3,5-dinitrobenzenesulfonamide	$-S(\rightarrow O)(\rightarrow O)-NH_3$	$-C_3H_7$	$-C_3H_7$	$-H$
Pendimethalin	*N*-(1-Ethylpropyl)-3,4-dimethyl-2,6-dinitrobenzenamine	$-CH_3$	$-CH(C_2H_5)_2$	$-H$	$-CH_3$
Trifluralin	2,6-Dinitro-*N*,*N*-dipropyl-4-(trifluoromethyl)benzanamine	$-CF_3$	$-C_3H_7$	$-C_3H_7$	$-H$

Benefin

Benefin is the common name for *N*-butyl-*N*-ethyl-2,6-dinitro-4-(trifluromethyl)benzenamine. The trade name is Balan®. Pure benefin is a yellow-orange crystalline solid and its water solubility is extremely low, 0.1 ppm at 25°C. It is formulated as an emulsifiable concentrate (1.5 lb/gal), dry flowable (60%), and granule (2.5%). Benefin has an acute oral LD_{50} >10 g/kg for adult rats.

Uses Benefin is primarily used as a preplant soil-incorporated treatment to control most annual grasses and many broadleaf weeds in several crops. It controls these weeds shortly after germination, but does not control established weeds. The crops are direct-seeded lettuce, peanuts, small seeded legumes [alfalfa, birdsfoot trefoil, clovers (alsike, ladino, red)] and transplanted tobacco (burley, dark). The use rate in these crops is 1.125 to 1.5 lb/acre; soil type and percent organic matter determine the actual use rate. In peanuts, it is also used in combination with metolachlor or vernolate as a tank mix. The granule formulation is used at 1.5 to 3.0 lb/acre to control annual grasses in established-perennial grass lawns or turf (bahiagrass, bermudagrass, bluegrass, centipede-grass, fescue, ryegrass, St. Agustinegrass, zoysiagrass).

Ethalfluralin

Ethalfluralin is the common name for *N*-ethyl-*N*-(2-methyl-2-propenyl)-2,6-dinitro-4-(trifluoromethyl)benzenamine. The trade name is Sonalan®. It is a yellow-orange crystalline solid with an extremely low water solubility of 0.3 ppm at 25°C. It is formulated as an emulsifiable concentrate (3 lb/gal). The acute oral LD_{50} is estimated to be > 10 g/kg for rats and mice.

Uses Ethalfluralin is preemergence or preplant soil-incorporated herbicide used to control most annual grass and many broadleaf weeds in a few crops; see Table A-0 for specific weeds controlled and Table 11-3 for detailed information on the crops.

Isopropalin

Isopropalin is the common name for 4-(1-methylethyl)-2,6-dinitro-*N*,*N*-dipropylbenzenamine. The trade name is Paarlan®. It is red-orange oil with an extremely low water solubility of 0.08 ppm at 25°C. It is formulated as an emulsifiable concentrate (6 lb/gal). The acute oral LD_{50} is >5 g/kg for rats.

Uses Isopropalin is used as a preplant soil-incorporated treatment to control many annual grass and broadleaf weeds in transplanted air-cured (burley, Maryland, dark) and flue-cured tobacco. The use rate is 1.5 to 2.0 lb/acre. See Table A-0 for specific weeds controlled.

TABLE 11-2. Trifluralin Uses[1]

Crop	Rate (lb/acre)	Combination	Remarks
Alfalfa	0.75–1.00	—	Established
Asparagus	0.50–2.00	—	Established, single or split Application, pre or postharvest
Beans, dry	0.50–0.75[2]	—	PPI[3] < 20 inches rain
	0.50–0.75[2]	EPTC	PPI, tank mix, < 20 inches rain
Corn, field	0.38–0.75	—	Post[3] > 8 inches tall, overtop or directed
Cotton	0.50–1.00	—	PPI, postplanting, or layby
	0.50–1.00	Fluometuron, prometryn	PPI, tank mix
	0.50–1.00	Diuron, fluometruon	PPI, sequential overlay
Fruit and nut crops and vineyards	0.50–1.00	—	PPI, new plantings, < 20 inches rain[4a]
	0.50–2.00	—	PPI, new plantings, < 20 inches rain[4b]
	0.50–1.00	—	PPI, new plantings, > 20 inches rain[4c]
	1.00–2.00	—	Established, nonbearing, > 20 inches rain[4d]
	1.00–2.00	—	Established, bearing, > 20 inches rain[4e]
	1.00–2.00	—	Established, nonbearing or bearing, < 20 inches rain[4f]
Hops	0.63–0.75	—	Dormant
Milo	0.38–0.75	—	Post > 8 inches tall, overtop or directed
Mint	0.50–0.75	—	Established, peppermint or spearment
Peas, dry and English	0.50–0.75	—	PPI
	0.38–0.50	Triallate	PPI, tank mix
Peanuts, spanish	0.50	—	PPI
	0.50	Vernolate	PPI, tank mix
Potatoes	0.50–0.75[2]	—	Postplant before emergence, or following dragoff, or after full emergence; < 20 inches rain
	0.38 + 0.38	—	Split application, PPI and after full emergence

	0.50–0.75[2]	EPTC	Postplant before emergence, or following dragoff; <20 inches rain
	0.38	EPTC	PPI
Safflower	0.50–0.75[2]	—	PPI, <20 inches rain
Soyabeans[5]	0.50–1.00	—	PPI
	0.50–1.00	Vernolate, chloramben plus metribuzin	PPI, tank mix
	0.50–1.00	Chloramben, metribuzin	PPI, tank mix or sequential overlay
Sugar beets	0.50–0.75	—	Post 2–6 inches tall, overtop
Sugarcane	1.00–2.00	—	Plant cane or ratoon
Sunflower	0.50–0.75[2]	—	PPI, <20 inches rain
	0.50–1.00	Chloramben	PPI, tank mix or sequential overlay
Tomatoes	0.50–0.75[2]	—	Post, direct seeded, directed spray at thinning, <20 inches rain
Wheat, winter	0.75–1.00	—	PPI
Wheat, spring	0.50–0.75	—	Postplant
	0.50–0.75	Triallate	Postplant, tank mix
Group 1[6]	0.50–0.75[2]	—	PPI, <20 inches rain
Group 2[7]	0.50–0.75	—	PPI

There are some limitations of some of the above uses relative to geographic location, variety, cultural practices, soil texture and organic matter, time of application, grazing and use of forage etc.; see product labels.

---

[1] This table does not include Supplemental Labeling for several crops or Speciality Uses, e.g., nursery stock, ornamentals, or under paved surfaces; see product labels.

[2] >20 inches rain highest rate = 1.00 lb/acre.

[3] PPI, preplant soil-incorporated; post, postemergence.

[4a] Almond, apricot, citrus, nectarine, peach, pecan, walnut; [4b] vineyards; [4c] citrus, pecan, vineyards; [4d] citrus, pecan; [4e] grapefruit, lemon, orange, pecan,tangelo, tangerine, [4f] almond, apricot, grapefruit, lemon, nectarine, orange, peach, plum, prune, tangelo, tangerine, walnut, vineyards.

[5] Special use instructions including different rates for specific weeds, organic soils, sequential applications, etc. are given in the label.

[6] Carrots, castor bean, celery (direct seeded and transplants), cole crops (broccoli, brussels sprout, cabbage, cauliflower-transplants), cucurbits (cantaloupe, cucumber, watermelon-postemergence), okra, pepper (transplants), southern peas.

[7] Beans (guar, lima, mung, snap), greens [collard, kale, mustard, turnip (processing)], mustard (seed or processing for food).

TABLE 11-3. Ethalfuralin Uses

Crop	Rate (lb/acre)	Combinations	Remarks
Beans, dry	0.56–1.13	—	PPI[1,2]
	1.14–1.70	—	PE[1]
	0.47–1.13	Alachlor, EPTC	PPI, tank mix[2]
	0.47–1.13	Chloramben, metolachlor	PPI, tank mix or sequential[2]
	0.47–1.13	Bentazon	PPI, bentazon post[1,2]
Soyabeans	0.60–1.13	—	PPI[3]
	0.47–1.13	Alachlor, chloramben, imazaquin, metolachlor, metribuzin	PPI, tank mix or sequential
	0.47–1.13	Clomazone, vernolate, clomazone plus metribuzin	PPI, tank mix
	0.47–1.13	Linuron	PPI, sequential
	0.56–1.13	Acifluorfen, bentazon	PPI, acifluorfen or bentazon post
Sunflower	0.56–1.13	—	PPI
	0.47–0.13	EPTC	PPI, tank mix
	0.47–1.13	Chloramben	PPI, tank mix or sequential

There are some limitations of some of the above uses relative to geographic location, grazing and use of forage, and crop variety; use rates vary with soil type; see product label.

[1] PPI, preplant soil-incorporated; PE, preemergence; post, postemergence.
[2] Higher rates (1.13–1.70 lb/acre) and more through soil incorporation reqired for annual groundcherry and nightshade control.
[3] Higher rates (1.13–1.30 lb/acre) and more through soil incorporation required for partial control of eastern black nightshade.

Oryzalin

Oryzalin is the common name for 4-(dipropylamino)-3,5-dinitrobenzenesulfonamide. The trade name is Surflan®. It is a yellow-orange crystalline solid with a very low water solubility of 2.5 ppm at 25°C. It is formulated as an aqueous suspension (4 lb/gal) and a wettable powder (75%). The acute oral LD_{50} is > 10 g/kg for rats.

Uses Oryzalin is a selective soil-applied herbicide used to control many annual grass and broadleaf weeds in numerous crops. In contrast to most dinitroanaline herbicides it is relatively nonvolatile and therefore does not require soil incorporation to prevent volatility losses. However, without overhead irrigation or rainfall within several days after application to the soil surface, shallow soil

Figure 11-1. General weed control in grapes with oryzalin. *Left*: After 3 consecutive years of treatment. *Right*: Control, no herbicide treatment. (Elanco Products Company, a division of Eli Lilly and Company.)

incorporation is required. See Table A-0 for specific weeds controlled and Table 11-4 for detailed information on the crops.

Pendimethalin

Pendimethalin is the common name for *N*-(1-ethylpropyl)-3,4-dimethyl-2,6-dinitrobenzenamine. The trade name is Prowl®. It is a crystalline orange-yellow solid with an extremely low water solubility of 0.5 ppm. It is formulated as an emulsifiable concentrate (4 lb/gal). The acute oral LD_{50} of the technical product for rats is 1050 (female) and 1250 (male) mg/kg.

Uses Pendimethalin controls most annual grass and certain broadleaf weeds in many crops. It has also controlled johnsongrass from rhizomes using a special 2-year program. It is applied preemergence, early postemergence, preplant soil-incorporated, or postemergence soil-incorporated depending on the crop. Like oryzalin and unlike the other dinitroaniline herbicides, it does not require soil incorporation with adequate rainfall or overhead irrigation because of its low

TABLE 11-4. Oryzalin Uses

Crop	Rate (lb/acre)	Combinations	Remarks
Barley	1.00–1.50	—	After fully tillered, jointing
	1.00–1.50	2,4-D	Initiated, but before heading
Cotton	1.00–1.50	—	At planting or within 2 days
Mint	1.00–1.50	—	To mint stubble
	1.00–1.50	Terbacil	Time varies with location
Peas	0.50	Trifluralin	Preplant, English or green peas
Potatoes	0.75–1.00	—	Postplant, before emergence
	0.75–1.00	Metribuzin	Postplant, before emergence
Small fruit	2.00–6.00	—	Bearing or nonbearing[1a]
	2.00–6.00	Several	See[1b–1h]
Soyabeans	0.75–1.50	—	From 4 weeks before planting
	0.50	Alachlor, alchlor plus metribuzin	After planting or within 2 days
	0.50–1.25	Chloramben	After planting or within 2 days
	0.50–1.25	Linuron, linuron plus paraquat	At planting or within 2 days
	0.50–1.25	Metribuzin	From 4 weeks before planting
	0.75–1.25	Metribuzin plus paraquat	At planting or within 2 days
Tobacco	0.50–1.00	—	Layby-directed
Tree fruits	2.00–6.00	—	Bearing or nonbearing[2a]
and nuts	2.00–6.00	Many	See[2b–2k]
Wheat, winter	1.00–1.50	—	After fully tillered, jointing
	1.00–1.50	2,4-D	Initiated, but before heading
Lawns/turf	2.00	—	> 2 weeks before grass-weed seeds germinate
Ornamentals			
Flowers	2.00–4.00	—	Field and container grown
Nurseries	2.00–4.00	—	Established, field grown
	2.00–4.00	Glyphosate	Field, burn down plus residual
Noncrop	2.00–6.00	Glyphosate	Burn down plus residual

There are some limitations of some of the above uses relative to geographic location, variety, cultural practices, soil texture and oganic matter, time of application, grazing and use of forage, etc.; see product labels.

[1a]Caneberries, currant, grape; [1b]diclobenil—blackberry, blueberry, grape, raspberry; [1c]diuron—caneberries, grape; [1d]glyphosate—grape; [1e]paraquat—blackberry, blueberry, boysenberry, raspberry paraquat; [1f]simazine—blackberry, boysenberry, grape, loganberry, raspberry; [1g]terbacil—blueberry, caneberries; [1h]oxyfluorfen plus paraquat—grape.

[2a]Almond, apple, apricot, avocado, cherry, English walnut, fig, filbert, grapefruit, kiwifruit, lemon, macadamia, nectarine, olive, orange, peach, pear, pecan, pistachio, plum, pomegranate, prune; [2b]bromacil—grapefruit, lemon, oange; [2c]diclobenil—almond, apple, cherry, English walnut, filbert, nectarine, peach, pear, pecan, plum; [2d]diuron—apple, grapefruit, lemon, orange, peach, pear; [2e]glyphosate—almond, English walnut, filbert, grapefruit, lemon, orange, pecan, pistachio; [2f]norflurfen—almond, apple, apricot, cherry, citrus, filbert, nectarine, peach, plum, prune, walnut; [2g]paraquat—almond, apple apricot, avocado, cherry, English walnut, fig, filbert, grapefruit, lemon, nectarine, olive, orange, peach, pear, pecan, plum, prune; [2h]simazine—almond, apple, avocado, cherry, English walnut, filbert, lemon, orange, peach, pear, pecan; [2i]terbacil—apple, grapefruit, lemon, orange, peach; [2j]bromacil plus diuron—grapefruit, lemon, orange; [2k]oxyfluorfen plus paraquat—apricot, nectarine, peach, plum, prune.

TABLE 11-5. Pendimethalin Uses

Crop	Rate (lb/acre)	Combinations	Remarks
Beans	0.50–1.50	—	PPI[1]; dry, lima, snap
	0.50–1.50	EPTC	PPI, tank mix; dry, snap
Cotton	0.50–1.50	—	PPI
	1.00–2.00	—	PPI, 2-year program for rhizome johnsongrass
	0.50–1.50	Diuron	PPI, diuron PE[1] sequential
	0.50–1.50	Fluometuron	PE, tank mix
	0.50–1.50	Fluometuron	PPI, fluometuron PE sequential
	0.50–1.50	Norflurazon	PPI or PE, tank mix
	0.50–1.50	Prometryn	PPI, tank mix
Corn, field	0.75–2.00	—	PE
	0.50–1.50	—	Post-1[1], > 4 inches high
	0.75–1.50	Atrazine	PE, tank mix
	0.50–1.50	Atrazine	Post-I, >4 inches high, tank mix
	0.75–1.50	Cyanazine, cyanazine plus atrazine	PE, tank mix
	1.50	Dicamba	PE,tank mix
Corn, sweet	0.75–2.00	—	PE
	1.25–2.00	Atrazine, cyanazine	PE or postemergence to 4-leaf
	0.75–1.50	Atrazine plus cyanazine	PE or postemergence to 4-leaf
Grapes	2.00–4.00	—	Nonbearing, dormant
Peanuts	0.50–1.00	—	PPI
	0.50–1.00	Metolachlor, vernolate	PPI, tank mix
Potatoes	0.75–1.50	—	Postplant; PE, post-I, after drag-off, or post[1] to 6 inches
	0.50–1.50	EPTC, linuron, metribuzin	Postplant; post-I or after drag-off
	0.50–1.50	Metribuzin	Post, emergence to 6 inches
Sorghum	0.50–1.50	—	Post-I, 4 inches to layby
	0.50–1.50	Atrazine	Post-I, 4 inches to layby
	0.75–1.00	Atrazine	Post, early, > 2-leaf
Soybeans	0.50–1.50	—	PPI or PE
	1.00–2.00	—	PPI; red rice, itchgrass suppression, and 2-year program for rhizome johnsongrass
	0.50–1.25	Alachlor, metolachlor	PPI or PE, tank mix
	0.75–1.25	Chloramben	PPI or PE, tank mix
	0.50–1.25	Linuron	PE, tank mix; PPI, sequential
	0.50–1.25	Metribuzin	PE, tank mix; PPI, tank mix or sequential
	0.75–1.25	Chloramben plus metribuzin	PPI or PE, tank mix

TABLE 11-5. (*Continued*)

Crop	Rate (lb/acre)	Combinations	Remarks
	0.50–1.25	Linuron plus paraquat, metribuzin plus paraquat	PE, tank mix
Sunflower	0.50–1.75	—	PPI, spring or fall
	0.75–1.25	Chloramben	PPI, tank mix or sequential
	0.50–1.50	EPTC	PPI, tank mix, fall
Tobacco	0.75–1.50	—	PPI, transplants
	0.50–1.00	—	Layby directed
Tree fruits and nuts[2]	2.00–4.00	—	Nonbearing

There are some limitations of some of the above uses relative to geographic location and soil type; use rates vary with soil texture and organic matter content; see product label.

[1] PPI, preplant soil-incorporated; PE, preemergence; post, postemergence; Post-I, postemergence soil-incorporated.

[2] Almonds, apples, apricots, cherries, citrus, English walnuts, nectarines, peaches, pistachlos, plums, prunes.

volatility. See Table A-0 for specific weeds controlled and Table 11-5 for detailed information on the crops.

Soil Influence on Dinitroanalines

In general, the soil influence on the dinitroaniline herbicides is similar, but they differ somewhat in detail. These differences will be discussed after presenting their common general characteristics.

The dinitroaniline herbicides are strongly adsorbed on clay and organic matter and not subject to significant leaching. The application rate may vary with the clay and organic matter content. They are subject to photodecomposition when allowed to remain on the soil surface for prolonged periods of time and this may reduce their effectiveness. Certain ones require immediate soil incorporation to prevent volatility losses while others can be used as a preemergence treatment. They appear to be primarily degraded in soil by microorganisms under normal field conditions.

Benefin, isopropalin, ethalfluralin, and trifluralin require soil incorporation due to their relatively high volatility, whereas oryzalin and pendimethalin can be used as a preemergence treatment. However, in the absence or limited rainfall or overhead irrigation the latter two herbicides also require soil incorporation.

Persistence of the dinitroaniline herbicides in soil is relatively long and must be taken into consideration for the subsequent planting of sensitive crops. At usual application rates and under normal field conditions, benefin persists 4 to 5 months and the others less than 12 months or one growing seasons (*Herbicide*

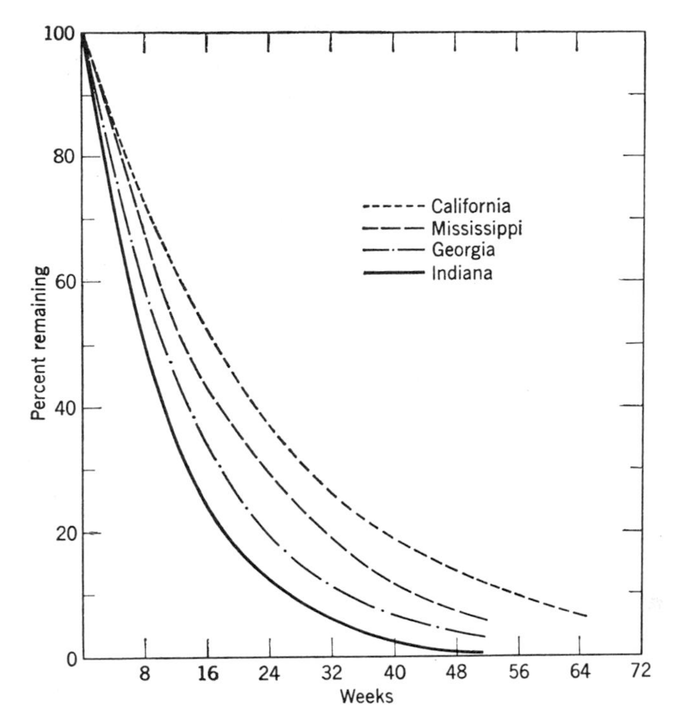

Figure 11-2. Field persistence of trifluralin at different geographic locations (different climatic conditions). In most location, almost all of the trifluralin disappeared within 48 weeks. However, it persisted longer under dryland conditions of California. (Lilly Research Laboratories, Greenfield, IN.)

Handbook, 1989). Trifluralin has been shown to persist longer in an arid environment than in areas receiving more rainfall (Figure11-2). Trifluralin has also been shown to be degraded rapidly under anaerobic conditions (Figure11-3) (Probst et al., 1967). Additional information on the degradation of trifluralin in soil has been published (Golab et al., 1967; Probst et al., 1975; Golab et al., 1979).

Mode of Action of Dinitroanilines

Most of the research on the mode of action of the dinitroaniline herbicides has been conducted with trifluralin. However, their similar phytotoxic symptoms suggest that they act in the same general manner although varying degrees of crop tolerance can be demonstrated.

These herbicides do not inhibit seed germination but rather inhibit early seedling growth shortly after seed germination. This is caused by the disruption of cell division. Characteristically, the root increases in diameter or swells in the active meristematic region near the root tip and shoot growth is also inhibited. Lateral or secondary root development is also inhibited, Figure11-4.

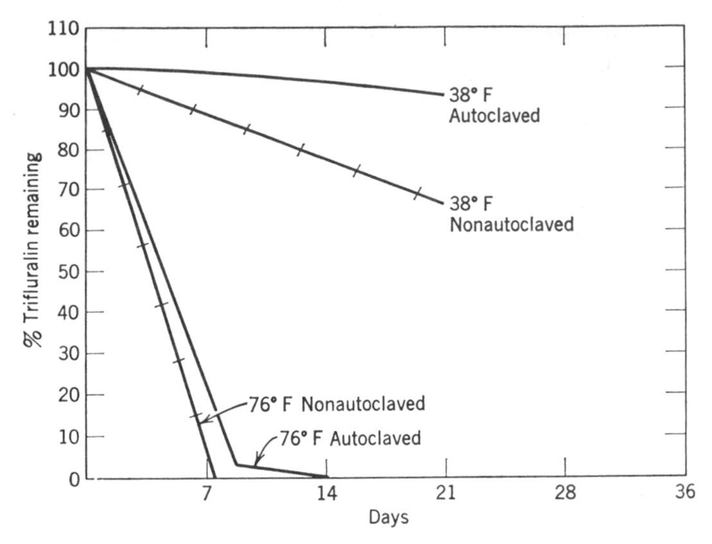

Figure 11-3. Anaerobic degradation of trifluralin in autoclaved and nonautoclaved soil as a function of temperature in soil saturated with water (200% field capacity). (Probst et al., 1967.)

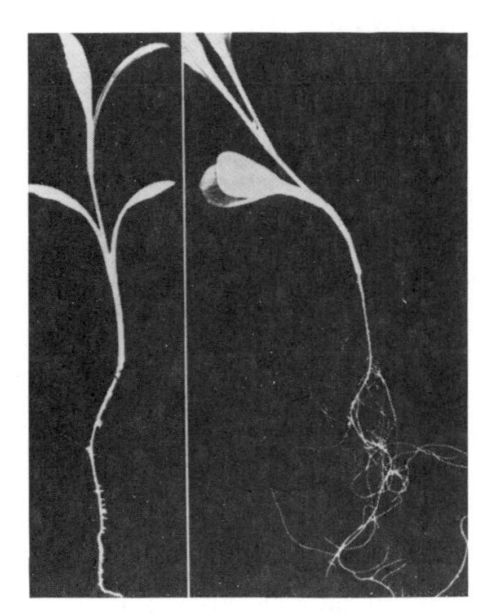

Figure 11-4. Inhibition of lateral root formation by trifluralin, a typical response of the dinitroaniline class of herbicides. *Left*: Treated. *Right*: Untreated. (D. E. Bayer, University of California, Davis.)

Trifluralin inhibits cell division by binding to tubulin and blocking the formation of spindle microtubules (Hess, 1982). Tubulin is a protein and an essential component of spindle microtubules. These microtubules are functional elements of the mitotic spindle apparatus and when they are not present cell division cannot proceed. Hess (1982) and Bartels (1985) present excellent reviews on the effect of trifluralin and other herbicides on cell division and plant development. Oryzalin also inhibits cell division and its effect appears to be similar to trifluralin (Bartles and Hilton, 1973). The herbicidal response of pendimethalin appears to involve an inhibition of cell division and cell elongation (*Herbicide Handbook*, 1989).

Most of these herbicides are absorbed by the developing seedling through the roots and shoot. There is very little translocation from the site of uptake. Pendimethalin is absorbed more readily by the foliage of very small broadleaf weeds than by grasses; very small amounts are absorbed by roots (*Herbicide Handbook*, 1989).

These herbicides are slowly degraded by higher plants and the degradation products appear to remain near the site of uptake. *N*-Dealkylation has been reported to be a major metabolic reaction in the degradation of trifluralin and other dinitroalinine herbicides in higher plants; reduction of the nitro groups has also been reported (Hatzios and Penner, 1982). The principal method of biological degradation of pendimethalin in animals, plants, and soil is the oxidation of the 4-methyl group on the benzene ring and oxidation of the *N*-ethylpropyl group in the amine moiety (*Herbicide Handbook*, 1989).

Trifluralin and/or other dinitroaniline herbicides have been reported to alter several biochemical processes and change the activity of several enzymes. Although some of these alterations may contribute to the demise of the plant, they do not appear to be primary considering the cytological and biophysical evidence of their action.

DIPHENYL ETHERS

Diphenyl ether herbicides control many annual broadleaf and grass weeds in soybeans and several other crops. They are particularly noted for their postemergence contact activity on emerged weeds, but they also are active as preemergence treatments to the soil. The first compound of this class of chemicals developed in the United States was nitrofen [2,4-dichloro-1-(4-nitrophenoxy)benzene]; however, it is no longer available in the United States. The current diphenyl ether herbicides available are acifluorfen, bifenox, fomesafen, lactofen, and oxyfluorfen. Their common name, chemical name, and chemical structure are given in Table11-6. They all consist of two ether-linked phenyl rings with a nitro (—NO_2) group attached to one of the phenyl rings.

Acifluorfen

Acifluorfen is the common name for 5-[2-chloro-4-(trifluoromethyl)phenoxy]-2-nitrobenzoic acid. The trade names are Blazer® and Tackle®; both are formulated as the sodium salt at 2 lb/gal. The technical product, a viscous sodium-salt solution, contains 45% of the active ingredient. It has a brown color and is infinitely soluble in water. The acute oral LD_{50} for rats is 1.3 g/kg for the technical product and 3.3 g/kg for the formulation.

TABLE 11-6. Common Name, Chemical Name, and Chemical Structure of Several Diphenyl Ether Herbicides

Common Name	Chemical Name	Chemical Structure
Acifluorfen[1]	5-[2-Chloro-4-(trifluoromethyl) phenoxyl-2-nitrobenzoic acid	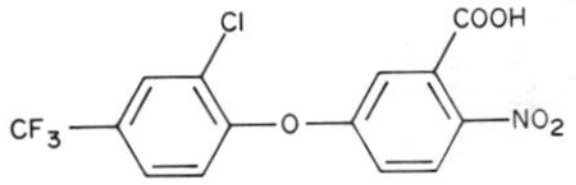 Cl, COOH, CF_3, O, NO_2
Bifenox	Methyl 5-(2, 4-dichlorophenoxy)-2-nitrobenzoate	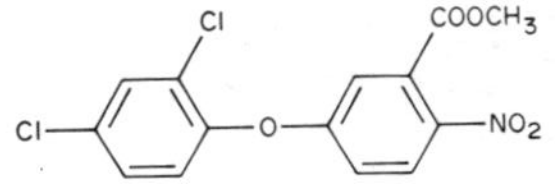 Cl, $COOCH_3$, Cl, O, NO_2
Fomesafen[1]	5-[2-Chloro-4-(triflurometyl)phenoxy]-*N*-(methylsulfonyl)-2-nitrobenzamide	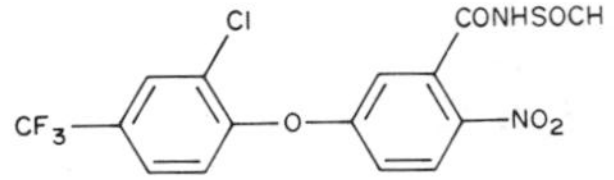 Cl, CONHSOCH, CF_3, O, NO_2
Lactofen	(±)-2-Ethoxy-1-methyl-2-oxoethyl 5-[2-chloro-4-(trifluromethyl)phenoxy]-2-nitrobenzoate	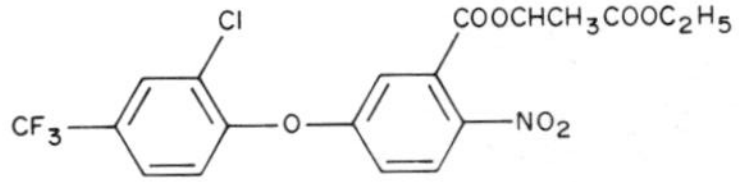 Cl, $COOCHCH_3COOC_2H_5$, CF_3, O, NO_2
Oxyfluorfen	2-Chloro-1-(3-ethoxy-4-nitrophenoxy)-4-trifluoromethyl) benzene	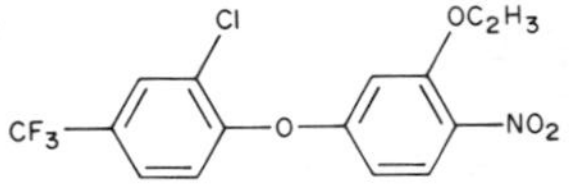 Cl, OC_2H_3, CF_3, O, NO_2

[1]Formulated as the sodium salt.

Uses Acifluorfen is primarily a selective postemergence contact herbicide used to control most annual broadleaf weeds, suppress the growth of certain young annual grasses, and suppress the growth of certain perennial weeds by killing the tops. It also has some preemergence soil activity but this is not its primary role. Tackle® is used only on rice and soybeans, whereas Blazer is used on peanuts, rice, and soybeans. The labels of these two products may differ slightly. Best weed control is obtained when the weeds are young and actively growing. A surfactant or nonphytotoxic oil increases its phytotoxcity, especially for certain relatively tolerant weeds. See Table A-0 for specific weeds controlled and Table11-7 for detailed information on the crops.

Soil Influence Acifluorfen is mainly applied to foliage; therefore, soil characteristics have little influence on its herbicidal effectiveness. It is strongly adsorbed on soil and not subject to leaching. It is readily photodecomposed into nonphytotoxic products. The half-life in soil is about 30 to 60 days or 2 to 4 weeks, depending on the source of the information.

Mode of Action Acifluorfen is readily absorbed by leaves. It is considered to be a

TABLE 11-7. Acifluorfen Uses[1]

Crop	Rate (lb/acre)	Combinations	Remarks
Peanuts	0.25–0.50	—	Early Post[2]
	0.38–0.50	2,4-DB	Post, tank mix, > 2 weeks old
Rice	0.13	—	Post, tillering to early boot
	0.25	Propanil	Post, tank mix, 3-leaf to end of tillering
Soybeans	0.38–0.75	—	Early Post
	0.50	UAN[3]	Post, 3-leaf to end of tillering
	0.38–0.50	Bentazon, chloramben, chlorimuron, 2,4-DB, imazaquin, bentazon plus UAN	Early post or post, tank mix
	0.38 0.50	Fluazifop, sethoxydim	Early post or post, tank mix or sequential[4]
	0.38–0.75	Bentazon plus sethoxydim	Post, tank mix or sequential[4]

There are some limitations on some of the above uses relative to geographic location, formulation, stage of crop growth, replanting, use for feed or forage, and use of adjvants, Use rate may vary with stage of crop growth and weed size and species.

[1]Blazer® and Tackle® labels differ somewhat for some of these uses.

[2]Post, postemergence.

[3]UAN, urea amonium nitrate.

[4]Sequential, time between applications varies depending on which herbicide is applied first.

contact herbicide since little translocation occurs and leaf injury is usually observed rather quickly, often within 24 to 48 hours. Light is required for these injury symptoms to develop. This injury includes chlorosis and necrosis. At the ultrastructural level, general membrane disruption including damage to the chloroplast envelope, tonoplast, and plasma membrane was observed (Orr and Hess, 1982a). Acifluorfen appears to be rapidly metabolized in plants with an initial conjugation with glutathione, which subsequently forms the *N*-malonylcysteine conjugate (Lamoureux and Rusness, 1983). However, exceedingly low or no radioactive was found in the crop seed at harvest (*Herbicide Handbook*, 1989). In an excellent review on the action of the diphenyl ether herbicides, Orr and Hess (1982b) proposed that these herbicides are activated in the light by carotenoids and then initiate radical chain reactions with membrane fatty acids. Recently, Abdallah et al. (1988) reported that acifluorfen greatly increased ethylene production and pretreatment of plants with compounds that interfere with ethylene production markedly reduced the phytotoxic symptoms. Also see discussion on the mechanisms of action of the diphenyl ether herbicides at end of this chapter.

Bifenox

Bifenox is the common name for methyl 5-(2,4-dichlorophenoxy)-2-nitrobenzoate. The trade name is Modown®. It is a yellow-tan crystalline solid with an extremely low water solubility of 0.35 ppm. It is available in a flowable formulation (4 lb/gal). The acute oral LD_{50} is 6.4 g/kg for rats.

Uses Bifenox is a selective preemergence, preplant soil-incorporated, or postemergence herbicide used to control annual weeds in several crops. See Table A-0 for specific weeds controlled and Table 11-8 for detailed information on the crops.

Soil Influence Bifenox is strongly adsorbed on soils and leaching is not a significant factor. Microorganisms are involved in its disappearance from soils. Photodecomposition or volatilization is not involved in these losses under normal field conditions. The half-life of bifenox is between 7 and 14 days with herbicidal activity lasting from 6 to 8 weeks at recommended application rates.

Mode of Action The phytotoxic symptoms of bifenox are foliar chlorosis and necrosis caused by rapid, light-dependent, disruption of cellular membranes. It is absorbed relatively rapidly by leaves and translocation is minimal from both leaves and roots. Studies with rice indicated that it was rapidly metabolized by ring hydroxylation and subsequent conjugation with or binding to normal plant constituents. Little information is available on its action at the molecular level; presumably its action is similar to other diphenyl ether herbicides; see discussion on the mechanisms of action of the diphenyl ether herbicides at end of this section.

Fomesafen

Fomesafen is the common name for 5-[2-chloro-4-(trifluromethyl)phenoxy]-*N*-(methylsulfonyl)-2-nitrobenzamide. The trade name is Reflex®. It is a white crystalline solid. The water solubility of the parent compound is very low (< 10 ppm) and the sodium salt water solubility is very high (600 g/liter). It is formulated as the sodium salt (2 lb/gal). The acute oral LD_{50} for rats is 8.7 g/kg (male) and 7.0 g/kg (female).

Uses Fomesafen is primarily a selective postemergence herbicide used to control a wide spectrum of broadleaf annual weeds in soybeans. Control is best when the weeds are young, actively growing, and not under stress. Certain weeds may be controlled by soil residual activity if rainfall occurs soon after application. See Table A-0 for the specific weeds controlled. The use rate is 0.19 to 0.38 lb/acre. It may be used alone or as a tank mix with bentazon, 2,4-DB, or fluazifop. Sequential treatments with fluazifop are also used. Limitations include geographic location and the rate used to control specific weeds of various stages of growth; see label.

Lactofen

Lactofen is the common name for (±)-2-ethoxy-1-methyl-2-oxoethyl 5-[2-chloro-4-(trifluoromethyl)phenoxy]-2-nitrobenzoate. The trade name is Cobra®. The technical material is a dark brown to tan material with an extremely low water solubility of < 1 ppm. It is formulated as an emulsifiable concentrate (2 lb/gal). The acute oral LD_{50} is > 5.0 g/kg for rats.

Uses Lactofen is used as a selective postemergence herbicide to control many broadleaf weeds in cotton and soybeans. See Table A-0 for specific weeds controlled. The use rate is 0.15 to 0.20 lb/acre for both of these crops. In cotton, it is used as a directed spray when the cotton is at least 6 inches tall. Cotton plants are not tolerant to over-the-top applications. In soybeans, it is used as an early postemergence treatment. There are grazing and use of forage for animals limitations; see label.

Oxyfluorfen

Oxyfluorfen is the common name for 2-chloro-1-(3-ethoxy-4-nitrophenoxy)-4-(trifluoromethyl)benzene. The trade name is Goal®. It has a deep red-brown color and an extremely low water solubility of 0.1 ppm. It is formulated as an emulsifiable concentrate (1.6 lb/gal). The acute oral $LD5_{50}$ of the technical material is > 5.0 g/kg for rats.

Uses Oxyfluorfen is a selective preemergence or postemergence herbicide used to control many annual broadleaf weeds and grasses in several crops. See

TABLE 11-8. Bifenox Uses

Crop	Rate (lb/acre)	Combinations	Remarks
Corn	1.50–2.00	—	Post[1], field, silage
	1.50–2.00	Alachlor, metolachlor	Post, field, silage, tank mix
Milo	1.50–2.00	—	PE[1]
	1.50–2.00	Alachlor	PE, tank mix, Screen® treated seed
	1.50–2.00	Metolachlor	PE, tank mix, Concept® treated seed
	1.50–2.00	Propachlor	PE, tank mix
	1.50–2.00	Glyphosate or paraquat	In conjunction with bifenox
	1.50–2.00	Glyphosate or paraquat plus alachlor	In conjunction with bifenox, Screen® treated seed
	1.50–2.00	Glyphosate or paraquat plus metolachlor	In conjunction with bifenox, Concept® treated seed
	1.50–2.00	Glyphosate or paraquat plus propachlor	In conjunction with bifenox
Rice	2.00–3.00	—	PE, drilled, dry-seeded
	2.00–3.00	—	Post, dry-seeded, water-seeded
	2.00–3.00	—	Post-flood, dry-seeded, water-seeded
	2.00–3.00	Propanil	Post, dry-seeded, water-seeded, tank mix

Small grain[2]	0.75–1.00	—	Post, 2–4 leaf stage
Soybeans	1.50–2.00	—	PE
	1.50–2.00	Alachlor, metolachlor	PE, PPI[1], tank mix
	1.25–1.50	Metribizin plus alachlor or metolachlor	PPI, tank mix
	1.50–2.00	Oryzalin	PE, tank mix
	1.50–2.00	Glyphosate or paraquat	In conjunction with bifenox
	1.50–2.00	Pendimethalin	Sequential, pendimethalin PPI followed by bifenox PE
	1.50–2.00	Trifluralin	Sequential, trifluralin PPI followed by bifenox PE
	1.25–1.50	Trifluralin plus metribuzin	Sequential, trifluralin plus metribuzin PPI followed by bifenox PE
Oranamentals	3.00	—	PE, field grown from seed
	3.00	—	Post, tree seedlings

There are limitations on some of the above uses relative to stage of crop growth, crop varities, and grazing and use of forage.

[1]PE, preemergence; post, postemergence; ppI, preplant soil-incorporated.
[2]Barley, oat, wheat (spring, spring-seeded and winter, fall-seeded).

TABLE 11-9. Oxyfluorfen Uses

Crop	Rate (lb/acre)	Combinations	Remarks
Artichokes	1.00–2.00	—	Post[1,2]-directed
Cotton	0.25–0.50	—	PP[1,2], > 14 days before planting
	0.25–0.50	Glyphosate, paraquat	PP, > 14 days before planting
	0.25–0.50	—	Post-directed
	0.25–0.50	Cyanazine, diuron, MSMA	Post-directed
Corn, field	0.50–0.75	—	Post-directed, > 24 inches, first application, witchweed
	0.25–0.50	—	Post-directed, > 24 inches repeat application, witchweed
Grapes	0.50–2.00	—	Dormant[2]
	0.50–2.00	Glyphosate, napropamide, norflurazon, oryzalin, paraquat, pronamide, simazine	Dormant
Jojoba	1.00–2.00	—	Post, > 6 inches
Mint	0.50–1.50	—	PE[1]
Onions	0.03–0.25	—	Post, > 2- or 3-true leaf stage
Tree fruits and nuts			
Citrus[3]	0.50–2.00	—	Nonbearing[2]
	0.50–2.00	Glyphosate, napropamide, norflurazon, oryzalin, paraquat, simazine	Nonbearing
Deciduous[4]	0.50–2.00	—	Dormant
	0.50–2.00	Glyphosate, napropamide, norflurazon, oryzalin, paraquat, pronamide, simazine	Dormant

Figs	0.50–2.00	—	Dormant
	0.50–2.00	Glyphosate, napropamide, oryzalin, paraquat	Dormant
Nuts[5]	0.50–2.00	—	Dormant
	0.50–2.00	Glyphosate, napropamide, oryzalin, paraquat, simazine	Dormant
Ornamental nurseries	0.25–1.00	—	Field grown, conifer seedbeds and transplants

There are some limitations on some of the above uses relative to geographic location, soil type, stage of crop or weed growth, grazing or use of forage, harvest time, surfactants, and method, time, or rate of application.

[1]PE, preemergence; post, postemergence; PP, preplant.
[2]Refers to crop, treatment may be PE or post to weeds.
[3]Citrus, citron, grapefruit, kumquat, lemon, lime, mandarin, orange, tangelos, tangerine.
[4]Apricot, cherry, pear, peach, plum, prune.
[5]Almond, pistachio, walnut.

Table A-0 for the specific weeds controlled and Table 11-9 for detailed information on the crops.

Soil Influence Oxyfluorfen is strongly adsorbed on soil, not readily desorbed, and shows negligible leaching. It undergoes detoxification in soils with a half-life of about 30 to 40 days under normal field conditions, but microbial degradation does not appear to be a major factor.

Mode of Action Oxyfluorfen is readily absorbed by both leaves and roots but translocation is limited. However, exposure of the shoot zone to the herbicide causes much more injury to plants than root exposure. This may be related to the light requirement for maximal injury in leaves. This injury appears to be caused by membrane destruction. It is not readily metabolized in plants.

Mechanism of Action of Diphenyl Ethers

In addition to the review discussed in the acifluorfen section of this chapter by Orr and Hess (1982b), Kunert et al. (1987) wrote a review paper on the diphenyl ether herbicides (DPEs) that compliments the earlier review. Kunert et al. (1987) considered the several responses that have been reported for the DPEs at the molecular level and the various substitutions on both phenyl rings. They concluded that the most important processes relative to herbicidal effects are lipid peroxidation (e.g., light-induced oxidative breakdown of cell constituents) and inhibition of carotene biosynthesis.

Recent research indicates that the DPEs cause an accumulation of tetrapyroles and these compounds may in turn induce lethal photooxidative reactions, e.g., lipid peroxidation (Maringe and Scalla, 1988; Witkowski and Halling, 1988).

IMIDAZOLINONES

Imidazolinones are a relatively new class of herbicides that control a broad spectrum of weeds. Imazapyr is nonselective and used on noncropland. Imazaquin, imazamethabenz, and imazethapyr are selective and used in crops. Los (1984) described the discovery and development of the imidazolinone herbicides by American Cyanamid Company.

Imazamethabenz

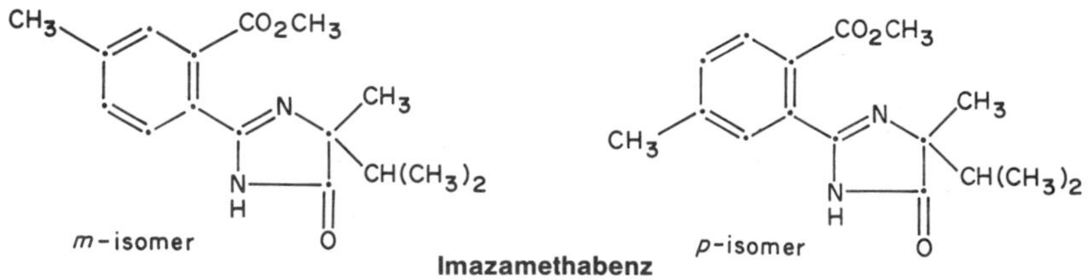

m-isomer *p*-isomer

Imazamethabenz

Imazamethabenz is the common name for a mixture of two positional isomers. The chemical name of the *m*-isomer is 6-(4-isopropyl-4-methyl-5-oxy-2-imidaxolin-2-yl) *m*-toluic acid and the *p*-isomer is 2-(4-isopropyl-4-methyl 5-oxo-2-imidaxolin-2-yl) *p*-toluic acid. Both isomers in the mixture are formulated as the methyl ester for herbicidal use and the following information applies to these methyl esters. The trade name is Assert®. The liquid concentrate formulation contains 2.5 lb/acre. They are off-white solids with water solubilities of 1370 ppm (*m*-isomer) and 857 ppm (*p*-isomer). The acute oral LD_{50} of the technical imazamethabenz is > 5.0 g/kg. The soil influence and mode of action of imazamethabenz are presented later in this chapter in sections covering these topics for all imidazolinone herbicides.

Uses Imazamethabenz is a selective postemergence herbicide used to control wild oats and certain annual broadleaf weeds in barley, wheat, and sunflower. See Table A-0 for the specific weeds controlled. It also has some residual soil activity. In barley and wheat, it is applied at 0.38 to 0.47 lb/acre. Wild oats should be at the one- to four-leaf stage of growth. Imazamethabenz can also be used as a tank mix with 2,4-D ester, MCPA ester, bromoxynil plus MCPA ester, chlorsulfuron, or metsulfuron to broaden the spectrum of weeds controlled in these crops.

In sunflower, it is used at 0.19 to 0.25 lb/acre to control wild mustard when the weed is in the rosette to prebloom stage. It is also used at 0.38 lb/acre to control wild oats when the weed is in the one- to four-leaf stage. Imazamethabenz can also be applied following application of a soil-applied herbicide registered for use in sunflowers, e.g., chloramben, EPTC, pendimethalin, or trifluralin to broaden the spectrum of weeds controlled.

Imazapyr

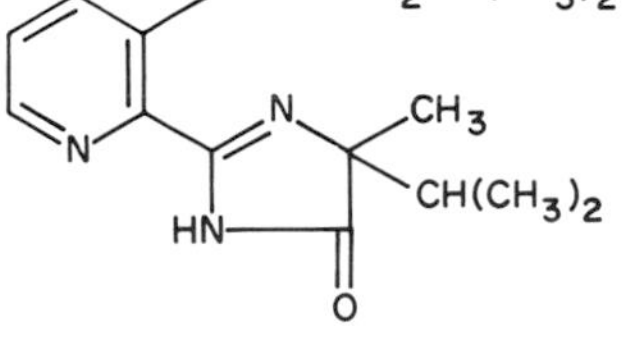

$COOH \cdot NH_2CH(CH_3)_2$ N CH_3 N $CH(CH_3)_2$ HN O

imazapyr

Imazapyr is the common name for (±)-2-[4,5-dihydro-4-methyl-4-(1-methylethyl)-5-oxy-1*H*-imidazol-2-yl]-3-pyridinecarboxylic acid. It is formulated as the water-soluble isopropylamine salt for herbicidal use but the application rates and concentration in the formulations refer to the acid equivalent. The trade names are Arsenal® (2 and 4 lb/gal) and Chopper® (2 lb/gal). The acid form is a white to tan solid and the isopropylamine salt is a pale yellow to dark green liquid. The isopropylamine salt has a very high water solubility, 10,000 to 15,000 ppm. The acute oral LD_{50} of technical imazapyr is > 5.0 g/kg for rats.

Uses Imazapyr is a nonselective herbicide used in noncropland land, forestry, and woody plant control. It controls most annual and perennial herbaceous grass and broadleaf weeds and many vines, brambles, brush species, and deciduous trees.

In noncropland land imazapyr can be applied either preemergence or postemergence to the weed. However, postemergence applications are generally superior, especially for perennial weeds. The preemergence activity of imazapyr provides residual control of most weed species following a postemergence application. The weeds should be vigorously growing at the time of application for maximum activity. The use rate in noncropland land ranges from 0.25 to 1.5 lb/acre depending on the species to be controlled.

In forestry imazapyr is used for site preparation at 0.75 to 1.25 lb/acre and release of loblolly pine stands at 0.5 to 1.0 lb/acre. Planting of loblolly pine should be delayed for 3 months following a site preparation application. Broadcast applications for loblolly pine release should not be used before the conifer is 3 years old. Treatments applied during periods of active growth may cause some minor growth inhibitions of the conifer. Treatments made after formation of final conifer resting bud formation in the fall minimize potential conifer injury. Directed spray applications may be made at all ages of loblolly pine.

For woody plant control imazapyr can be applied as a cut stump, tree injection, frill, girdle, or low volume basal bark treatment. The use rate is 8.0 to 12.0 fluid ounces of product (Chopper®, 2 lb ae/gal) in 1 gallon of water, diesel oil, or a penetrating oil.

Imazaquin

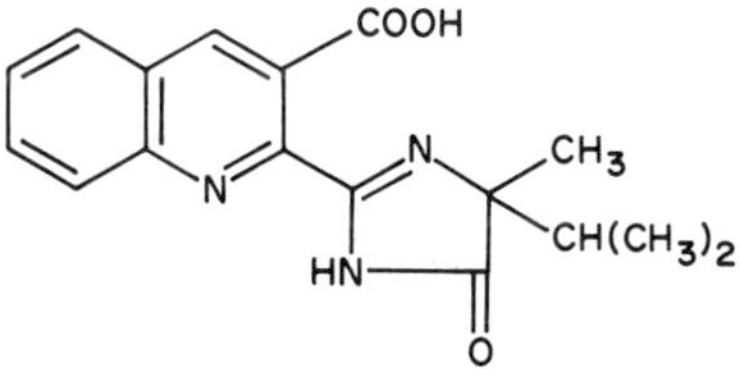

COOH

N

N

CH_3

$CH(CH_3)_2$

HN

O

imazaquin

Imazaquin is the common name for 2-[4,5-dihydro-4-methyl-4-(1-methylethyl)-5-oxo-1*H*-imidazol-2-yl]-3-quinolinecarboxylic acid. The trade name is Scepter®. It is formulated as an aqueous solution (1.5 lb/gal). Imazaquin is a tan solid with a low to moderate water solubility of 60 to 120 ppm at 25°C. The acute oral LD_{50} of imazaquin is > 5.0 g/kg for male and female rats. The soil influence and mode of action of imazaquin are presented later in this chapter in sections covering these topics for all imidazolinone herbicides.

Uses Imazaquin is a selective soybean herbicide and can be applied preplant soil-incorporated, preemergence, or postemergence at 0.125 lb/acre. Preemergence applications can also be used in minimum or no-till systems. It controls a broad spectrum of annual broadleaf weeds. It also controls certain annual grasses and reduces competition from certain other annual grasses and yellow nutsedge. However, some weeds are only controlled adequately by certain types of application; see label. Also see Table A-0 for specific weeds controlled by imazaquin. Imazaquin can also be used in combination with certain other herbicides to broaden the spectrum of weeds controlled. Imazaquin can be applied as a sequential preemergence of postemergence treatment following the application of alachlor, metholachlor, pendimethalin, or trifluralin for maximum grass control. Imazaquin may be followed by a postemergence application of flamprop or sethoxydim; do not tank mix. Imazaquin may be used in combination with a registered broadleaf herbicide for weeds not controlled by imazaquin. For no-till uses, imazaquin may be tank mixed with paraquat, glyphosate, or glyphosate plus alachlor. There are some limitations on some of the above uses including geographic and irrigation; see label.

Imazethapyr

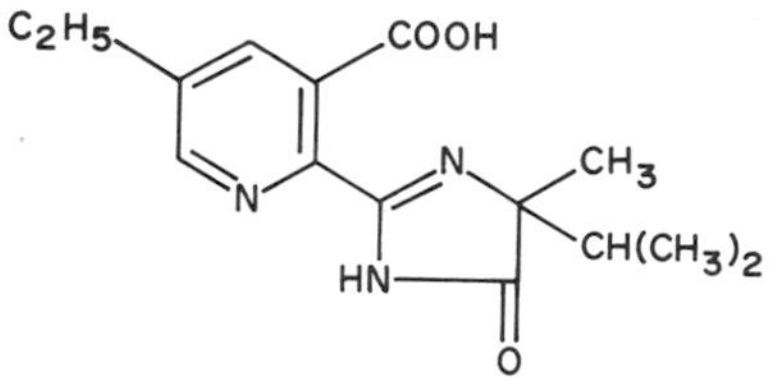

Imazethapyr is the common name for (±)-2-[4,5-dihydro-4-methyl-4-(1-methylethyl)-5-oxo-1*H*-imidazol-2-yl]-5-ethyl-3-pyridinecarboxylic acid. It is formulated as the ammonium salt for herbicidal use but the application rates and concentration in the formulation refer to the acid equivalent. The trade name is Pursuit®. Imazethapyr is an off-white to tan solid with a high water solubility of 1400 ppm. The acute oral LD_{50} of technical imazethapyr is > 5.0 g/kg for male and female rats. The soil influence and mode of action of imazethapyr are presented later in this chapter in sections covering these topics for all imidazolinone herbicides.

Uses Imazethapyr is a broad spectrum herbicide for use in soybeans and other leguminous crops. It has provided excellent weed control when applied preplant soil-incorporated, preemergence, at cracking, and early postemergence. The suggested rates applied are 0.032 to 0.125 lb/acre. Imazethapyr may be used in combination with other herbicides to broaden the spectrum of weeds controlled.

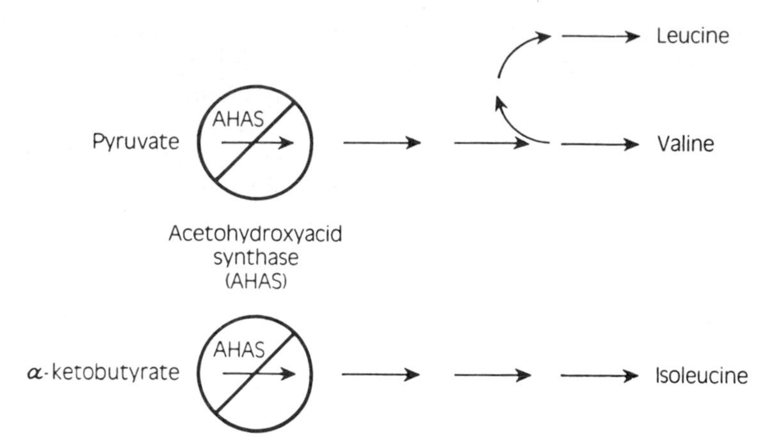

Figure 11-5. Inhibition of branched-chain amino acid synthesis by the imidazolinone class herbicides. (American Cyanamid Company, Wayne, NJ.)

Soil Influence on Imidazolinones

Imidazolinone herbicides are similar in regard to the soil influence on their performance and persistence. They are adsorbed on soil colloids and leaching is minimal. Their persistence at phytotoxic levels appears to be from several weeks to several months under temperate conditions. Persistence is shorter under tropical conditions. Although degradation in soils with 5.0 and 7.0 pH is relatively slow, they undergo rapid hydrolysis at pH 9.0. Their loss from the soil is primarily attributed to microbial activity and uptake by tolerant plants. They are also subject to photolytic modification by sunlight, but this is probably not a significant factor under field conditions.

Mode of Action of Imidazolinones

Imidazolinone herbicides are readily absorbed by roots and leaves, translocated in both the symplast (phloem) and apoplast (xylem), and accumulate in meristematic tissue. The phytotoxic symptoms in herbaceous species include inhibition of growth, especially growing points, followed by purple discoloration or chlorosis and necrosis of the leaves within 2 to 4 weeks. In woody plants, growth inhibition is usually followed by redding or browning of the foliage within 1 month and defoliation within 3 months. However, some susceptible species may retain green foliage for a few months. After preemergence or preplant soil incorporated treatments, susceptible annual weeds may germinate and emerge; however, normal growth ceases at the cotyledon stage in broadleaf weeds and before the two-leaf stage in grasses.

Rapid metabolic degradation of these herbicides in tolerant crops is their basis of selectivity. Susceptible weeds either cannot metabolize the herbicide or metabolize it too slowly for detoxification. For example, the half-life of imazaquin in the tolerant crop soybeans is 3 days whereas in the susceptible weed common cocklebur it is 30 days.

Imidazolinone herbicides act by inhibiting the enzyme acetohydroxy acid synthase (AHAS) (see Figure 11-5). This enzyme is common to the biosynthetic pathway for three branched-chain aliphatic acids, namely isoleucine, leucine, and valine. The reduction in the levels of the three amino acids causes a disruption in protein synthesis and other subsequent biochemical reaction, which, in turn, inhibit plant growth.

SUGGESTED ADDITIONAL READING

Abdullah, M. M. F., F. M. Ashton, and C. L. Elmore, 1988, *Proc. Western Soc. Weed Sci.*, **41**, 162.

Alder, E. F., W. L. Wright, and Q. R. Soper, 1960, *Proc. 17th North Central Weed Control Conf.*, p. 23.

Ashton, F. M., and A. S. Crafts, 1981, *Mode of Action of Herbicides*, Wiley, New York.

Bartels, P. G., 1985, Effects of herbicides on chloroplast and cellular development, in S. O. Duke, Ed., *Weed Physiology*, Vol. II, *Herbicide Physiology*, pp. 63–90, CRC Press, Boca Raton, FL.

Bartels, P. G., and J. L. Hilton, 1973, *Pestic. Biochem. Physiol.* **3**, 462.

Corbett, J. R., K. Wright, and A. C. Baillie, 1984, *The Biochemical Mode of Action of Pesticides*, Academic Press, New York.

Crop Protection Chemical Reference, 1989, Wiley, New York.

Duke, S. O., Ed., 1985, *Weed Physiology*, Vol. II, *Herbicide Physiology*, CRC Press, Boca Raton, FL.

Farm Chemicals Handbook, 1987, Meister, Willough, OH.

Fedtke, C., 1982, *Biochemistry and Physiology of Herbicide Action*, Springer-Verlag, New York.

Golab, T., W. A. Althaus, and H. L. Wooten, 1979, *J. Agric. Food Chem.* **27**, 163.

Golab, T., R. J. Herberg, S. J. Parka, and J. B. Tepe, 1967, *J. Agric, Food Chem.* **15**, 638.

Hatzios, K. K., and D. Penner, 1982. *Metabolism of Herbicides in Higher Plants*, Burgess, Minneapolis.

Herbicide Handbook, 1989, Weed Science Society of America, Champaign, IL.

Hess, F. D., 1982, Determining causes and categorizing types of growth inhibitions induced by herbicides, in D. E. Moreland, J. B. St. John, and F. D. Hess, Eds., *Biochemical Responses Induced by Herbicides*, p. 207, ACS Symposium Series 181, American Chemical Society, Washington, D.C.

Kunert, K-J., G. Sandmann, and P. Böger, 1987, *Rev. Weed Sci.* **3**, 35.

Lamoureux, G. L., and D. G. Rusness, 1981, Catabolism of glutathione conjugates of pesticides in higher plants, in J. D. Rosen, P. S. Magee, and J. E. Casida, Eds., *Sulfur in Pesticide Action and Metabolism*, p. 133, ACS Symposium Series 158, American Chemical Society, Washington, D.C.

Los, M., 1984, o-(5-Oxo-2-imidazolin-2-yl)arylcarboxylates: A new class of herbicides, in

P. S. Magee, G. K. Kohn, and J. J. Menn, Eds., *Pesticide Synthesis through Rational Approaches*, ACS Symposium Series 255, American Chemical Society, Washington, D.C.

Matringe, M., and R. Scalla, 1988, *Plant Physiol.* **86**, 619.

Matsunaka, S., 1976, Diphenyl ethers, in P. C. Kearney and D. D. Kaufman, Eds., *Herbicides*, Vol. 2, pp. 709–739, Dekker, New York.

Muhitch, M. J., D. L. Shaner, and M. A. Stidham, 1987, *Plant Physiol.* **83**, 451.

Orr, G. L. and F. D. Hess, 1982a, *Plant Physiol.* **69**, 502.

Orr, G. L. and F. D. Hess, 1982b, Proposed site(s) of action of new diphenyl ether herbicides, in C. E. Moreland, J. B. St. John, and F. D. Hess, Eds., *Biochemical Responses Induced by Herbicides*, pp. 131–152, ACS Symposium Series 181, American Chemical Society, Washington, D.C.

Parka, S. J., and Q. F. Sopar, 1977, *Weed Sci.* **25**, 79.

Probst, G. W., T. Golab, R. J. Herberg, F. J. Holzer, S. J. Parka, C. Van der Schans, and J. B. Tepe, 1967, *J. Agric. Food Chem.* **15**, 592.

Probst, G. W., T. Golab, and W. L. Wright, 1975, Dinitroanilines, in P. C. Kearney and D. D. Kaufman, Eds., *Herbicides*, Vol. 1, pp. 453–500, Dekker, New York.

Shaner, D. L., P. C. Anderson, and M. A. Stidham, 1984, *Plant Physiol.* **76**, 545.

Shaner, D. L., and P. C. Anderson, 1985, *Biotechnol. Plant Sci.* 287–299.

Shaner, D. L., and P. A. Robson, 1985, *Weed Sci.* **33**, 469.

Van Ellis, M. R., and D. L. Shaner, 1988, *Pestic. Sci.* **23**, 25.

Weed Control Manual and Herbicide Guide, 1988, Meister, Willough, OH.

Witkowski, D. A., and B. P. Halling, 1988, *Plant Physiol.* **87**, 632.

For herbicide use, see the manufacturer's or supplier's label and follow these directions. Also see preface.

12 Nitriles, Phenoxys, and Pyradazinones

NITRILES

Nitriles are organic compounds containing a —C≡N group. Those used as herbicides are benzonitriles with —OH, —Cl, and/or —Br substitutions on the benzene ring. Dichlobenil has two —Cl substitutions and bromoxynil has one —OH and two —Br substitutions. Herbicidal properties of these compounds were discovered in the late 1950s and early 1960s. Dichlobenil is applied to the soil, whereas bromoxynil is applied to the foliage of the weeds to be controlled.

Dichlobenil

Dichlobenil is the common name for 2,6-dichlorobenzonitrile. The trade names are Casoron®, Dyclomec®, and Norosac®. It is a white crystalline solid with a low water solubility of 18 ppm at 20°C. It is available only as granule formulations since it is volatile and volatility losses are less from granule formulations than from the previously available wettable powder and dispersible powder formulations. These granule formulations are 4 and 10% for Casoron® and Norosac® and 4% for Dyclomec®. The acute oral LD_{50} for dichlobenil is 3.1 g/kg in rats.

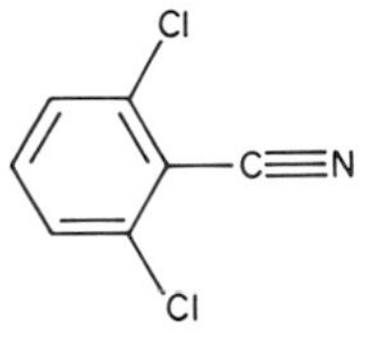

Cl

C≡N

Cl

dichlobenil

Uses Dichlobenil is a potent inhibitor of seed germination and actively dividing meristems, e.g., growing points and root tips. Therefore, it can control both broadleaf and grassy weeds at several stages of growth when it comes in contact with actively growing organs in the soil. These stages include seed germination, early seedling growth, older shallow-rooted weeds, and emerging shoots of perennial weeds. In general, it can be used selectively in several established crops whose roots do not come in contact with dichlobenil that is located in the upper layers of the soil. It is also used on noncrop land and for aquatic weed control. See Table A-0 for specific weeds controlled and Table 12-1 for detailed information on the crops.

Soil Influence Dichlobenil is tightly adsorbed on soil colloids, particularly organic matter. Therefore, it is not subject to leaching in most agricultural soils. However, it is volatile and its vapors are subject to some movement in soils. Its volatility and codistillation with water can cause its rapid loss from the soil surface. This loss is accelerated under high temperatures, wet soil, and low relative air humidity. This loss is minimized when it is applied to dry soil just prior to rainfall, overhead irrigation, or mechanical soil incorporation. In laboratory studies, dichlobenil was shown to be slowly degraded into 2,6-dichlorobenzamide by soil microorganisms with a half-life of 1.5 to 12 months depending on soil type. Under field conditions, it persists at herbicidal levels from 2 to 12 months at usual application rates.

Mode of Action Dichlobenil acts primarily on apical growing points and root tips. This inhibition of growth is followed by a gross disruption of tissues, mostly

TABLE 12-1. Dichlobenil Uses

Crop	Rate (lb/acre)	Remarks
Grapes	4.0–6.0	>4 weeks after transplant
Small fruit[1]	4.0–6.0	>4 weeks after transplant
Cranberries	4.0–6.0	Prebloom or postharvest
Tree fruit and nuts		
Citrus[2]	3.0–4.0	New groves: >1 year old
	4.0–6.0	Nurseries: >4 weeks after transplant
Others[3]	4.0–6.0	>4 weeks after transplant
Ornamental trees	4.0–6.0	Field grown, >4 weeks after transplant, annual weeds
	6.0–8.0	Field grown, >1 year old, perennial weeds
	10.0–20.0	Container pad, nutsedge
Still water		
Marshes	7.0–10.0	Before weed emergence
Ponds/lakes	10.0–15.0	Early spring
Noncrop	12.0–20.0	Air temperature <21°C

There are limitations on some of the above uses relative to soil incorporation, time and rate of application, harvest time, grazing, and use of water and fish; see product label.

[1] Blackberry, blueberry, raspberry.
[2] Grapefruit, lemon, lime, orange, tangerine.
[3] Deciduous: apple, cherry, nectarine, peach, pear, prune plum; subtropical: mango; nuts: almond, filbert, peacan, English walnut.

in the meristems and phloem. This may cause swelling or collapse of stems, roots, and leaf petioles. Apical meristems and leaves may show dark discoloration (Milborrow, 1964).

Because dichlobenil is applied to soil, its rapid absorption by roots and seeds is of particular significance. It is also absorbed by leaves in vapor form (Massini, 1961). Translocation upward from roots to shoots in the apoplast is rapid, but transport downward from leaves to roots is limited. Dichlobenil has been reported to greatly restrict the translocation of photosynthate from leaves to other plant organs.

Dichlobenil is metabolized by higher plants by hydroxylation of the ring. These hydroxylation products are phytotoxic. Glucose conjugates and insoluble residues have also been detected. Dichlobenil has been reported to inhibit cellulose synthesis and extractable glutamine synthetase activity as well as increase membrane leakage. Although dichlobenil appears to have little effect on electron transport and phosphorylation in chloroplasts and mitochondria, the hydroxy degradation products are strong inhibitors of these reactions.

Bromoxynil

Bromoxynil is the common name for 3,5-dibromo-4-hydroxybenzonitrile. It is a light buff to creamy powder with a moderate water solubility of 130 ppm at 20 to 25°C. When formulated as the octanoic acid ester of bromoxynil at 2 lb/gal the trade name is Buctril®. Buctril® 4EC is the trade name for a combination of the heptanoic acid and butyric acid esters of bromoxynil at 4 lb/gal. When formulated as the octanoic acid ester of bromoxynil (2 lb/gal) in combination with the isooctyl ester of MCPA (2 lb/gal) the trade name is Bronate®. A package mix of the octanoic acid ester of bromoxynil (1 lb/gal) plus atrazine (2 lb/gal) is also available. The acute oral LD_{50} of bromoxynil is 440 mg/kg for adult rats.

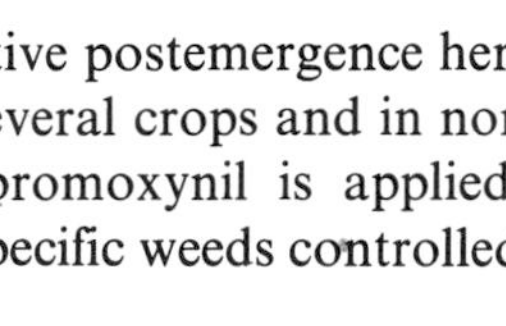

Br
HO— —C≡N
Br
bromoxynil

Uses Bromoxynil is a selective postemergence herbicide used to control many annual broadleaf weeds in several crops and in noncrop areas. Optimum weed control is obtained when bromoxynil is applied to actively growing weed seedling. See Table A-0 for specific weeds controlled and Table 12-2 for detailed information on the crops.

Soil Influence Little research has been conducted on the influence of soil on bromoxynil performance since it is applied postemergence to the target species.

TABLE 12-2. Bromoxynil Uses

Crop	Rate (lb/acre)	Combinations	Remarks
Corn	0.25–0.38	—	Field or pop, post[1]
	0.25–0.38	Atrazine	Field or pop, post, tank mix
	0.25–0.38	2,4-D, dicamba	Field only, post, tank mix
Flax	0.25	—	Post, 2 to 8 inches prior to bud stage
Garlic	0.50	—	Post, to 12 inches
Milo	0.25–0.38	—	Post, 3- to 10-leaf stage or boot
	0.25–0.38	Atrazine	Post, 3- to 10-leaf stage or boot
Mint[2]	1.00	—	Dormant
Onion	0.25–0.38	—	Post, 2- to 5-true-leaf stage
Small grain			
barley and	0.38–0.50	—	Post, emergence to boot stage
wheat	0.38–0.50	Diclofop, difenzoquat	Post, emergence to boot stage
	0.38–0.50	2,4-D, MCPA	Post, after tillering before boot
	0.19–0.38	Chlorsulfuron, chlorsulfuron plus metsulfuron	Post, > 2- to 3-leaf stage, before boot
oat and rye	0.38–0.50	—	Post, emergence to boot stage
	0.38–0.50	2,4-D, MCPA	Post, after tillering before boot
Canarygrass, annual	0.25–0.50	MCPA	Post, > 3-leaf stage
Turfgrasses	0.38–0.50	—	Post, seedling and established[3]
	1.00–2.00	—	Post, established[3]
	0.50–1.00	Mecoprop, dicamba, mecoprop plus dicamba	Post, established[3]
	0.50	2,4-D plus mecoprop	Post, established[3]
Covercrops on CRP[4] areas	0.38–0.50	—	Post, grasses > 2- to 3-leaf stage; alfalfa > 3-trifoliate leaf stage
Noncrop	0.50–1.00	—	Post, plus surfactant or diesel oil

There are limitations of some of the above uses relative to formulation, stage of crop and weed growth, method, rate, and time of application, geographic location, grazing, and harvest time; see product labels.

[1] Post postemergence.
[2] Pepperment, spearment.
[3] See label for grass types tolerant.
[4] Conservation Reserve Program areas, set-aside land.

However, recent observations suggest that it may also be active through the soil and this may account for some of its herbicidal activity.

Mode of Action The phytotoxic symptoms of bromoxynil action appear as blistered or necrotic spots on leaves within 24 hours, essentially a contact action. Later, these spots coalesce with extensive leaf damage. The ester formulations are readily absorbed by leaves and rapidly hydrolyzed to bromoxynil, the active form. Complete spray coverage of the leaves is essential for weed control since translocation is limited.

Bromoxynil is metabolized in higher plants by hydrolysis of the cyano group to an amide and then to a carboxylic acid (Hatzios and Penner, 1982). The presence of benzoic acid and the decarboxylation of the carboxylic acid formed after hydrolysis has also been reported. It may also undergo ring hydroxylation and subsequent conjugation. Bromoxynil is both a photosynthetic and respiratory inhibitor.

PHENOXYS

The introduction of 2,4-D and MCPA in the mid-1940s, immediately after World War II, revolutionized weed control. They demonstrated that synthetic organic compounds could be developed and used to selectively control weeds in crops economically. Following their introduction, the chemical industry began major synthesis and evaluation programs, which lead to the development of the wide array of herbicides that are available today.

The discovery of the phytotoxic properties of the phenoxy herbicides came directly from basic research on plant growth regulators. This early research on these herbicides was not reported as it progressed because of World War II security regulations; see *Botanical Gazette*, volume 107 (1946) for several of these papers. Slade et al. (1945) reported that the plant growth regulator α-

TABLE 12-3. Common Name, Chemical Name, and Chemical Structure of Several Phenoxy Herbicides

O—R_1 ; R_2 ; Cl

Common Name	Chemical Name	R_1	R_2
2,4-D	(2,4-Dichlorophenoxy) acetic acid	$-CH_2-COOH$	—Cl
MCPA	(4-Chloro-2-methylphenoxy) acetic acid	$-CH_2-COOH$	$-CH_3$
2,4-DB	4-(2,4-Dichlorophenoxy) butanoic acid	$-(CH_2)_3-COOH$	—Cl
MCPB	4-(4-Chloro-2-methylphenoxyl) butanoic acid	$-(CH_2)_3-COOH$	$-CH_3$
Dichlorprop (2,4-DP)	(±)-2-(2,4-Dichlorophenoxy)-propanoic acid	$-CH(CH_3)-COOH$	—Cl
Mecoprop (MCPP)	(±)-2-(4-Chloro-2-methylphenoxy)-propanoic acid	$-CH(CH_3)-COOH$	$-CH_3$

naphthaleneacetic acid controlled yellow charlock (wild mustard) in oats with only slight injury. The phenoxy herbicides are often referred to as auxin-like herbicides because they induce twisting and curvature (epinasty) of petioles and stems of broadleaf plants reminiscent of high rates of the native auxin, indole-3-acetic acid.

Six phenoxy herbicides are currently used in the United States. In addition to the phenoxy "mainframe," all have a chlorine atom on the 4-position of the ring and an aliphatic acid attached to the oxygen atom. These aliphatic acids are acetic, butyric, or propionic acid. Each of these acids has a chlorine atom or methyl group on the 2-position of the ring. The common name, chemical name, and chemical structure of these compounds are given in Table 12-3. Although these six phenoxy herbicides have many characteristics in common, each has its own unique selective and phytotoxic uses. In general, the acetic acid forms are used in grass crops, lawns and turf, and noncrop land and the butyric acid forms are used in legume crops. The propionic acid form dichlorprop (2,4-DP) is used for woody plant control and the propionic form mecoprop (MCPP) is used in lawns and turf.

The previously used phenoxy herbicides 2,4,5-T and silvex have been withdrawn from the market because of the presence of teratogenic contaminant, namely 2,3,7,8-tetrachlorodibenzo-*p*-dioxin (TCDD). The synthesis methods for the six phenoxy herbicides currently used in the United States preclude the formation of TCDD.

2,4-D

2,4-D is the common name for (2,4-dichlorophenoxy)acetic acid. As the acid, it is a white crystalline solid with a moderate water solubility of about 900 ppm at 25°C. Other molecular forms of 2,4-D have much higher or much lower water solubilities. There are numerous trade names and formulations; the latter are discussed in detail below. The acute oral LD_{50} of the various formulations fall in the range of 0.3 to 1.0 g/kg for rats, guinea pigs, and rabbits.

The first reference to 2,4-D in the literature is an article by Pokorny (1941). He described a simple method for its synthesis but made no reference to its use. In 1942, Zimmerman and Hitchcock of Boyce Thompson Insitute first described the use of 2,4-D as a plant growth regulator. In 1944, Marth and Mitchell of the USDA reported that 2,4-D killed dandelion, plantain, and other weeds in a bluegrass lawn and Hamner and Tukey described successful field trials using 2,4-D as a herbicide. An excellent paper on the discovery and development of 2,4-D has been written by Peterson (1967).

Common Forms 2,4-D is used as the parent acid, amine salts, esters, and inorganic salts. These various forms involve the substitution of another chemical group for the terminal hydrogen atom of the acetic side chain of the parent molecule. These substitutions alter the physical and biological characteristics of the parent molecule and thereby facilitate the use and/or increase the effectiveness

TABLE 12-4. General Characteristics of Different Forms of 2,4-D

Form	Solubility in Water	Solubility in oil	Appearance When Mixed with Water	Precipates Formed in Hard Water	Volatility[1] Hazard
Acid	Low	Low	Milky	Yes	Low
Amine salts					
Water-soluble	High	Low	Clear	Yes	None
Oil-soluble	Low	High	Milky	Yes	None
Esters					
Low-volatile	Low	High	Milky	No	Medium
High-volatile	Low	High	Milky	No	High
Inorganic salts	Medium	Low	Clear	Yes	None

[1] The tendency to form volatile fumes or gases that can injure susceptible plants.

of 2,4-D in the field. The relative effectiveness of the various forms usually refers to their different degrees of phytotoxicity at equal rates of application. Increased effectiveness of a particular form is usually associated with increased absorption but volatility of the compound usually also increases with increased absorption. The general characteristics of these different forms are given in Table 12-4. However, regardless of the substitution, it is the parent molecule that acts as the herbicide at its site of action. Some formulations may contain more than one molecular form of 2,4-D in a given formulation, e.g., two different amines or ester plus acid. The concentration of essentially all phenoxy herbicide formulations is expressed as acid equivalent in pounds per gallon. Acid equivalent refers to that part of a formulation that theoretically can be converted to the acid. Recommendations are also made on this basis.

Acid The acid form is not commonly used because it is only moderately soluble in water, slightly volatile, and relatively expensive to formulate. Other less expensive formulations are equally effective for many purposes. However, it is more effective on certain hard-to-kill weeds than the amine forms. It is available as emulsifiable concentrates, alone or in combination with other forms of 2,4-D or other herbicides. One formulation contains a 2,4-D ester and is particularly effective for the control of field bindweed, Russian knapweed, Canada thistle, leafyspurge, cattails, tules, and nutsedge.

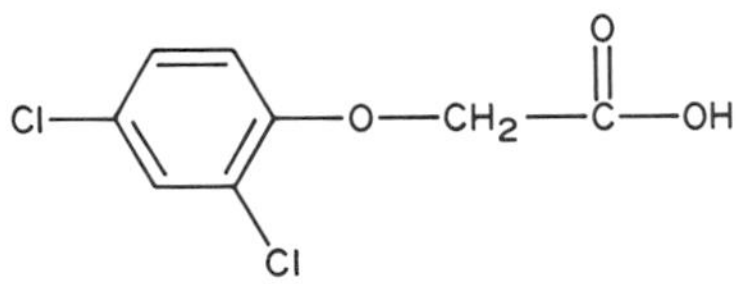

Cl—C₆H₃(Cl)—O—CH₂—C(=O)—OH

2,4-D or (2,4-dichlorophenoxy)acetic acid

Amines Amines are produced by reacting the 2,4-D acid with an amine forming an amine salt of 2,4-D. Several different amines are used. There are two types of

2,4-D amine salts, namely water-soluble amines and oil-soluble amines. They are distinctly different in their physical and biological properties.

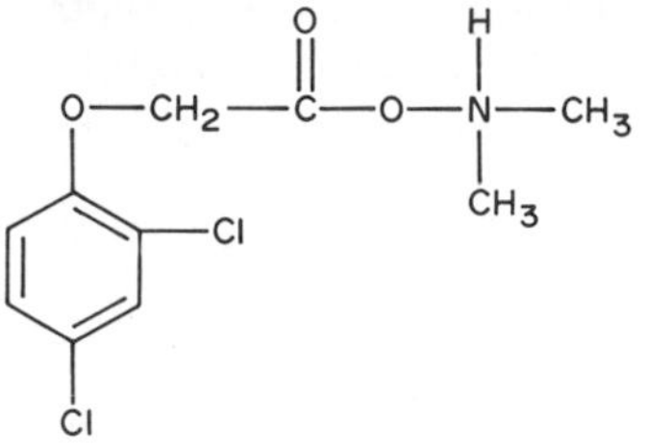

dimethylamine salt of 2,4-D (water soluble)

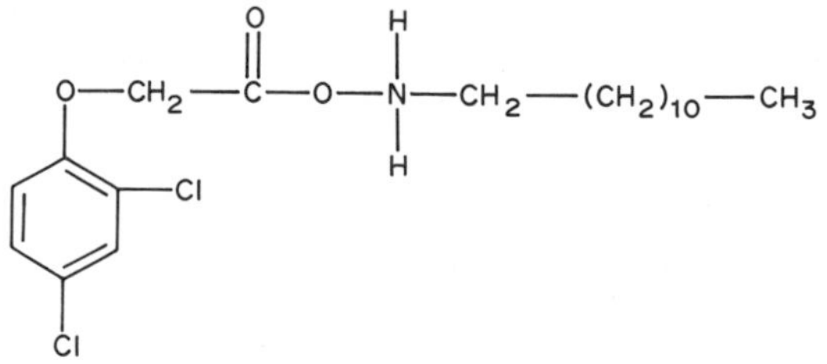

dodecylamine salt of 2,4-D (oil soluble)

Water-soluble amines are the most commonly used form of 2,4-D because of their high water solubility, very low volatility, ease of handling in the field, and overall low cost. They are formulated as water-soluble concentrates. They are somewhat less effective than most other forms but provide effective weed control for many purposes at a minimal cost.

Oil-soluble amines are essentially insoluble in water and used as emulsifiable concentrates. Their major advantage is that they approach the effectiveness of low-volatile esters of 2,4-D with minimal volatility hazard, especially at high temperatures.

Esters Esters are produced by reacting the 2,4-D acid with an alcohol forming an ester of 2,4-D. A number of different esters are used. Increasing the length of the alcohol side chain reduces the volatility of the compound and generally reduces its absorption. However, esters in general are absorbed more readily than any of the other forms of 2,4-D. There are three types of 2,4-D esters, namely low-volatile esters, high-volatile esters, and invert esters. They are distinctly different in their physical and biological properties.

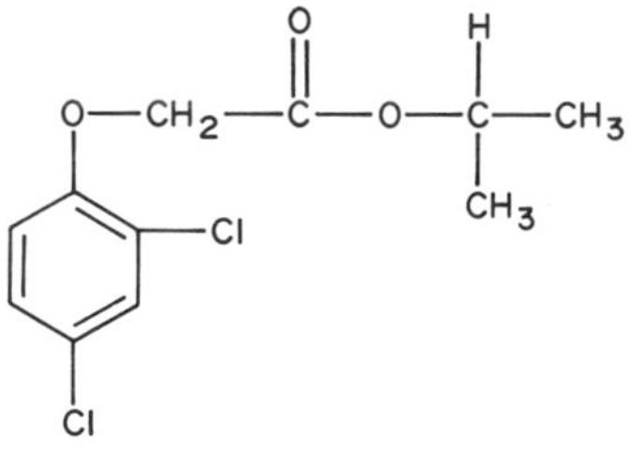

isopropyl ester of 2,4-D (a volatile form)

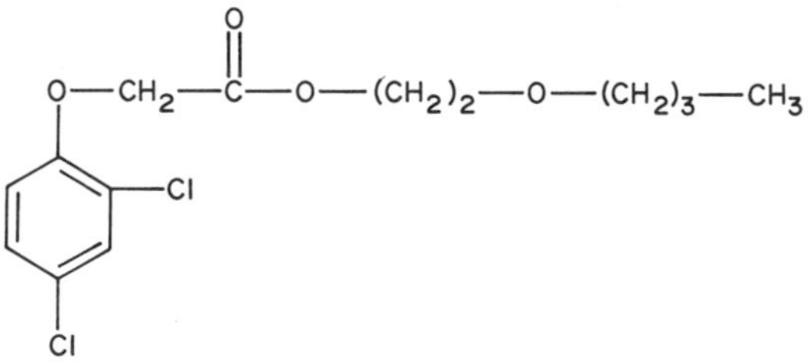

butoxyethyl ester of 2,4-D (a low volatile form)

Low-volatile esters are essentially insoluble in water and used as emulsifiable concentrates. For some uses, they are dissolved in kerosene or diesel oil. They are somewhat volatile and present a volatility hazard, particularly under hot conditions. They are more effective than amines for controlling certain hard-to-kill weeds, e.g., bindweed, thistles, smartweeds, wild garlic, curled dock, tansy ragwort, and wild onion.

High-volatile esters are essentially insoluble in water and used as emulsifiable concentrates. They are very volatile and present a serious volatility hazard. Therefore, they are used only in isolated areas where volatility drift will not cause injury to desirable species. Their use is prohibited in many areas.

Invert esters are unique formulation that produce an invert emulsion (water-in-oil, W/O) when mixed with water. This is in contrast to the oil-in-water (O/W) type of emulsion commonly used in herbicidal sprays. A detailed discussion of emulsions is presented in Chapter 7. Invert emulsions produce a more viscous solution than oil-in-water emulsions and therefore less subject to spray drift. Special application equipment is required because of their viscous nature. They are commonly used for brush control in narrow noncrop areas, e.g., rights-of-way, roadsides, and ditchbanks.

Inorganic Salts Inorganic salts are produced by reacting the 2,4-D acid with a basic inorganic molecule, usually the hydroxide, to form the inorganic salt of 2,4-D. They are water soluble and nonvolatile but the least effective of all forms of 2,4-D. The lithium salt is relatively effective and still available. The once commonly used ammonium, potassium, and sodium salts are no longer available, but the sodium salt was available in the recent past.

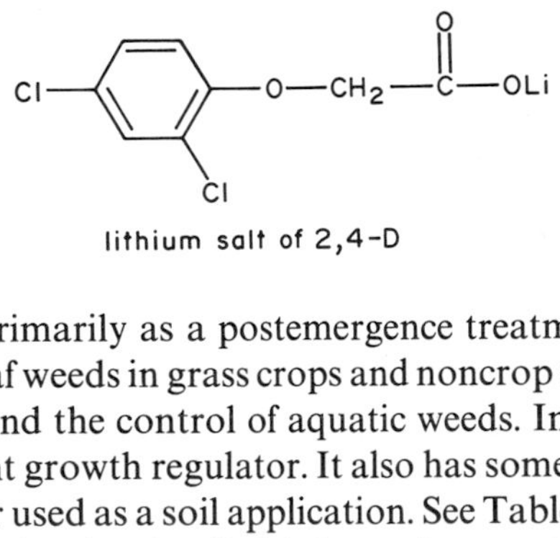

Cl—C_6H_3(Cl)—O—CH_2—C(=O)—OLi

lithium salt of 2,4-D

Uses 2,4-D is used primarily as a postemergence treatment to control annual and perennial broadleaf weeds in grass crops and noncrop areas. It is also used for woody plant control and the control of aquatic weeds. In low concentrations it has been used as a plant growth regulator. It also has some herbicidal activity via the soil but is no longer used as a soil application. See Table A-0 for specific weeds controlled and Table 12-5 for detailed information on the crops.

Soil Influence Soil type and formulation of 2,4-D influence its leaching in soil. It is adsorbed onto soil colloids and less leaching occurs in clay and organic soils than in sandy soils. Microorganisms are of major importance in its disappearance from soil. It persists at phytotoxic levels from 1 to 4 weeks in warm-moist loam soil at usual application rates. Although it does not generally reduce the total number of microorganisms in the soil, it may reduce nodulation of legume species (Payne and Fultz, 1947).

Mode of Action Epinasty is among the most obvious effect of 2,4-D and other

TABLE 12-5. 2,4-D Uses[1]

Crop	Rate[2] (lb/acre)	Combinations	Remarks[2]
Corn[3]	1.00–1.50	—	Pre,[4] just before emergence
	0.50–0.75	—	Post,[4] after emergence to 5–8 inches tall
	0.25–0.50	—	Post, use drop nozzles after 10–12 inches tall, not in tassel
Milo	0.25–0.50	—	Post, 4–12 inches tall, secondary roots established, drop nozzles > 10 inches, not from flowering to dough stage
Rice	0.75–1.50	—	Post, after fully tillered, not during boot stage
Small grains[5]	0.25–1.00	—	Post, after fully tillered, not during boot thru dough stage
	0.25–1.00	Bromoxynil	as above
Turf/lawns	0.50–2.00	—	Post, established
	0.50–2.00	AMA, dicamba, dichlorprop MCPA, mecoprop, dicamba plus dichlorprop or mecoprop or mecoprop plus MSMA	As above
Pastures, rangeland	0.50–3.00	—	Post, established
	0.50–1.00	Picloram, triclopyr	As above
Noncrop	1.00–4.00	—	Post, usually for herbaceous weeds, some woody species
	1.00–9.00	Dicamba, picloram, triclopyr, dicamba plus dichlorprop or mecoprop	Post, herbaceous weeds and/or woody species, the latter may require basal, stem, cut-surface, or stump treatment
CRP[6] areas	0.25–2.00	—	Post
	0.25–2.00	Triclopyr	Post
Water			
Still (lakes and ponds)	1.25–2.25	—	Post
	1.25–2.25	Dicamba	Post
Still (marshes, shoreline)	1.25–2.25	—	Post
	1.25–2.25	Dicamba, dicamba plus dalapon	Post
Moving (irrigation and drainage canals)	1.25–4.00	—	Post
	1.25–2.25	Dicamba, dicamba plus dalapon	Post
Moving (ditchbanks)	1.00–2.00	—	Post
	1.00–2.00	Dicamba plus dichlorprop	Post
Quiescent or slow-moving	2.00–4.00	—	Post

There are some limitations on some of the above uses relative to formulation, variety or species, soil type and soil moisture, grazing, geographic location, days to harvest, and water use; see product labels.

[1]Formulations not given. In some cases, the rates are an average from several labels. All formulations are not suitable for all uses. [2]Stage of crop and/or weed growth may alter the rate used and the rate may vary for different formulations. [3]Field, silage, sweet. [4]Pre, preemergence; post, postemergence. [5]Barley, oat, rye, wheat. [6]Conservation Reserve Program areas, set-aside land.

phenoxy-type herbicides on broadleaf plants. These plants usually develop grotesque and malformed leaves and stems when treated with 2,4-D (Figures 12-1, 12-2, and 12-3). Brace roots of corn also develop abnormally (Figure 12-4). The herbicide concentrates in young embryonic or meristematic tissues that are growing rapidly. It affects these tissues more than mature or relatively inactive young tissues.

Histological studies with red kidney bean showed that the cambium, endodermis, embryonic pericycle, phloem parenchyma, and phloem rays were grossly altered by 2,4-D. The cortex and xylem parenchyma showed little response and the epidermis, pith, mature xylem, mature sieve tubes, and differentiated pericycle showed no response (Swanson, 1946). These results suggest that active cell division is essential for the development of 2,4-D toxicity symptoms. The types of tissue affected by 2,4-D in field bindweed and sow thistle were much the same as those in bean (Tukey et al., 1946). Dunlap (1948), Zussman (1949), Rodgers (1952), and others have reported on related abnormalities induced by 2,4-D in several plant species.

Leaves readily absorb nonpolar forms (acid, esters, oil-soluble amines) of 2,4-D, whereas polar forms (inorganic salts, water soluble amines) are absorbed more slowly. The use of surfactants usually increases foliar absorption. Absorption increases with increasing temperature and humidity. Rainfall shortly after application may reduce its effectiveness, but rainfall 6 to 12 hours later has little effect. Nonpolar forms have a tendency to resist removal by rainfall. Plant stems and roots also absorb 2,4-D.

After 2,4-D is absorbed by the foliage, it is translocated both up and down the plant via the symplastic system. It moves from the leaves (source) with the photosynthate in the phloem, but more slowly than the photosynthate. It accumulates in those areas of high photosynthate utilization (sink), e.g., rapidly developing organs and meristems. Limited translocation of 2,4-D in grasses relative to broadleaf plants may partially explain their tolerance to this herbicide (Ashton, 1958).

Translocation of a foliar applied herbicide to underground roots and rhizomes is essential for the control of perennial weeds. Therefore, periods of maximum growth and photosynthate accumulation in these underground organs and minimum growth of the aboveground organs favor the control of perennial weeds. This usually occurs in the fall for most perennial species but may occur at other time of the year for certain perennial species. Excessive rates of application may damage the phloem and reduce translocation, therefore sequential low rates usually give better control of perennial weeds than a single high application rate.

Plant age and associated rate of growth influence its susceptibility to 2,4-D. In general, young plants are more susceptible than older plants of the same species. However, some plants are tolerant while still small and others never gain more than a slight tolerance. Some plants may develop a second period of susceptibility. Small grains are very susceptible to 2-4-D in the germinating and small-seedling stages. They become tolerant in the fully tillered stage, susceptible again in the jointing, heading, and flowering stages, and tolerant again in the

Figure 12-1. Scientists studying the selective killing of giant ragweed in corn in Henderson County, Kentucky, in 1947. This was one of the first large-scale tests made with 2,4-D in the United States. (Sherwin-Williams Company.)

Figure 12-2. A common burdock plant twisted and curled following treatment with 2,4-D.

Figure 12-3. Bean leaves showing the effect of 2,4-D. The leaf second from the right is normal, not treated with 2,4-D.

Figure 12-4. Abnormal corn brace roots induced by a high rate of 2,4-D applied during a susceptible stage of growth.

"soft-dough" stage (Figure 15-2). The periods of susceptibility coincide with periods of rapid growth. At this time, the cells of the meristems are dividing rapidly, have a high level of metabolic activity, and are very susceptible to 2,4-D.

The degradation of the phenoxy herbicides by higher plants has been reviewed by Loos (1975) and Hatios and Penner (1982). 2,4-D is degraded to nonphytotoxic forms in higher plants, undergoing decarboxylation and hydroxylation of the side chain as well as dechlorination and hydroxylation of the ring. 2,4-D also forms conjugates with glucose and certain amino acids; 2,4-D metabolites also conjugate with glucose. The resistance of certain species to 2,4-D has been in part attributed to their ability to rapidly degrade 2,4-D to nontoxic molecules.

2,4-D appears to be acting in a manner similar to the native auxin, indole-3-acetic acid (IAA). However, IAA has endogenous control mechanisms that control its concentration within the physiological range. 2,4-D has no such control mechanisms. 2,4-D stimulates and inhibits plant growth and related metabolic processes depending on the level of 2,4-D present in the tissue. In general, low concentrations stimulate and high concentrations inhibit these processes. The level of 2,4-D in the meristems and developing organs of the intact plant increases with time after the application, initially low and later high. Thus, first there is a stimulation of these processes causing uncontrolled growth and later an inhibition of these processes and growth.

Two plant processes are most relevant to the biochemical mechanism of action of the phenoxy herbicides, nucleic acid metabolism and cell-wall plasticity. Low levels of 2,4-D stimulate RNA polymerase, which results in an increase in RNA and protein synthesis. Low levels of 2,4-D also induce cell enlargement by increasing the activity of certain enzymes responsible for "loosening" cell walls and the formation of new cell-wall material. Abnormal stimulation of these processes leads to uncontrolled growth. High levels of 2,4-D act just the opposite, they inhibit these processes and growth. 2,4-D also increases ethylene production and this is considered to be involved in the development of some of the formative effects including epinasty.

Hanson and Slife (1969) proposed that the immediate cause of death is physiological dysfunction of the plant brought about by abnormal growth. In turn, the abnormal growth is believed to be based on an abnormal nucleic acid metabolism. Penner and Ashton (1966) reviewed the biochemical and metabolic changes in plants induced by the chlorophenoxy herbicides.

MCPA

MCPA is the common name for (4-chloro-2-methylphenoxy)acetic acid. The trade names are Weedar® MCPA (amine), Weedone® MCPA (ester), and Weedone® Sodium MCPA. MCPA has a chemical structure identical to 2,4-D except it has a methyl (—CH_3) group at the number 2-position of the ring instead of a chlorine atom. It is a light-brown solid and essentially insoluble in water. The acute oral LD_{50} is 800 mg/kg for mice.

Uses MCPA was one of the first hormone-type herbicides discovered in England. It is a postemergence herbicide. Its herbicidal characteristics are similar to 2,4-D except it is more selective on cereals, legumes, and flax at equal rates and may be more effective than 2,4-D on certain broadleaf weed species. It can be formulated as amines, esters, and inorganic salts. The water-soluble dimethylamine salt is used in flax, rice, peas, and small grains. The sodium salt is used in rice, peas, and small grains. The low-volatile 2-butoxyethyl ester is used in small grains. Application rates vary from 0.25 to 1.25 or 1.50 lb/acre (acid equivalent) depending on formulation, weed type (annual, biennial, perennial, and stage of crop growth and variety; see product labels. It is also formulated as a package mix with bromoxynil.

The soil influence on MCPA is similar to 2,4-D. The mode of action of MCPA also appears to be similar to 2,4-D.

2,4-DB

2,4-DB is the common name for 4-(2,4-dichlorophenoxy)butanoic acid. The trade name is Butyrac®. It is a white crystalline solid and practically insoluble in water. It is formulated as the water-soluble dimethylamine salt and low-volatile butoxyethanol ester. The acute oral LD_{50} is about 2.0 g/kg for rats.

Uses 2,4-DB is especially selective on small-seeded and certain other legume crops as a postemergence treatment to control broadleaf weeds. The amine is used in alfalfa clovers, birdsfoot trefoil, peanuts, and soybeans and the ester is used in alfalfa and birdsfoot trefoil. See Table A-O for specific weeds controlled and Table 12-6 for detailed information on the crops. The soil influence on 2,4-DB is similar to 2,4-D.

Mode of Action 2,4-DB is not highly phytotoxic per se; however, it undergoes β-oxidation in plants and soil to form 2,4-D, which is relatively phytotoxic. This reaction is more rapid in susceptible plants than in tolerant plants (e.g., small-seeded legumes). Therefore, many broadleaf weeds are controlled by 2,4-DB, whereas the small-seeded legumes are less subject to injury.

O—CH_2—CH_2—CH_2—C(=O)—OH
Cl
Cl
2,4-DB
Beta-oxidation
—(—CH_2—CH_2—)
O—CH_2—C(=O)—OH
Cl
Cl
2,4-D

MCPB

MCPB is the common name for 4-(4-chloro-2-methylphenoxy)butanoic acid. The trade name is Thistrol®. MCPB has a chemical structure identical to 2,4-DB

TABLE 12-6. 2,4-DB Uses[1]

Crop	Rate[2] (lb/acre)	Combinations	Remarks[2]
Peanuts	0.20–0.40	—	Post,[3] 2nd application O.K.
Small-seeded legumes[4]	0.50–1.50	—	Post, established or seedlings (1–4 trifoliate leaf stage)
Soybeans	0.20–0.40	—	Post, 7–10 days before bloom through midbloom, directed sprays safer
	See remarks	Acifluorfen, bentazon, fomesafen, linuron, metribuzin	Post; rate of 2,4-DB, method and time of application, etc. vary for each combination
CRP[5] areas	0.50–1.50	—	Post

There are some limitations on some of the above uses relative to formulation, species or cultivars, geographic location, stage of crop and/or weed growth, grazing, and harvest date; see product label.

[1] Formulations not given. All formulations are not suitable for all uses.
[2] Stage of crop and/or weed growth may alter the rate used and the rate may vary for different formulations.
[3] Post, postemergence.
[4] Alfalfa, birdsfoot trefoil, lespedeza, clovers (alsike, ladino, red). All species not on all labels.
[5] Conservation Reserve Program areas, set-aside land.

except it has a methyl (—CH_3) group at the number 2-position of the ring instead of a chlorine atom. It is a white solid and practically insoluble in water, however, its sodium salt is very soluble in water. It is formulated as the sodium salt. The acute oral LD_{50} is 600 mg/kg for rats.

Uses MCPB is used as a postemergence treatment to control Canada thistle and other broadleaf weeds in peas. The use rate is 0.50 to 1.50 lb/acre depending on the weed species to be controlled and their size. The soil influence on MCPB is similar to 2,4-D and the mode of action of MCPB is similar to 2,4-DB.

Dichlorprop

Dichlorprop is the common name for (±)-2-(2,4-dichlorophenoxy)propanoic acid. It has also been referred to as 2,4-DP. The trade name is Weedone® 2,4-DP. It is a white to tan crystalline solid with a moderate water solubility of 710 ppm at 28°C. It is formulated as the low-volatile butoxyethyl ester. The acute oral LD_{50} is 800 mg/kg for rats.

Uses Dichlorprop is used for woody plant control. It controls mixed brush on highways, railroads, and utility rights-of-way. It is also used for pine release and to control solid stands of post, blackjack, and sand shinnery oaks and sandsage. Retreatment the following year may be required to control some species.

Endangered species restrictions may limit its use in some areas. The soil influence and mode of action are considered to be similar to 2,4-D.

Mecoprop

Mecoprop is the common name for (±)-2-(4-chloro-2-methylphenoxy)propanoic acid. It has also been referred to as MCPP. The trade names are Mecomec® and Cleary's® MCPP. It is also available as a package mix with 2,4-D with other trade names. It is a colorless crystalline solid with a moderate water solubility of 620 ppm at 20°C. It is usually formulated as a water-soluble amine salt. The acute oral LD_{50} is 650 mg/kg for mice and 1.1 g/kg for the technical material with rats.

Uses Mecoprop is used to control 2,4-D tolerant weeds in lawns and turf. The spectrum of weed species controlled is broadened when it is used in combination with 2,4-D. There are some restrictions for certain grass species, environmental conditions, and time of mowing; see product labels. The soil influence and mode of action are considered to be similar to 2,4-D.

PYRIDAZINONES

Pyridazinones basic chemical structure consists of two adjacent nitrogen atoms and four carbon atoms in a six-membered ring with an oxygen atom attached to a carbon atom. The herbicides of this class have various substitutions in the R_1, R_2, and R_3 positions of the basic structure. Norflurazon and pyrazon are the only two herbicides of this class available in the United States and are used to control annual weeds in a number of crops. Norflurazon is a Sandoz, Incorporated product and pryazon a BASF Corporation product.

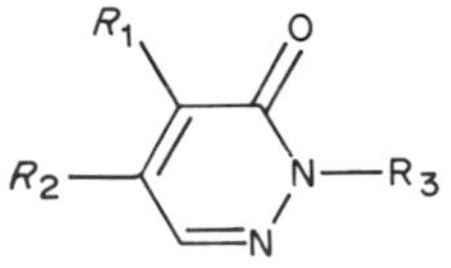

pyridazinone

Norflurazon

Norflurazon is the common name for 4-chloro-5-(methylamino)-2-[3-(trifluromethyl)phenyl]-3(2*H*)-pyridazinone. The trade names are Evital®, Solicam® and Zorial®. The former is a granule formulation (5%) and the latter two are dry flowable formulations (80%). Each product is for use in particular crops. Norflurazon is a white to brownish-gray crystalline solid with a low water solubility of 28 ppm at 25°C. The acute oral LD_{50} of technical norflurazon is > 8.0 g/kg for and > 10.0 g/kg for the 80% formulation.

norflurazon

Uses Norflurazon is a selective soil applied herbicide used to control many annual grass and broadleaf weeds and suppress the growth of certain perennials in several crops. It is applied before weed emergence or weed growth resumes. Evital® is used in cranberries, Solicam® in small fruits and tree crops, and Zorial® in cotton and soybeans. See Table A-0 for the specific weeds controlled and Table 12-7 for detailed information on the crops.

Soil Influence Norflurazon is readily adsorbed by the clay and organic matter in soils. Their content determines the rate of application and the degree of leaching. However, it is not leached appreciably through most soils. Norflurazon is degraded by soil microorganisms, but this is considered to be only partially

TABLE 12-7. Norflurazon Uses

Crop	Rate (lb/acre)	Combinations	Remarks[1]
Cotton	1.0–2.0	—	PPI, pre; or both at 1/2 rate
Nuts[2]	2.0–4.0	—	Established
	2.0–4.0	Paraquat	Established, tank mix
	2.0–3.0	Simazine	Established, tank mix
Small fruits			
Cranberries	4.0–8.0	—	Early spring or fall after harvest
Others[3]	2.0–4.0	—	Fall to early spring
Soybeans	1.0–2.0	—	Pre, PPI + pre at 1/2 rate each
	1.0–1.5	Metribuzin	Pre, tank mix
Tree fruits			
Citrus	2.0–4.0	—	Newly planted or established
	2.0–4.0	Paraquat	Tank mix
Citrus[4]	2.0–3.0	Simazine	Tank mix
Deciduous[5]	2.0–4.0	—	Established
	2.0–4.0	Glyphosate or paraquat	Established, tank mix
Deciduous[6]	2.0–3.0	Simazine	Established, tank mix

There are some limitations on some of the above uses relative to geographic locations and soil types; see product labels.

[1] Apply norflurazon before annual weed emergence or growth of other weeds resumes.
[2] Almonds, filberts, pecans, walnuts.
[3] Bluberries, blackberries, raspberries.
[4] Grapefruit, lemons, oranges.
[5] Apples, apricots, cherries, nectarines, peaches, pears, plums, prunes.
[6] Apricots, cherries, peaches, pears, plums.

responsible for its disappearance from soil. Volatilization and photodecomposition also contribute to its loss when it remains on the surface of the soil. The average half-life of norflurazon is 45 to 180 days in soils of the Mississippi Delta and Southeastern United States. The colloidal content of the soil influences the persistence period.

Mode of Action Norflurazon causes light-grown seedlings to emerge with chlorotic leaves and die following exhaustion of food reserves (Bartels and Watson, 1978). It inhibits the biosynthesis of carotenoid pigments, and since carotenoid pigments protect chlorophyll from photodegradation, chlorophyll is destroyed. The original research on this phenomenon has been reviewed by Ashton and Crafts (1981).

Norflurazon is absorbed by roots and translocated to shoots, presumably via the apoplastic system. It is metbolized by higher plants. In cotton, tolerance appears to be associated with limited transport rather than degradation (Strang and Rogers, 1974). However in cranberries, degradation rather than limited transport may have the dominant role in tolerance (Yaklich et al., 1974).

Pyrazon

Pyrazon is the common name for 5-amino-4-chloro-2-phenyl-3-(2*H*)-pryidazinone. The trade name is Pyramin®. It is a tan to brown solid with a moderate water solubility of 400 ppm at 20°C. It is formulated as a flowable liquid (4 lb/gal). Pyrazon has an acute oral LD_{50} of 3.0 g/kg for rats.

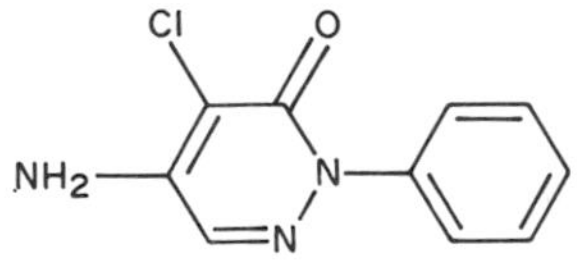

pyrazon

Uses Pyrazon is a selective herbicide used to control annual broadleaf weeds in sugar beets and red (table) beets. When used alone it is applied as a preemergence or early postemergence treatment. In combination with certain other herbicides it is also applied as preplant or preplant soil-incorporated treatment. These combinations are used to broaden the spectrum of weeds controlled, particularly grasses, but are not registered for use in red beets. See Table A-0 for the specific weeds controlled and Table 12-8 for detailed information on the crops.

Soil Influence Soil type has a considerable influence on the effectiveness and selectivity of pyrazon. Soil applications are not recommended on sands or loamy sands because of leaching and possible crop injury. In contrast, adsorption on soils containing more than 5% organic matter precludes adequate weed control. Pyrazon degradation by microorganisms is moderately rapid and its persistence

TABLE 12-8. Pyrazon Uses

Crop	Rate (lb/acre)	Combinations	Remarks
Red beets	3.2–3.7	—	Pre
	3.7	—	Post, crop > 2 true leaves
Sugar beets	3.2–3.7	—	Pre, post; crop > 2 true leaves
	2.6–3.2	Diethatyl	PPI, pre; tank mix
	3.2	Ethofumesate	PP, pre; tank mix
	3.2–3.7	TCA	Pre, tank mix
	3.2–3.7	Dalapon	Post, early; tank mix

There are some limitations on some of the above uses relative to geographic location and soil type; see product label.

in warm-moist soils is 1 to 3 months. The principal soil degradation product is dephenylated pyrazon, which is not significantly phytotoxic.

Mode of Action Pyrazon progressively induces wilting, chlorosis, and necrosis in leaves as well as acting as a general growth inhibitor in susceptible species (Frank and Switzer, 1969; Rodebush and Anderson, 1970). Microscopic studies have shown that chloroplast structure is also altered (Anderson and Schaelling, 1970). Although pyrazon is absorbed by leaves, it is not translocated from them to a significant degree. It is readily absorbed by roots and distributed throughout the plant. These translocation patterns suggest that pyrazon is primarily translocated in the apoplast.

Although pyrazon is degraded in plants, the primary basis of tolerance of sugar beets appears to be associated with the conjugation of pyrazon with glucose to form a nonphytotoxic molecule, *N*-glucosyl pyrazon (Hatzios and Penner, 1982). The biochemical mechanism of action of pyrazon has been shown to be the inhibition of photosynthesis (Eshel, 1969; Hilton et al., 1969) by binding to an electron acceptor in the electron transport chain between photosystem I and photosystem II (Tischer and Strotmann, 1977).

SUGGESTED ADDITIONAL READING

Anderson, J. L., and J. P. Schaelling, 1970, *Weed Sci.* **18**, 455.

Ashton, F. M., 1958, *Weeds* **6**, 257.

Ashton, F. M., and A. S. Crafts, 1981, *Mode of Action of Herbicides*, Wiley, New York.

Bartels, P. G., and C. W. Watson, 1978, *Weed Sci.* **26**, 198.

Corbett, J. R., K. Wright, and A. C. Baillie, 1984, *The Biochemical Mode of Action of Pesticides*, Academic Press, New York.

Crop Protection Chemical Reference, 1989, Wiley, New York (revised annually).

Duke, S. O., Ed., 1985, *Weed Physiology*, Vol II, *Herbicide Physiology*, CRC Press, Boca Raton, FL.

Dunlap, A. A., 1948, *Phytopathology* **38**, 638.

Eshel, Y., 1969, *Weed Res.* **10**, 196.

Farm Chemicals Handbook, 1987, Meister, Willough, OH.

Fedtke, C., 1982, *Biochemistry and Physiology of Herbicide Action*, Springer-Verlag, New York.

Frank, R., and C. M. Switzer, 1969, *Weed Sci.* **17**, 344.

Frear, C. S., 1976, The benzoic acid herbicides, in P. C. Kearney and D. D. Kaufman, Eds., *Herbicides*, Vol 2, pp. 541–607, Dekker, New York.

Hammer, C. L., and H. B. Tukey, 1944, *Bot. Gaz.* **106**, 232.

Hanson, J. B., and F. W. Slife, 1969, *Residue Rev.* **25**, 59.

Hatzios, K. K., and D. Penner, 1982, *Metabolism of Herbicides in Higher Plants*, Burgess, Minneapolis.

Herbicide Handbook, 1989, Weed Science Society of America, Champaign, IL.

Hilton, J. L., A. L. Scharen, J. B. St. John, D. E. Moreland, and K. H. Norris, 1969, *Weed Sci.* **17**, 541.

Loos, M. A., 1975, Phenoxyalkanoic acids, in P. C. Kearney and D. D. Kaufman, Eds., *Herbicides*, Vol 1, pp. 1–128, Dekker, New York.

Marth, P. C., and J. W. Mitchell, 1944, *Bot. Gaz.* **126**, 224.

Massini, P., 1961, *Weed Res.* **1**, 142.

Milborrow, B. V., 1964, *J. Exp. Bot.* **15**, 515.

Mullison, W. R., and R. W. Hummer, 1949, *Bot. Gaz.* **111**, 77.

Payne, M. G., and J. L. Fults, 1947, *J. Am. Soc. Agron.* **39**, 52.

Peterson, G. E., 1967, *Agric. History* **41**, 243.

Penner, D., and F. M. Ashton, 1966, *Residue Rev.* **14**, 39.

Pokorny, R., 1941, *J. Am. Chem. Soc.* **63**, 1768.

Rodebush, J. E., and J. L. Anderson, 1970, *Weed Sci.* **18**, 443.

Rodgers, E. G., 1952, *Plant Physiol.* **27**, 153.

Slade, R. E., W. G. Templeman, and W. A. Sexton, 1945, *Nature* (*London*) **155**, 497.

Strang, R. H., and R. L. Rogers, 1974, *J. Agric. Food Chem.* **22**, 1119.

Swanson, C. P., 1946, *Bot. Gaz.* **107**, 522.

Tischer, W., and H. Strotmann, 1977, *Biochim. Biophys. Acta* **460**, 113.

Tukey, H. B., C. L. Hamner, and B. Imkoffe, 1946, *Bot. Gaz.* **107**, 62.

Weed Control Manual and Herbicide Guide, 1988, Meister, Willough, OH (revised annually).

Yaklich, R. W., S. J. Karczmarczyk, and R. M. Devlin, 1974, *Weed Sci.* **22**, 595.

Yamaguchi, S., and A. S. Crafts, 1958, *Hilgardia* **28**, 161.

Zimmerman, P. W., and A. E. Hitchcock, 1942, *Contr. Boyce Thompson Inst.* **12**, 321.

Zussman, H. W., 1949, *Agric. Chem.* **4**, 27–29, 73.

For herbicide use, see the manufacturer's or supplier's label and follow these directions. Also see the Preface.

13 Pyridines, Sulfonylureas, and Triazines

PYRIDINES

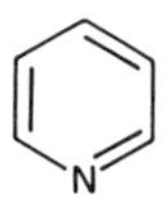

pyridine

Pyridines basic chemical structure consists of one nitrogen atom and five carbon atoms in a six-membered ring. The three herbicides in this class in order of their development are picloram, triclopyr, and clopyralid. They are similar to 2,4-D in some ways such as selectivity and phytotoxic symptoms, but differ from 2,4-D in their relative phytotoxicity and persistence in both plants and in the soil. They are particularly effective on broadleaf weeds and many woody plants but most grasses are tolerant. Their phytotoxic symptoms are auxin-like: epinasty, cuplike leaves, and tissue proliferation. However, in general, they are more phytotoxic and more persistent in both plants and soils than 2,4-D. Within the pyridine class, picloram is more phytotoxic and more persistent than triclopyr and clopyralid. Dow Chemical has developed the pyridine-type herbicides.

Picloram

Picloram is the common name for 4-amino-3,5,6-trichloro-2-pyridinecarboxylic acid. The trade name is Tordon®. Tordon® 22K is a potassium salt formulation (2 lb, ae/gal). Tordon® RTU is a mixture of triisopropanol-amine salts of picloram (3.0%, ae) and 2,4-D (11.2%, ae). Picloram is a white solid with a moderate water solubility of 430 ppm at 25°C. The acute oral LD_{50} of picloram is 8.2 g/kg for rats. Picloram is a restricted use pesticide. It is highly potent and very persistent. Very small amounts can kill or injure many broadleaf plants. Extreme care should be taken to prevent it from escaping the target site.

Figure 13-1. Picloram-induced injury to wheat. *Left*: Untreated. *Right*: Treated. (C. L. Elmore, University of California, Davis.)

Uses Picloram is relatively nonselective but can be used selectively in small grains. It is particularly effective on many woody and perennial broadleaf weeds. It also controls most annual broadleaf weeds but not grasses. It is applied to both the foliage and the soil and is also used as a basal or cut surface treatment for unwanted tree control. See Table A-0 for specific weeds controlled and Table 13-1 for specific uses.

Soil Influence Picloram is adsorbed by organic matter and certain clays (*Herbicide Handbook*, 1989). It is readily leached through sandy and montmorillonitc clay soils low in organic matter, but not through soils high in organic matter or lateritic soils. Salts of picloram appear to be leached more readily than the parent acid form.

Picloram is very persistent in soils, which is one of the reasons it is a restricted use herbicide. It is slowly degraded by microorganisms. Conditions that favor microbial growth, warm-moist soil and organic matter, reduce its period of persistence. The application rate also influcnces its period of phytotoxicity in soils. Phytotoxicity may often be detected well over 1 year after application.

Mode of Action Picloram is highly phytotoxic to most woody and broadleaf species. Its phytotoxic symptoms include epinasty, cuplike leaves, and tissue proliferation. It is readily absorbed by leaves, stems, and roots and appears to be truly systemic, being readily translocated in both the symplast and the apoplast. It accumulates in areas of rapid growth and metabolic sinks. Picloram persists in plant tissue since its detoxification by plant metabolism is very slow. Relative to 2,4-D, picloram is more phytotoxic, more readily translocated, and degraded

TABLE 13-1. Picloram Uses

Crop	Rate (lb/acre)	Combinations	Remarks
Barley,	0.016–0.023	—	Post, crop 3–5 leaf stage
spring	0.016–0.023	2,4-D or MCPA	As above, tank mix
Oats	0.016–0.023	—	Post, crop 3–5 leaf stage
	0.016–0.023	MCPA	As above, tank mix
Wheat	0.016–0.023	—	Post, crop 3–5 leaf stage
spring	0.016–0.023	2,4-D or MCPA	As above, tank mix
Wheat, winter	0.016–0.023	—	Post, spring, after growth resumption to early boot stage
	0.016–0.023	2,4-D or MCPA	As above, tank mix
Pastures, established	0.125–2.000	—	Spring or weeds actively growing
Rangeland, established	0.250–2.000	—	Weeds and/or brush actively growing
Brush control[1]	0.250–1.000	—	Low volume, actively growing
	1.000–2.000	—	High volume, actively growing
Noncropland	2.000–3.000	—	See label
	RTU-undiluted	2,4-D	Stem injection, frills, girdles; freshly cut stumps or stems
Fallowland	0.125–0.250	—	Apply between small grain crops
Set-aside land	0.250	—	Apply >4 months before seedling Certain perennial grasses
	0.063–0.500	—	Postseeding treatment

There are some limitations on some of the above uses relative to geographic location, formulation, rate for species, and grazing and use of forage; see product labels.

[1] Rangelands and permanent grass pastures.

more slowly. However, the biochemical mechanism of action of picloram is probably similar to 2,4-D, namely, an interference with nucleic acid metabolism. These are briefly discussed in Chapter 12 and more extensively in Ashton and Crafts (1981) and Fedtke (1982).

Clopyralid

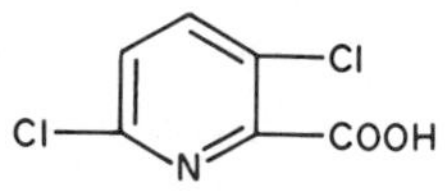

clopyralid

Clopyralid is the common name for 3,6-dichloro-2-pyridinecarboxylic acid. The trade name Curtail® is a mixture of the alkanolamine salts of clopyralid (0.38 lb, ae/gal) and 2,4-D (2.0 lb ae/gal). Clopyralid is a white crystalline solid with a high to moderate water solubility of 1000 ppm. The acute oral LD_{50} of clopyralid for rats is 4.3 to > 5.0 g/kg. In general, the mode of action of clopyralid appears to be similar to that of picloram discussed above. However, the basis for the greater selectivity of clopyralid has not been determined.

Uses Clopyralid is a postemergence herbicide applied to the foliage of plants, but it can also affect susceptible species by root uptake. It controls many annual and perennial broadleaf weeds and certain woody species, e.g., mesquite and associated species. It is particularly effective on members of the Polygonaceae, Compositae, and Legumnosae families but it does not control grasses. Clopyralid is used in barley, oats, wheat, fallow cropland, rangeland, permanent grass pastures, turf, noncropland, and conservation reserve programs (CRP). See Table A-0 for the specific weeds controlled and Table 13-2 for detailed information on the crops.

TABLE 13-2. Clopyralid Uses

Crop	Rate (lb/acre)	Combinations	Remarks
Small grains[1]	0.094–0.125	—	Post, 3-leaf to early boot stage
	0.094–0.125	MCPA, bromoxynil, chlorsulfuron, dicamba, diuron, bromoxynil plus MCPA	As above
Small grains[2]	0.094–0.125	2,4-D, terbutryn, metribuzin	As above
Fallow cropland	0.094–0.188	—	Post
Noncropland	0.094–0.188	—	Post
	0.094–0.188	2,4-D, triclopyr	Post
Turf	0.094–0.188	—	Post
	0.094–0.188	Turflon®	Post
Rangeland, pastures[3]	0.250–0.500	—	Post
CPR[4]	0.250–0.500	—	Post
	0.250	2,4-D	Post

There are some limitations on some of the above uses relative to endangered species, spray drift, and grazing and use of hay.

[1] Barley, oats, wheat.
[2] Barley, wheat.
[3] Permanent grass pastures.
[4] Conservation Reserve Program.

Soil Influence Clopyralid in not strongly adsorbed by soil colloids (*Herbicide Handbook*, 1989). It exists in the soil primarily in the salt form and therefore is subject to leaching. It is degraded by microorganisms at a medium to fast rate and has an average half-life of 12 to 70 days in a wide range of soils. No injury to susceptible broadleaf crops was observed the following year from a field application of 0.5 lb/acre.

Triclopyr

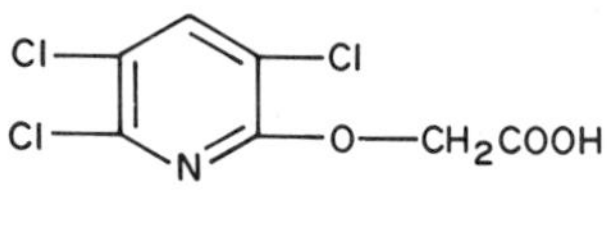

triclopyr

Triclopyr is the common name for [(3,5,6-trichloro-2-pyridinyl)oxy]acetic acid. The trade names of the various formulations are Turflon® Amine, Turfloan® II Amine, Turflon® D, Turflon® Ester, and Crossbow®. Turflon® Amine and Turflon® Ester contain only triclopyr and the others are combinations of triclopyr and 2,4-D. The form and concentration triclopyr and 2,4-D vary with the various formulations. Triclopyr is a white crystalline solid with a moderate water solubility of 430 ppm. The acute oral LD_{50} of triclopyr is 713 mg/kg for rats. The mode of action of triclopyr appears to be similar to picloram discussed above.

Uses Triclopyr is used to control many woody and broadleaf weeds. It controls ash, oaks, and other root sprouting species better than other auxin-type herbicides. Most grass species are tolerant. The Turflon® formulations are used on turf and Crossbow® on trees and brush, rangeland, permanent grass pastures, fence rows, nonirrigated ditch banks, other noncrop areas, and industrial sites. The ester form is usually more effective than the amine form on certain hard-to-kill species; see labels.

Soil Influence Organic matter content and pH influence triclopyr adsorption but it is not considered to be strongly adsorbed on soil colloids (*Herbicide Handbook*, 1989). Some leaching may occur in light soils under high rainfall conditions. It is degraded in soils by microorganisms at a rate that is considered to be relatively rapid. The average half-life is 46 days and is influenced by soil type and climatic conditions.

SULFONYLUREAS

The sulfonylurea class herbicides have been developed relatively recently. They are particularly attractive because of the high phytotoxicity that allows them to

TABLE 13-3. Common Name, Chemical Name, and Chemical Structure of the Sulfonylurea Herbicides

Common Name	Chemical Name	Chemical Structure
Bensulfuron methyl	Methyl 2-[[[(4,6-dimethoxy-pyrimidin-2-yl)aminocarbonyl]-aminosulfonyl]methyl]benzoate	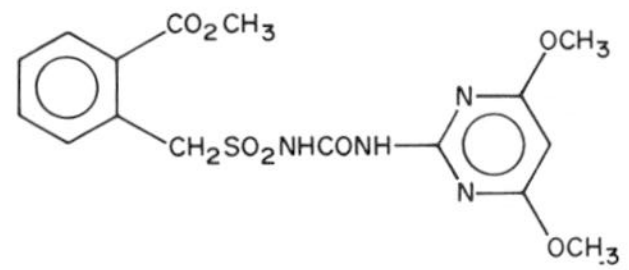 CO_2CH_3; $CH_2SO_2NHCONH$; N; N; OCH_3; OCH_3
Chlorimuron ethyl	Ethyl 2-[[(4-chloro-6-methoxy-pyrimidin-2-yl)aminocarbonyl]-aminosulfonyl]benzoate	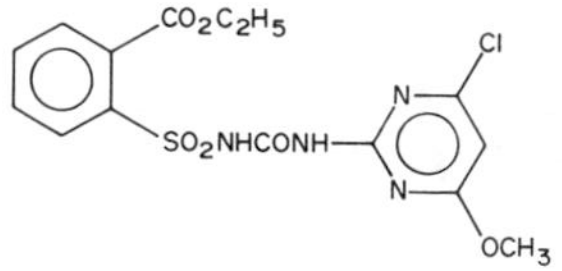 $CO_2C_2H_5$; $SO_2NHCONH$; N; N; Cl; OCH_3
Chlorsulfuron	2-Chloro-*N*-[4-methoxy-6-methyl-1,3,5,-triazin-2-yl)aminocarbonyl]-benzenesulfonamide	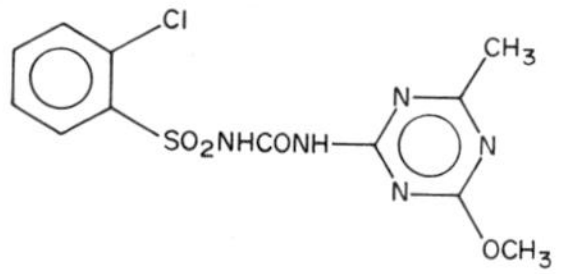 Cl; $SO_2NHCONH$; N; N; N; CH_3; OCH_3
Metsulfuron methyl	Methyl 2-[[(4-methoxy-6-methyl-1,3,5-triazin-2-yl)aminocarbonyl]-aminosulfonyl]benzoate	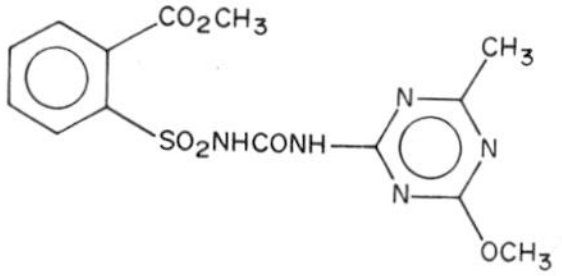 CO_2CH_3; $SO_2NHCONH$; N; N; N; CH_3; OCH_3
Sulfometuron methyl	Methyl 2-[[(4,6-dimethyl-pyrimidin-2-yl)aminocarbonyl]-Aminosulfonyl]benzoate	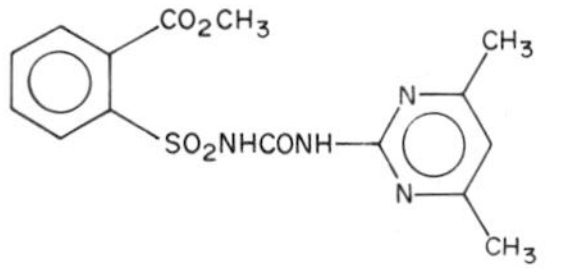 CO_2CH_3; $SO_2NHCONH$; N; N; CH_3; CH_3
Thiameturon methyl	Methyl 3-[[(4-methoxy-6-methyl-1,3,5-triazin-2-yl]aminocarbonyl]-aminosulfonyl]-2-thiophene-carboxylate	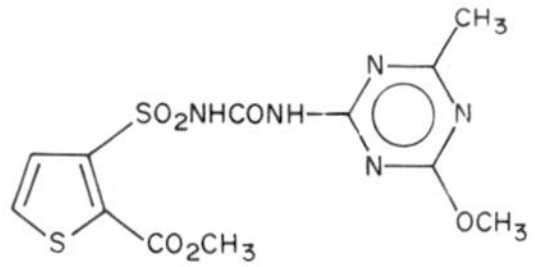 $SO_2NHCONH$; N; N; N; CH_3; OCH_3; S; CO_2CH_3

be used at very low rates and their low mammalian toxicity. These reduce any potential environmental hazard, an increasingly important goal in the development of all pesticides. Following several years of research and development, E. I. du Pont de Nemours & Company, Inc. commercialized chlorsulfuron and sulfometuron in 1982. Additional sulfonylurea type herbicides subsequently developed include bensulfuron, chlorimuron, metsulfuron, and thiameturon. All of these herbicides except chlorsulfuron are either methyl or ethyl derivatives and are often referred to by these respective common names, e.g., sulfometuron methyl. Table 13-3 gives the common name, chemical name, and chemical structure of each of these six sulfonylurea herbicides. A comprehensive review of the history, chemistry, biology, mode of action, and degradation in plants, animals, and soils of the sulfonylurea herbicides has recently been published (Beyer et al., 1988).

In many applications, the addition of a nonionic surfactant improves the postemergence performance of the sulfonylurea herbicides.

Bensulfuron

Bensulfuron is the common name for 2-[[[(4,6-dimethoxy-2-pyrimidinyl)amino]sulfonyl]methyl]benzoic acid. The methyl derivative, bensulfuron methyl, is the form used as a herbicide and the following information refers to this form. The trade name is Londax®. It is a white crystalline solid and its water solubility at 25°C is 3.0 ppm at pH 5.0 and 120 ppm at pH 7.0. It is available as a dry flowable formulation (60%). The acute oral LD_{50} of bensulfuron methyl is > 5.0 g/kg. The soil influence and mode of action of bensulfuron methyl are presented later in this chapter in sections covering these topics for all sulfonylurea herbicides.

Uses Bensulfuron methyl was introduced in 1984 for use in direct-seeded and transplanted rice in Thailand in 1985 (Beyer et al., 1988). In the United States it is proposed for use in direct-seeded and transplanted rice at 0.5 to 1.0 oz/acre before emergence to early post of the weeds; full registration is expected soon. It is particularly effective on many broadleaf weeds and sedges; see Table A-0 for specific weeds controlled. A minimum holding period of 5 days is required for maximum efficacy when applied to standing water. It can be used in combination with most rice herbicides, especially molinate, for grass control.

Chlorimuron

Chlorimuron is the common name for 2-[[[(4-chloro-6-methoxy-2-pyrimidinyl)amino]carbonyl]amino]sulfonyl]benzoic acid. The ethyl derivative, chlorimuron ethyl, is the form used as a herbicide and the following information refers to this form. The trade name is Classic®; as a package mix with metsulfuron, the trade name is Finesse®. It is a white crystalline solid and its water solubility at

25°C is 11 ppm at pH 5.0 and 1200 ppm at pH 7.0. It is formulated as a dry dispersible granule (25%). The acute oral LD_{50} of chlorimuron ethyl is 4.1 and 4.2 g/kg for male and female rats, respectively. The soil influence and mode of action of chlorimuron ethyl are presented later in this chapter in sections covering these topics for all sulfonylurea herbicides.

Uses Chlorimuron ethyl was introduced under an experimental use permit (EUP) in 1984 for postemergence use in soybeans and commercialized in 1986 in the United States (Beyer et al., 1988). It is a selective postemergence herbicide for this crop used for the control of many broadleaf weeds and yellow nutsedge. See Table A-0 for specific weeds controlled. The use rate is 0.125 to 0.20 oz/acre. It is applied after crop emergence but not later than 60 days before soybean maturity. Best postemergence results are obtained when applied to young, actively growing weeds. Do not graze treated fields or harvest for forage or hay. Chlorimuron can be tank mixed and applied with acifluorfen or applied after an application of linuron or metribuzin to broaden the spectrum of weeds controlled.

Chlorsulfuron

Chlorsulfuron is the common name for 2-chloro-*N*-[[(4-methoxy-6-methyl-1,3,5-triazin-2-yl)amino]carbonyl]benzenesulfonamide. The trade name is Glean®. Chlorsulfuron is a white crystalline solid and it water solubility at 25°C is 60 ppm at pH 5.0 and 7000 ppm at pH 7.0. It is available as a dry flowable formulation (75%). The acute oral LD_{50} of chlorsulfuron is 5.5 and 6.3 g/kg for male and female rats, respectively. The soil influence and mode of action of chlorsulfuron are presented later in this chapter in sections covering these topics for all sulfonylurea herbicides.

Uses Chlorsulfuron was first sold in 1982 for use in small grains to control or suppress the growth of many weeds common to these crops. See Table A-0 for specific weeds controlled. It is primarily for use on land dedicated to continuous barley/wheat/fallow rotations. It can also be used in oats. It is applied postemergence to these crops at 0.125 to 0.375 oz/acre after the crop is in the two-leaf stage but before the boot stage. Since wheat is somewhat more tolerant than barley or oats to chlorsulfuron, it can be applied in wheat at times or methods not suitable for the other two crops including preplant soil-incorporated, preemergence, split-treatment (preemergence plus postemergence). Finesse®, a package mix of chlorsulfuron plus metsulfuron, is available for use in wheat and barley. Finesse® can also be used as a tank mix with diuron or linuron for use in wheat and barley. Special techniques are required for the control of specific weeds (e.g., Canada thistle, kochia) including a combination with certain other herbicides in some cases, see labels. There are numerous other limitations or suggestions on the Glean® and Finesse® labels including soil pH, geographic location, and crop variety.

Metsulfuron

Metsulfuron is the common name for 2-[[[[(4-methoxy-6-methyl-1,3,5-triazin-2-yl)amino]carbonyl]amino]sulfonyl]benzoic acid. The methyl derivative, metsulfuron methyl, is the form used as a herbicide and the following information refers to this form. It is available as a dry flowable formulation (60%). The trade names are Ally® (barley and wheat use) and Escort® (noncrop land use); as a package mix with chlorsulfuron, the trade name is Finesse®. Metsulfuron methyl is a white crystalline solid and its water solubility at 25°C is 1100 ppm at pH 5.0 and 9500 ppm at pH 7.0. The acute oral LD_{50} of metsulfuron methyl is > 5.0 g/kg for rats. The soil influence and mode of action of metsulfuron are presented later in this chapter in sections covering these topics for all sulfonylurea herbicides.

Uses Metsulfuron was introduced in Europe and the United States from 1984 to 1986 for the control of broadleaf weeds in barley and wheat and/or noncrop land. It is also used for brush control. See Table A-0 for the specific weeds controlled. In barley and wheat, it is applied postemergence at the two-leaf stage to the boot stage of the crop at 0.06 oz/acre. However, these uses are limited to certain geographic locations; see label. The label also lists other limitations and suggests several tank mixes with other herbicides to broaden the spectrum of weeds controlled. Finesse® (metsulfuron plus chlorsulfuron) can also be used in barley and wheat. On noncrop land, the use rate is 0.15 to 1.2 oz/acre; the specific rate depends on the weeds species present. For brush control, the application rate depends on the brush species and density. Low volume sprays utilize 0.3 to 0.45 oz in 5 to 25 gallon of water and high volume sprays utilize 0.3 to 2.4 oz in 100 gallon of water.

Sulfometuron

Sulfometuron is the common name for 2-[[[[(4,6-dimethyl-2-pyrimidinyl)amino]carbonyl]amino]sulfonyl]benzoic acid. The methyl derivative, sulfometuron methyl, is the form used as a herbicide and the following information refers to this form. It is available as a dispersible granule formulation (75%). The trade name is Oust®. Sulfometuron methyl is a white crystalline solid and its water solubility at 25°C is 8 ppm at pH 5.0 and 70 ppm at pH 7.0. The acute oral LD_{50} of sulfometuron methyl is > 5.0 g/kg for rats. The soil influence and mode of action of sulfometuron are presented later in this chapter in sections covering these topics for all sulfonylurea herbicides.

Uses Sulfometuron is usually used on noncrop land to control many annual and perennial grass and broadleaf weeds. It is applied preemergence or postemergence to weeds. It is generally considered to be nonselective but can be used selectively for release and suppression of certain established grasses and for conifer site preparation and release at low rates. For nonselective weed control, it is applied at 2.25 to 9.0 oz/acre depending the size and species of the weeds to be controlled.

Thiameturon

Thiameturon is the common name for 3-[[[[(4-methoxy-6-methyl-1,3,5-triazin-2-yl)amino]carbonyl]amino]sulfonyl]-2-thiophenecarboxylic acid. The methyl derivative, thiameturon methyl, is the form used as a herbicide and the following information refers to this form. It is available as a dispersible granule formulation (75%). The trade name is Harmony®. Thiameturon methyl is a white crystalline solid and its water solubility is pH dependent; at 25°C, 260 ppm at pH 5.0 and 2400 ppm at pH 7.0. The acute oral LD_{50} is > 5.0 g/kg for rats. The soil influence and mode of action of thiameturon are presented later in this chapter in sections covering these topics for all sulfonylurea herbicides.

Uses Thiameturon is used as an early postemergence treatment to control many broadleaf weeds and wild garlic in barley and spring wheat. It is applied after the crop is in the two-leaf stage but before flag leaf emergence at 0.25 to 0.50 oz/acre. Thiameturon is degraded rapidly in soil and any crop can be planted in thiameturon-treated soil 60 days after application. This provides considerable rotational crop flexibility, especially when compared to chlorsulfuron, which may cause injury to sensitive crops for a year or longer depending on soil, climate, and use rate.

SOIL INFLUENCE

The mobility of a particular sulfonylurea herbicide in soil generally increases with increasing soil pH and decreasing soil organic matter (Beyer et al., 1988; Mersie and Foy, 1986; Nicholls and Evans, 1985; Priester, 1988). Priester's data from experiments using two sandy loam and two silt loam soils suggest that bensulfuron methyl is the least mobile, chlorimuron ethyl and sulfometuron methyl are intermediate, and chlorsulfuron, metsulfuron methyl, and thiameturon methyl are the most mobile. As a class, the sulfonylurea herbicides are considered to be moderately mobile in soil. Chemical hydrolysis and microbial breakdown are the most important methods of sulfonylurea degradation/dissipation in soil (Beyer et al., 1988). The rate of degradation of sulfonylurea herbicides in soil is greatest in warm, moist, light-textured, low pH soils and slowest in cold, dry, heavy, high pH soils. Summarizing worldwide field experience, Palm et al. (1980) reported the half-life of chlorsulfuron in soil is usually 1 to 2 months. Under field conditions in Europe, Canada, and the United States, Doig et al. (1983) reported the half-life of metsulfuron methyl in soil ranged from 1 week to 1 month. However, under adverse conditions the half-life of the sulfonylurea herbicides may be considerably longer.

MODE OF ACTION

Ray (1980, 1982) and Rost (1984) showed that the sulfonylurea herbicides are rapid and potent inhibitors of cell division. They inhibit growth of susceptible species after either preemergence or postemergence application. The susceptible weed seedlings may emerge following a preemergence application. If they do emerge, and in postemergence applications, the leaves become chlorotic in many species and this is followed by death of the growing points. However, in some susceptible weed species, the leaves may remain green but they are stunted and noncompetitive. The sulfonylurea herbicides are readily absorbed by both leaves and roots and translocated in both the symplast and apoplast.

Beyer et al. (1988) reviewed and proposed metabolic pathways for the degradation of most sulfonylurea herbicides in higher plants. The pathways vary for each herbicide of this class and among species for each herbicide. Although several different types of reactions occur within these many pathways, hydroxylation of one of the side chains and a separation of the two ring are common to most of these herbicides. Conjugation with glucose (chlorsulfuron, metsulfuron) and homoglutathione (chlorimuron) also occurs in some species. The rate of metabolism is the principal factor for the differential tolerance or selectivity of various plant species to the sulfonylurea herbicides. Tolerant species render the herbicide nonphytotoxic by rapid degradation whereas its degradation is slow in susceptible species.

The sulfonylurea herbicides are tightly but reversibly bound to acetolactate synthase (ALS). This binding inhibits ALS activity and the synthesis of branched chain amino acids, e.g., valine and isoleucine. ALS is considered to be the primary biochemical site of action of the sulfonylurea herbicides. A detailed discussion of the elegant research leading to this concept has been published (Beyer et al., 1988).

TRIAZINES

In 1952, J. R. Geigy, S. A., currently CIBA-GEIGY, started investigations with triazine derivatives as potential herbicides (Gysin, 1974). Herbicidal properties of chlorazine were reported by Gast et al. (1955) and Antognini and Day (1955). Since that time numerous triazine derivatives have been synthesized and screened for their herbicidal properties. Eight triazine herbicides are now being used in the United States. However, some of these herbicides and/or some of their specific uses are in the processes of being reregistrated with EPA. During this procedure some registrations discussed below may be canceled. Therefore, consultation with company representatives and regulatory agencies is recommended before they are used; also see Preface.

Chemically, the triazines are heterocyclic nitrogen derivatives. The word *heterocyclic* is used to designate a ring structure composed of atoms of different kinds. In this case the ring is composed of nitrogen and carbon atoms. Most

TABLE 13-4. Common Name, Chemical Name, and Chemical Structure of the Symmetrical Triazine Herbicides[1,2]

R1, R2—C, N, C, N, C—R3, N

Common Name	Chemical Name	R_1	R_2	R_3
Ametryn	*N*-Ethyl-*N*′-(methylethyl)-6-(methylthio)triazine-2,4-diamine	—SCH_3	—$NH \cdot iso\text{-}C_3H_7$	—NHC_2H_5
Atrazine	6-Chloro-*N*-ethyl-*N*′-(1-methylethyl)-1,3,5-triazine-2,4-diamine	—Cl	—$NH \cdot iso\text{-}C_3H_7$	—NHC_2H_5
Prometon	6-Methoxy-*N*,*N*′-bis(1-methylethyl)-1,3,5-triazine-2,4-diamine	—OCH_3	—$NH \cdot iso\text{-}C_3H_7$	—$NH \cdot iso\text{-}C_3H_7$
Prometryn	*N*,*N*′-Bis(methylethyl)-6-(methylthio)-1,3,5-triazine-2,4-diamine	—SCH_3	—$NH \cdot iso\text{-}C_3H_7$	—$NH \cdot iso\text{-}C_3H_7$
Propazine	6-Chloro-*N*,*N*′-bis(1-methylethyl)-1,3,5-triazine-2,4-diamine	—Cl	—$NH \cdot iso\text{-}C_3H_7$	—$NH \cdot iso\text{-}C_3H_7$
Simazine	6-Chloro-*N*,*N*′-diethyl-1,3,5-triazine-2,4-diamine	—Cl	—NHC_2H_5	—NHC_2H_5

[1] With a symmetrical molecule, the numbering can start from any one of the nitrogen atoms in the ring.

[2] The chemical name and chemical structure of the symmetrical triazinc hexazinone and the asymmetrical triazine metribuzin are given in the text.

triazine herbicides are symmetrical, that is, they have alternating carbon and nitrogen atoms in the ring. However, one exception is metribuzin, which is asymmetrical.

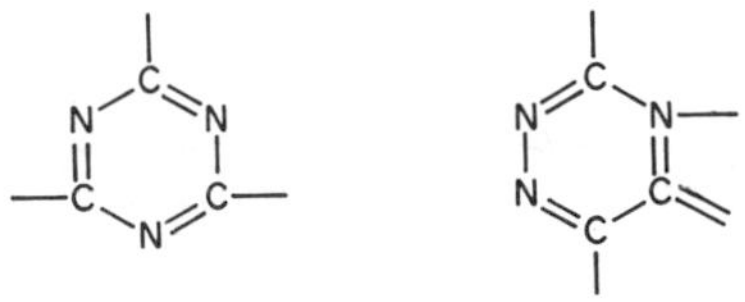

symmetrical triazine   asymmetrical triazine

Structures of symmetrical triazine herbicides are given in Table 13-4. The substitution on the R_1 position of the ring determines the ending of the common name: *-azine* = chlorine atom; *-tryn* = methylthio group ($—SCH_3$); and *-ton* = methoxy group ($—OCH_3$). Water solubility of a series of triazines is markedly influenced by the substitution of R_1: prometon ($—OCH_3$), 750 ppm; prometryn ($—SCH_3$), 48 ppm; and propazine (—Cl), 8.6 ppm.

The greatest use of the triazine herbicides has been as a selective herbicide on cropland. Several excellent reviews have been written on various aspects of triazine herbicides, including their history (Gysin, 1974), uses (Gast, 1970), degradation (Esser et al., 1975), molecular structure and function (Gysin, 1960, 1974), mode of action (Ashton, 1965; Ashton and Crafts, 1981; Esser et al., 1975; Ebert and Dumford, 1976), comprehensive (Gysin and Knüsli, 1960); and biotype resistance (LeBaron and Gressel, 1982). Volume 32 of *Residue Reviews* contains 15 chapters dealing with triazine–soil interactions.

Sections on soil influence and mode of action of all of the triazine herbicides are presented following their individual coverages since they have common characteristics and relative comparisons can be made.

Ametryn

Ametryn is the common name for *N*-ethyl-*N′*-(methylethyl)-6-(methylthio)-triazine-2,4-diamine. The trade name is Evik®. It is a white crystalline solid with a moderate water solubility of 185 ppm at 20°C. It is formulated as a wettable powder (80%). The acute oral LD_{50} of the formulation is 1.1 g/kg.

Uses Ametryn is a selective herbicide that can be used either preemergence or postemergence to the weeds. It can be used in corn, sugarcane, certain citrus and subtropical fruits, and noncropland. It is also used in potatoes for preharvest desiccation of both crop and weeds. Detailed information on the crops is given in Table 13-5 and Table A-0 lists the specific weeds controlled.

Atrazine

Atrazine is the common name for 6-chloro-*N*-ethyl-*N′*-(1-methylethyl)-1,3,5-triazine-2,4-diamine. The trade names are AAtrex®, Atratol®, and Atrazine.

TABLE 13-5. Ametryn Uses[1]

Crop	Rate (lb/acre)	Remarks
Bananas and plantains	3.2–8.0[2]	Pre and/or post to weeds, directed spray
Citrus[3]	1.6–6.4[2]	Post to weeds, established crop
Corn[4]	0.6–2.0	Post, directed spray, corn > 12 inches high
Pineapple	2.0–8.0[2]	Pre and/or post to weeds
Potatoes	1.0–2.4	Post, preharvest desiccant
Sugarcane[5]	0.4–8.0[2]	Pre and/or post to weeds

There are some limitations for some of these treatments relative to geographic location, crop rotation, harvest date, and grazing and feeding forage; see product label.

[1] Addition of surfactant or crop oil suggested for some of these treatments.
[2] Rates vary greatly with location and time of application. Most treatments can be repeated as necessary.
[3] Grapefruit, oranges.
[4] Field, Sweet, or pop.
[5] Tank mixes of 2,4-D or diuron with ametryn are also suggested for some weeds in some locations.

Certain package mix combinations of atrazine with other herbicides have specific trade names. Atrazine is a white crystalline solid with a low water solubility of 33 ppm at 27°C. It is formulated as a liquid concentrate (4 lb/gal), wettable powder (80%), and a dispersible granule (85.5%). The acute oral LD_{50} of the 80% wettable powder formulation is 5.1 g/kg.

Uses Atrazine is widely used to control broadleaf weeds and some grasses and in corn, sorghum, rangeland, fallow land, noncropland, and certain tropical plantations, evergreen tree nurseries, and lawns. It is applied as preplant soil-incorporated, at planting, preemergence, or postemergence treatments. The specific type of treatment used depends on the crop, weeds, and herbicide combinations. See Table A-0 for specific weeds controlled and Table 13-6 for detailed information on the crops.

Hexazinone

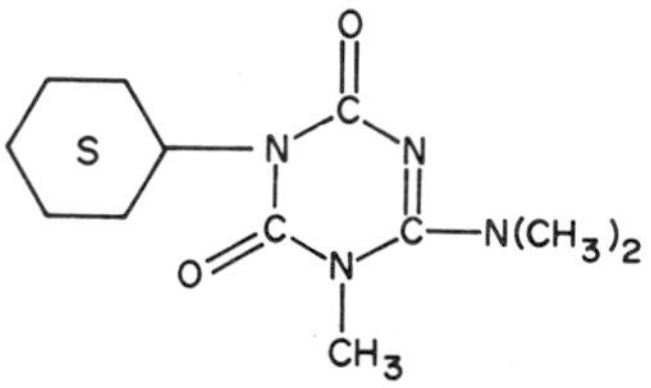

hexazinone

TABLE 13-6. Atrazine Uses

Crop	Rate (1b/acre)	Combinations	Remarks
Corn[1]	2.0–4.0	—	PP, PPI, pre, post, LBY; rates total for split applications
	1.2	—	Post, plus emulsifiable oil
	2.0–3.0	Paraquat	PPI, pre; emerged vegetation and residual control
	1.0–1.6	Alachlor[2]	PPI, pre, post
	1.0–1.6	Butylate(Sutan®+)	PPI
	1.0–2.0	Metolachlor[2]	PPI, pre, post
	1.0–1.6	Propachlor	Pre
	0.7–1.4	Simazine[2]	PP, pre
	0.6–1.0	Simazine plus metolachlor[2]	Pre
Guava	2.0–4.0	—	Established, directed spray[3]
Macadamia	2.0–4.0	—	Before harvest, prior to weed emergence[3]
Pineapple	4.6	—	Immediately after planting or following harvest[3]
Sorghum[4]	1.6–2.4	—	PP, PPI, pre
	2.0–3.0	—	Post, sorghum > 6 inches high
	1.2	—	Post, plus surfactant, sorghum > 6 inches high
	1.2–1.6	Metolachlor	PPI, pre
	0.8	Terbutryn	PPI
Sugarcane	2.0–4.0	—	Pre, post[3]
	0.4–2.0	—	Post, directed, plus surfactant
Turf/lawns	1.0–2.0	—	Established, pre to weeds
Rangeland	0.5–2.0	—	Established

Fallow land	0.5–3.0	—	Crop stubble after harvest
	0.5–1.0	Paraquat	Crop stubble after harvest
	0.5–3.0	Terbutryn	Crop stubble after harvest
Noncropland	4.8–10.0	—	Pre to early post, annuals,
	10.0–20.0	—	Pre to early post, annuals and certain perennials
	20.0–40.0	—	Pre to early post, hard-to-kill biennials and perennials
	4.0–10.0	Diuron	Pre to early post, added surfactant > post activity
	4.0–10.0	Imazapyr	Post, higher rates on established stands
	2.6–10.0	Simazine	Pre to early post
	4.0–16.0	Sulfometruon	Pre to early post; added surfactant > post activity
	8.0–32.0	Chlorate–borate mixture	Pre, post

There are some restrictions on some of the above uses relative to specific rate within the range given, geographic location, soil type, crop rotation, varieties, and grazing and feeding forage; see product labels.

[1] Field, silage, seed (except combinations). Most treatments also for sweet and pop; see labels.
[2] Glyphosate or paraquat added to control emerged vegetation.
[3] Most of these treatments can be repeated as necessary; see lables.
[4] Grain forage.

Hexazinone is the common name for 3-cyclohexyl-6-(dimethylamino)-1-methyl-1,3,5-triazine-2,4(1*H*,3*H*)-dione. It is a symmetrical triazine but the ring substitutions and bonds in the ring vary considerably from the other symmetrical triazines. Compare the structure of hexazinone given above with the other symmetrical triazines structures given in Table 13-4. The trade name of hexazinone is Velpar®. It is a white crystalline solid with a very high water solubility of 33,000 ppm. This very high water solubility permits it to be formulated as a water-soluble powder (90%). It is also formulated as a miscible liquid (2 lb/gal). The acute oral LD_{50} of hexazinone is 1.7 g/kg for rats.

Uses Hexazinone can be used as a preemergence herbicide, or a postemergence herbicide during active plant growth. It controls many annual grass and broadleaf weeds at selective rates and in addition certain perennial weed on noncropland at higher rates. It is used in alfalfa, blueberries, sugarcane, and noncropland and for pine release. In alfalfa, it is used at 0.45 to 1.35 lb/acre. It is applied in the fall or winter to the dormant crop or in the spring before alfalfa begins to grow. In certain areas it can be applied postdormancy and after cuttings. In blueberries, it can be applied at 0.45 to 2.7 lb/acre in the spring before budbreak in some areas; however, some clones are susceptible to injury. In sugarcane, hexazinone is used in the continental United States and the rate and time of application vary for different states. In general, the use rate is 0.45 to 0.9 lb/acre and split applications are common. For field-grown pine release, it is applied to 0.9 to 1.8 lb/acre in the spring before conifer budbreak. On noncropland, hexazinone is used at 2 to 12 lb/acre as a foliar spray to control many brush species, certain perennial weeds, and annual weeds. Its preemergence activity will also control weeds subsequently germinating from seeds. See Table A-0 for specific weeds controlled.

Metribuzin

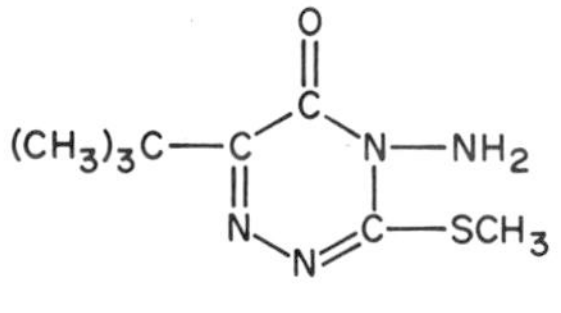

metribuzin

Metribuzin is an asymmetrical triazine with the chemical structure shown above. Its chemical name is 4-amino-6-(1,1-dimethylethyl)-3-(methylthio)-1,2,4-triazin-5(4*H*)-one. The trade names are Lexone®, Preview®, and Sencor®. As a package mix with chlorimuron, the trade name is Canopy®. It is a white crystalline solid with a high water solubility of 1220 ppm. It is available as a liquid suspension or a dry flowable formation. The acute oral LD_{50} of technical metribuzin is 1.1 and 1.2 g/kg for male and female rats, respectively.

Uses Metribuzin is a selective herbicide used to control annual broadleaf weeds in several crops. It is used as a preemergence or postemergence herbicide and frequently in combination with other herbicides to also control annual grass weeds. Some of these combinations are applied as a preplant soil-incorporated treatment. The crops include alfalfa, asparagus, barley, potatoes, soybeans, sugarcane, tomatoes, wheat, and turf; see Table 13-7 for details on the crops. The specific weeds controlled are given in Table A-0.

Prometon

Prometon is the common name for 6-methoxy-*N*,*N'*-bis(1-methylethyl)-1,3,5-triazine-2,4-diamine. The trade name is Pramitol® 25E. Pramitol® 5Ps is a mixture of prometon, simazine, sodium chlorate, and sodium metaborate and Vegemec® is a mixture of prometon and 2,4-D. Prometon is a white crystalline solid with a moderate water solubility of 750 ppm at 20°C. Pramitol® 25E is a liquid formulation (2 lb/gal). Pramitol® 5PS is a solid formulation containing 5.0% prometon, 0.75% simazine, 40.0% sodium chlorate, and 50.0% sodium metaborate. Vegemec® is available in two liquid formulations containing prometon and 2,4-D. Prometon (25EC) has an acute oral LD_{50} of 2.3 g/kg for rats.

Uses Prometon is a nonselective preemergence and postemergence herbicide used to control most annual and broadleaf weeds and certain perennial weeds on noncrop land. Although it has considerable activity through the foliage, much of its activity is through the roots. Therefore, maximum activity requires sufficient rainfall to move it into the root zone. The application rate is 10 to 60 lb/acre depending on climatic conditions, soil type, weed species present, stage of weed growth, and period of residual control desired. In general, lower rates are used for annual weed control and relatively short control periods; higher rates are used for the control of hard-to-kill perennial weeds and extended periods of control. It can be applied in water or oil before or after the weeds begin to grow. It is also used to control weeds under asphalt pavement and mixed with cutback asphalt for weed control in areas being stabilized. The combination of prometon with the herbicides mentioned above broadens the spectrum of weeds controlled and/or extends the period of weed control.

Prometryn Prometryn is the common name for *N*,*N'*-bis(1-methylethyl)-6-(methylthio)-1,3,5-triazine-2,4-diamine. The trade names are Caparol® and Cotton-Pro®. It is a white crystalline solid with a low water solubility of 48 ppm at 20°C and is formulated as a liquid suspension (4 lb/gal). Prometryn (80% WP) has an acute oral LD_{50} of 3.8 g/kg for rats.

Uses Prometryn is a selective herbicide used to control many annual grass and broadleaf weeds in celery, cotton, and peas. See Table A-0 for specific weeds controlled and Table 13-8 for detailed information on the crops.

TABLE 13-7. Metribuzin Uses

Crop	Rate[1] (lb/acre)	Combinations	Remarks[1]
Alfalfa, sainfoin	0.25–1.00	—	PP, post; dormant, spring or fall
Asparagus	1.00–2.00	—	Pre, established, before spear emergence
	0.50–1.00 + 1.00–1.50	—	Pre, established, before spear emergence + postharvest
Barley	0.25–0.50	—	Post, after fully tillered but before jointing
Potato	0.50–1.00	—	Pre
	0.25–0.50	—	Post
	0.50–1.00 + 0.25–0.50	—	Pre + post, not more than 1.00 lb/acre total
	0.38–1.00	Metolachlor	Pre, tank mix
	0.50–1.00	Pendimethalin	Pre, tank mix
Soybean	0.13–1.00	—	Pre
	0.25–0.50	—	Post; directed, crop > 12 inches
	0.13–1.00	Alachlor, chloramben	Pre, tank mix
	0.13–0.63	Oryzalin	Pre, tank mix
	0.25–0.63	Alachlor	PPI, tank mix
	0.25–0.50	Pendimethalin	PP, pre; tank mix
	0.38–1.00	Paraquat	PP, pre; tank mix
	0.25–0.50	Trifluralin	PP, PPI; tank mix
	0.25–0.50	Dimethazone	PP, PPI (shallow); tank mix
	0.13–0.75	Metolachlor	PPI, pre; tank mix
	0.25–0.50	2,4-DB	Post; directed, tank mix
	0.25–0.75	Alachlor + glyphosate	Pre; tank mix
	0.38–1.00	Alachlor + paraquat	PP, pre; tank mix
	0.13–0.50	Linuron + alachlor or metolachlor	Pre, tank mix
	0.33–0.50	Oryzalin + paraquat	PP, pre; tank mix
	0.25–0.50	2,4-D + alachlor, metolachlor, oryzalin, or pendimethalin	Pre; tank mix

Crop	Rate	Herbicide combination	Application
	0.25–1.00	Pendimethalin	PPI (pendimethalin alone) + pre (metribuzin alone)
	0.38–1.00	Trifluralin	PPI (trifluralin alone) + pre (metribuzin alone)
	0.00 + 0.38–0.75	Metolachlor, pendimethalin, trifluralin	PPI (other herbicide alone) + pre (metribuzin alone)
	0.13–0.25 + 0.25–0.50	Trifluralin	PPI (metribuzin alone) + pre (combination), tank mix
	0.13–0.38 + 0.25–0.50	Alachlor, pendimethalin	PPI (metribuzin alone) + pre (combination), tank mix
	0.25–0.50 + 0.13–0.38	Metolachlor	PPI (combination), tank mix + pre (metribuzin alone)
	0.25–0.63 + 0.13–0.50	Alachlor, metolachlor	PP (15–30 days before planting) + pre; tank mix combination both times
Sugarcane	1.00–4.00	—	Pre, post
	0.75–2.00	Atrazine	Pre, post
Tomato	0.25–0.50	—	PPI, transplants
	0.25–0.38	—	Post; established, broadcast
	0.50–1.00	—	Post; established, directed
	0.25–0.50	Trifluralin	PPI, transplants
Wheat (winter)	0.25–0.75	—	Post; fall or spring, after fully tillered but before jointing
Lawns, turf, sod	0.25–0.50	—	Post, actively growing established bermudagrass
	0.50	—	Post, dormant established bermudagrass

There are some restrictions on some of the above uses relative to geographic location, soil type, crop variety, grazing and forage use, and harvest date; see product labels; all labels do not list all above uses.

[1] Split applications are indicated by a + sign.

Table 13-8. Prometryn Uses

Crop	Rate (lb/acre)	Combinations	Remarks
Celery, direct-seeded	1.2–1.6	—	Pre
	0.6–1.0	—	Post; crop 2–5 true leaves, weeds before 2 inches
Celery, transplants	0.8–3.2	—	Post; crop 2–6 weeks after transplanting, weeds before 2 inches
Cotton	0.6–0.8	—	PP, weeds pre to 4 inches
	0.5–2.8	—	PPI
	0.8–2.8	—	Pre
	0.5–2.8	—	Post, directed
	1.2–2.4	Pendimethalin	PPI, tank mix
	1.2–2.0	Trifluralin	PPI, tank mix
	0.5–0.8	DSMA, MSMA	Post; directed, tank mix
Peas, pigeon	2.0–3.0	—	Pre

There are some restrictions on some of the above uses relative to geographic location, soil type, grazing, and use of forage; see product labels.

Propazine

Propazine is the common name for 6-chloro-*N*, *N*′-bis(1-methylethyl)-1,3,5-triazine-2,4-diamine. The trade names are Milogard®, Milogard®, Maax™, and Propazine. It is a colorless crystalline solid with a very low water solubility of 8.6 ppm at 20°C. It is formulated as a liquid at 4 lb/gal, Milogard® and Propazine, and as a 90% water dispersible granule. Milogard® and Maax™. The acute oral LD_{50} of propazine (80% WP) is greater than 5.0 g/kg.

Uses Propazine is used as a selective preplant or preemergence herbicide to control annual grass and broadleaf weeds in milo and sweet sorghum. It is used at 2.0 to 3.2 lb/acre within 2 weeks prior to planting or at 1.2 to 3.2 lb/acre at planting or immediately after planting. There are geographic and soil type limitations: see product labels.

Simazine

Simazine is the common name for 6-chloro-*N*,*N*′-diethyl-1,35-triazine-2,4-diamine. The trade names are Princep®, Princep®, Caliber® 90, Simazine, Sim-Trol®, and Aquazine®. It is a white crystalline solid with a very low water solubility of 3.5 ppm at 20°C and a low water solubility of 85 ppm at 85°C. It is formulated as an emulsifiable concentrate (4 lb/gal), wettable powder (80%), water dispersible granule (90%), and granule (4%). The acute oral LD_{50} of technical simazine is > 5.0 g/kg for rats.

Figure 13-2. Weed-free plots were treated with simazine, applied preemergence to both weeds and corn. The corn was not injured. (Ciba-Geigy Corporation.)

Uses Simazine was the first widely used triazine herbicide. It is used in many crops and on noncropland to control annual grass and broadleaf weeds and certain perennial weeds. It is also used for aquatic weed control. It is applied as a preemergence or preplant soil-incorporated treatment. See Table A-0 for specific weeds controlled and Table 13-9 for detailed information on the crops.

SOIL INFLUENCE ON THE TRIAZINE HERBICIDES

Triazine herbicides are reversibly absorbed by clay and organic colloids. They are not subject to excessive leaching in most soil types. In a study of five triazine herbicides on 25 soil types, adsorption almost always increased in the following order: propazine > atrazine > simazine > prometon > prometryn (Talbert and Fletchall, 1965). Correlation analysis indicated that adsorption of the methylthio-(prometryn) and methoxy-(prometon) triazines was more highly related to clay content, whereas adsorption of chloro-triazines (simazine, atrazine, propazine) was more highly related to organic matter. The following relative leachability in Lakeland fine sand has been reported:

Table 13-9. Simazine Uses

Crop	Rate (1b/acre)	Combinations	Remarks
Artichokes	2.0–4.0	—	Pre; directed, after last fall tillage
Asparagus	2.0–4.0	—	Pre; established, spring or after harvest
Corn, field	2.0–4.0	—[1]	PP, pre; tank mix
	1.0–1.9	Atrazine[1]	PP, pre; tank mix
	1.0–3.0	Butylate, EPTC + antidote	PPI, tank mix
	0.6–1.1	Atrazine + metolachlor[1]	Pre, tank mix
	2.0–3.0	Glyphosate, paraquat[2]	Pre; cover crop, established sod, or previous crop residues
	1.0–2.0	Atrazine + glyphosate or paraquat	Pre; cover crop, established sod, or previous crop residues
Grapes	2.0–4.8	—	Established, between harvest and early spring
	2.0–4.8	Glyphosate, paraquat	As above
	2.0–4.8	Oryzalin	As above
Nuts[3]	2.0–4.0[3]	—	Established
	2.0–4.0[3]	Glyphosate, paraquat	Established, tank mix
	2.0–4.0[3]	Oryzalin[4]	Established, tank mix
Small fruits[5]	2.0–4.0	—	< 6 months old or split application use 1/2 rate
Cranberries	2.0–4.0	—	After fall harvest or before spring growth
Strawberries	1.0	—	October through November, > 4 months after transplanting
Sugarcane	2.0–4.0	—	At planting, after harvest, or inter line-directed
Tree fruits			
Citrus[6]	1.6–9.6	—	Established
	3.2–4.0	Ametryn,[7] bromacil[7]	Established, tank mix
	1.6–9.6	Glyphosate, paraquat, oryzalin	Established, tank mix

Deciduous[8]	2.0–4.0	—	Established
	2.0–4.0	Paraquat, oryzalin	Established, tank mix
Subtropical[9]	2.0–4.0	—	Established
	2.0–4.0	Paraquat, oryzalin[10]	Established, tank mix
Nurseries,	2.0–4.0	—	Established, field grown
woody plants	2.0–4.0	Oryzalin, metolachlor	Established, field grown
Lawns and sod	1.0–2.0	—	Species limitations
Noncropland	4.8–18.0	—	Higher rates for perennials or longer residual
	4.8–18.0	Glyphosate	Also emerged weeds
	4.0–16.0	Sulfometuron	Also actively growing weeds, higher rates on > 2.5% soil organic matter
Aquatic weeds			
Algae	0.1–1.0 ppm	—	Still water (lakes and ponds)
Submerged	1.3–2.5 ppm	—	Still water (lakes and ponds)

There are some restrictions on some of the above uses relative to geographic location, soil type, cultural practices, climatic conditions, species, plant stresses, and application and harvest times; see product labels; all labels do not list all above uses.

[1] Also for sweet corn and popcorn.
[2] Also for sweet corn and popcorn except a PP application.
[3] Almonds, filberts, macadamias, pecans, and walnuts; use lower rates on almonds (1.0–2.0 lb/acre).
[4] Not for macadamias.
[5] Blackberries, boysenberries, loganberries, raspberries, and blueberries.
[6] Grapefruit, lemons, and oranges.
[7] Not for lemons.
[8] Apples, cherries, peaches, pears, and plums.
[9] Avocados and olives.
[10] Not for olives.

atraton > propazine > atrazine > simazine > ipazine > ametryn > prometryn (Rodgers, 1968).

Note that the order of triazine herbicides common to these two studies is the same, indicating that the leachability of triazine herbicides is directly related to their adsorption to soil colloids. These two studies also indicate that adsorption and leachability have little or no relationship to water solubility of the compounds. A reduction in phytotoxicity of triazine herbicides is associated with increasing amounts of clay and organic matter in soil (Weber, 1970).

Triazine herbicides vary widely in their persistence in soils. Soil type and environmental conditions have considerable influence on the actual period of persistence. Methoxytriazines are generally more persistent than methylthio- or chlorotriazines. Prometon, the most persistent, can remain at phytotoxic levels for several years. Atrazine, propazine, and simazine are less persistent but can still injure sensitive plants the next season. Ametryn, cyprazine, prometryn, and terbutryn are usually even less persistent, but may last from six to nine months. Cyanazine and metribuzin appear to be least persistent of the triazines; their half-life is two to four weeks under most conditions. At these rates of disappearance, less than 10% of that applied would remain after two to four months.

A monograph contains several review papers on interaction of triazine herbicides and soil (Gunther, 1970).

MODE OF ACTION OF TRIAZINE HERBICIDES

The triazine herbicides inhibit plant growth, but this is considered to be a secondary effect caused by an inhibition of photosynthesis (Ashton and Crafts, 1981). At herbicidal concentration, triazine herbicides cause foliar chlorosis followed by death of the leaf (Figure 13-3). Other leaf effects include loss of membrane integrity and chloroplast destruction. At sublethal levels, however, increased greening of leaves may occur.

Triazine herbicides are absorbed by leaves, but translocation from them is essentially nil. The amount of foliar absorption varies for various compounds. Propazine and simazine are poorly absorbed by leaves, whereas ametryn and prometryn are readily absorbed. The others appear to be intermediate. All triazine herbicides are rapidly absorbed by roots and readily translocated throughout the plant by the transpiration stream. They are considered to be translocated almost exclusively in the apoplast system.

The distribution of [^{14}C]simazine and/or ^{14}C-labeled degradation products in three species is shown in autoradiographs (Figure 13-4) (Davis et al., 1959). The roots of the plants were treated 4 days earlier with [^{14}C]simazine via culture solution. The ^{14}C-patterns may reflect the relative susceptibility and possible degree of degradation in these three species. Cucumber is quite susceptible and the accumulation of ^{14}C in the leaf margins is typical of [^{14}C] simazine per se. The general distribution of radioactivity in tolerant corn suggests it is tolerant because of extensive metabolism of simazine. Cotton, the species of intermediate

Figure 13-3. Leaf chlorosis induced by triazine herbicides. *Upper*: Chlorotriazines (e.g., simazine) almost always show interveinal chlorosis. *Lower*: Methylthiotriazines (e.g., ametryn) usually show veinal chlorosis. (C. L. Elmore, University of California, Davis.)

tolerance, shows an accumulation of radioactivity in the lysigneous glands, which may account for its intermediate degree of tolerance. A metabolism study indicated that simazine was degraded most rapidly in corn, intermediately in cotton, and least in cucumber.

The rate of degradation of triazine herbicides in higher plants varies greatly with different species. In resistant species they are rapidly degraded, whereas in susceptible species, the herbicides are degraded slowly; thus the rate of degradation appears to be the primary basis of selectivity. This process occurs by hydroxylation, dechlorination, demethoxylation, or demethythiolation, depending on the parent substitution. Conjugation with glutathione and perhaps

Figure 13-4. Autoradiographs showing the distribution of [^{14}C]simazine and/or ^{14}C-labeled degradation products 4 days after root treatment via culture solution in (*A*) susceptible cucumber, (*B*) moderately susceptible cotton, and (*C*) tolerant corn. (Davis et al., 1959.)

other peptides is also an important inactivation reaction in certain species. Dealkylation of the alkyl side chains also occurs.

The mechanism of action of triazine herbicides involves a severe inhibition of the Hill reaction of photosynthesis (Gast, 1958; Moreland et al., 1959). Total herbicidal effect, however, must be more complex than this because the plants do not merely starve to death. It has been postulated that the action involves the interaction of light, chlorophyll, and triazine to produce a secondary phytotoxic substance (Ashton et al., 1963).

Resistance of biotypes of certain weeds usually controlled by a given herbicide has become increasingly recognized (LeBaron and Gressel, 1982). They reported that 30 common annual weed species in 18 genera, including 23 dicots and 7 monocots, previously susceptible to the triazine herbicides have been found to be resistant. An increase of a resistant weed in a field can usually be prevented or controlled by not using the same herbicide or a herbicide of the same class on the same field year after year.

These triazine-resistant biotypes differ from the susceptible biotypes of the same species at an adsorption site on the thylakoid membrane of the chloroplast (Pfister et al., 1979). This adsorption site is a proteinaceous electron acceptor in the photosynthetic electron transport chain between photosystem-I and photosystem-II. When the triazine molecule is adsorbed at this active site in the susceptible biotype photosynthesis is inhibited, but in the resistant biotype the triazine molecule is not adsorbed at this active site and photosynthesis is not inhibited.

Considerable research has been conducted over the past several years on the transfer of the gene responsible for herbicide resistance of a given herbicide from a resistant plant to a susceptible crop plant by conventional breeding and modern technology of genetic engineering, including cell and tissue cultures and transfer of recombinant DNA. Although progress in the laboratory has been noteworthy, field application has been limited. Conventional plant breeding has been successful in transferring triazine resistance from wild mustard to oil seed rape (canola). In Canada, over 50% of the canola acreage is planted to the triazine resistant cultivar "Triton."

SUGGESTED ADDITIONAL READING

Antognini, J., and B. E. Day, 1955, *Proc. 8th Southern Weed Conf.*, pp. 92–98.

Ashton, F. M., 1965, *Proc. 18th Southern Weed Conf.*, pp. 596–602.

Ashton, F. M., and A. S. Crafts, 1981. *Mode of Action of Herbicides*, Wiley, New York.

Ashton, F. M., E. M. Gifford, and T. Bisalputra, 1963, *Bot. Gaz.*, **124**, 329.

Beyer, E. M., Jr., M. J. Duffy, J. V. Hay, and D. D. Schlueter, 1988, Sulfonylureas, in P. C. Kearney and D. D. Kaufman, Eds., *Herbicides*, Vol. 3, pp. 117–189, Dekker, New York.

Corbett, J. R., K. Wright, and A. C. Baillie, 1984, *The Biochemical Mode of Action of Pesticides*, Academic Press, New York.

Crop Protection Chemical Reference, 1989, Wiley, New York (revised annually).

Davis, D. E., H. H. Funderburk, Jr., and N. G. Sansing, 1959, *Weeds*, **7**, 300.

Doig, R. I., G. A. Carraro, and N. D. McKinley, 1983, *Proc. Br. Crop Prot. Conf.*, **3**, 20.

Duke, S. O., Ed., 1985, *Weed Physiology*, Vol II, *Herbicide Physiology*, CRC Press, Boca Raton, FL.

Farm Chemicals Handbook, 1987, Meister, Willough, OH (revised annually).

Fedtke, C., 1982, *Biochemistry and Physiology of Herbicide Action*, Springer-Verlag, New York.

Gast, A., 1958, *Experientia*, **13**, 134.

Gast, A., 1970, *Residue Rev.*, **32**, 11.

Gast, A., E. Knüsli, and H. Gysin, 1955, *Experientia*, **11**, 107.

Gunther, F. A., Ed., 1970, *Residue Rev.*, **32**, pp. 1–413.

Gysin, H., 1960, *Weeds*, **4**, 541.

Gysin, H., 1974, *Weed Sci.*, **22**, 523.

Gysin, H., and E. Knüsli, 1960, *Adv. Pest Control Res.*, **3**, 289.

Hatzios, K. K., and D. Penner, 1982, *Metabolism of Herbicides in Higher Plants*, Burgess, Minneapolis.

Herbicide Handbook, 1989, Weed Science Society of America, Champaign, IL.

Hess, F. D., 1985, Herbicide absorption and translocation and their relationship to plant tolerances and susceptibility, in S. O. Duke, Ed., *Weed Physiology*, Vol. II, *Herbicide Physiology*, pp. 191–214, CRC Press, Boca Raton, FL.

LeBaron, H. M., and J. Gressel, 1982, *Herbicide Resistance in Plants*, Wiley, New York.

Mersie, W. and C. L. Foy, 1986, *J. Agric. Food Chem.*, **34**, 89.

Moreland, D. E., W. A. Genter, J. L. Hilton, and K. L. Hill, 1959, *Plant Physiol.*, **34**, 432.

Nicholls, P., and A. A. Evans, 1985, *Proc. Br. Crop Prot. Conf. Weeds*, p. 333.

Palm, H. L., J. D. Riffleman, and D. A. Allison, 1980, *Proc. Br. Crop Prot. Conf. Weeds*, p. 1.

Pfister, K., S. R. Radosevich, and C. J. Arntzen, 1979, *Plant Physiol.*, **64**, 995.

Priester, T. M., 1988, unpublished, see Beyer et al. (1988).

Ray, T. B., 1980, *Proc. Br. Crop Prot. Conf. Weeds*, p. 7.

Ray, T. B., 1982, *Pestic, Biochem. Physiol.*, **17**, 10.

Ray, T. B., 1984, *Plant Physiol.*, **75**, 827.

Rodgers, E. G., 1968, *Weed Sci.*, **16**, 117.

Rost, T. L., 1984, *J. Plant Growth Regul.*, **3**, 51.

Smith, L. L., Jr., and J. Geronimo, 1984, *Down to Earth*, **40**, 25.

Talbert, R. E., and O. H. Fletchall, 1965, *Weeds*, **13**, 46.

Weber, J. B., 1970, *Residue Rey.*, **32**, 93.

Weed Control Manual and Herbicide Guide, 1988, Meister, Willough, OH (revised annually).

For herbicide use, see the manufacturer's or supplier's label and follow these directions. Also see the Preface.

14 Ureas and Uracils, Other Organic Herbicides, and Inorganic Herbicides

UREAS AND URACILS

Substituted urea and uracil-type herbicides are included in the same section because they have many features in common. The substituted ureas are discussed first and the individual herbicides are arranged alphabetically within each class. Pioneering development of both classes of herbicides was conducted by E. I. duPont de Nemours and Company.

UREAS

$$(H)_2N-\overset{\overset{\displaystyle O}{\|}}{C}-N(H)_2$$

urea

Urea is a common nitrogen fertilizer. Effective herbicides are produced by substituting three of the hydrogen atoms of urea with other chemical groups. Common substitutions include a phenyl, methyl, and/or methoxy group. Other groups occur in certain molecules. The phenyl group may also have chlorine, bromine, or other substitutions. The common name, chemical name, and chemical structure of the urea-type herbicides are given in Table 14-1.

Most urea herbicides are relatively nonselective at high rates of usage, and they are usually applied to the soil. Certain ones, however, have foliar activity, which may be increased by the addition of surfactants to the spray solution. At low rates of application, they may be selective. Selectivity may result from lack of leaching of the herbicide into the absorption zone of the crop plant or inherent plant tolerance.

TABLE 14-1. Common Name, Chemical Name, and Chemical Structure of the Substituted Urea Herbicides

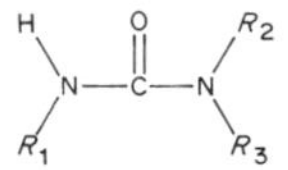

H, O, R_2, N—C—N, R_1, R_3

Common Name	Chemical Name	R_1	R_2	R_3
Diuron[1]	*N'*-(3,4-Dichlorophenyl)-*N,N*-dimethylurea	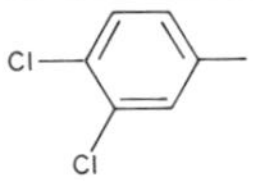 Cl, Cl	CH_3—	CH_3—
Fluometuron	*N,N*-Dimethyl-*N'*-[3-(trifluoromethyl)phenyl]urea	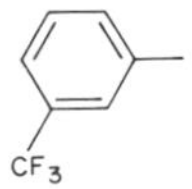 CF_3	CH_3—	CH_3—
Linuron	*N'*-(3,4-Dichlorophenyl)-*N*-methoxy-*N*-methylurea	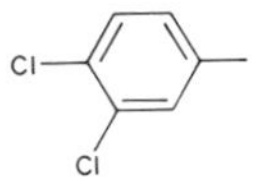 Cl, Cl	CH_3—	CH_3O—
Monuron[1]	*N'*-(4-Chlorophenyl)-*N,N*-dimethylurea	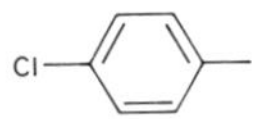 Cl	CH_3—	CH_3—
Siduron	*N*-(Methylcyclohexyl)-*N'*-phenylurea	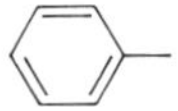	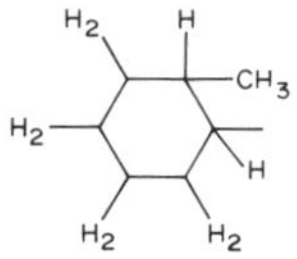 H_2, H, CH_3, H_2, H, H_2, H_2	H
Tebuthiuron	*N*-[5-(1,1-Dimethylethyl)-1,3,4-thiadiazol-2-yl]-*N,N'*-dimethylurea	CH_3—	CH_3	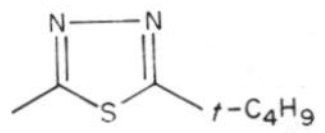 N—N, S, *t*-C_4H_9

[1] Monuron and diuron were previously designated as CMU and DCMU, respectively; the photosynthetic biologists continue to use the older terminology.

Diuron

Diuron is the common name for *N*′-(3,4-dichlorophenyl)-*N*,*N*-dimethylurea. The trade names are Karmex®, Direx®, and Diuron. It is a white crystalline solid with a low water solubility of 42 ppm at 25°C. It is formulated as a liquid concentrate (4 lb/gal) and wettable powder (80%). It has an acute oral LD_{50} of 3.4 g/kg for rats.

Uses Diuron is widely used at relatively low rates as a selective soil-applied herbicide in many crops to control many annual broadleaf and grass weeds. At higher rates, it is used on noncropland to control most annual and perennial weeds as a soil treatment before weed emergence or as a foliar-soil treatment for emerged weeds. Diuron alone has very little foliar activity on most plants; however, the addition of certain surfactants to the spray solution enhances foliar activity. See Table A-0 for specific weeds controlled and Table 14-2 for detailed information on crop.

Fluometuron

Flumeturon is the common name for *N*,*N*-dimethyl-*N*′-[3-(trifluoromethyl)-phenyl]urea. The trade names are Cotoran® and Meturon®. It is formulated as a liquid concentrate (4 lb/gal) and wettable powder (80%). It is also available as a liquid package mix with MSMA that has the trade name of Croak®. Fluometuron is a white crystalline solid with a low water solubility of 90 ppm at 20°C. The acute oral LD_{50} of fluometuron (80% WP) is 1.8 g/kg for rats.

Uses Fluometuron is used to control many annual broadleaf and grass weeds in cotton and sugarcane. In cotton, it is applied as a preplant soil-incorporated, preemergence, and/or postemergence treatment at 1.0 to 2.0 lb/acre depending on soil type. It is also applied preplant soil-incorporated with trifluralin, preemergence with metolachlor, or postemergence with DSMA or MSMA. A sequential preplant soil-incorporated treatment of pendimethalin or trifluralin followed by a preemergence treatment of fluometuron is also used. In sugarcane, it is applied preemergence and/or postemergence at 2.0 to 4.0 lb/acre. Following a preemergence treatment, a postemergence treatment can be applied directed, semidirected, or over-the-top when the sugarcane is 12 to 18 inches tall. Do not apply more than 8.0 lb/year or graze treated fields. See Table A-0 for the specific weeds controlled.

Linuron

Linuron is the common name for *N*′-(3,4-dichlorophenyl)-*N*-methoxy-*N*-methylurea. The trade names are Lorox®, Linex®, and Linuron. It is a white crystalline solid with a low water solubility of 75 ppm at 25°C. It is formulated as a liquid concentrate (4 lb/gal) and dry flowable (50%). A dispersible granule package mix of linuron and chlorimuron known as Gemini® is also available. The acute oral LD_{50} of linuron is approximately 1.5 g/kg for rats.

TABLE 14-2. Diuron Uses

Crop	Rate (lb/acre)	Combinations	Remarks
Alfalfa	0.8–2.5	—	Established, dormant
Bermudagrass, forage	0.8–2.4	—	Pre, post; newly sprigged
Birdsfoot trefoil	1.5–1.6	—	Established, dormant
Artichokes	1.6–3.2	—	Directed, late fall or early winter after last cultivation
Asparagus	0.8–3.2	—	Pre and/or postharvest, established
Barley, winter	1.1–1.6	—	Pre
Cotton	0.8–2.0	—	PP
	0.5–1.6	—	Pre
	0.5–1.6	Trifluralin	PPI (trifluralin) followed by pre (diuron), sequential
	0.4	Surfactant	Post; directed, crop > 6 inches
	0.8–1.2	± Surfactant	Post; directed, crop > 12 inches
	0.4	DSMA + surfactant	Post; directed, crop > 12 inches
Grapes	1.6–4.8	—	Established, late fall or winter
Nuts			
Macadamias	1.6–4.8	± Surfactant	Established; directed, after harvest
Pecans	1.6–3.2	—	Established; directed, spring or early summer
Walnuts	2.4–4.0/ 1.6–2.4	—	Established; directed, initial rate/annual rate
Oats			
Spring	0.8–1.2	—	Pre, post; < 6 weeks of plant
Winter	1.2–1.6	—	Pre
Red clover	1.6	—	Established, dormant
Small fruits[1]	1.6–3.2	—	Established; directed, spring and/or fall
Sorghum	0.1–0.4	Surfactant	Post; directed, crop > 15 inches
Sugarcane	1.6–3.2	—	Pre; after planting or harvest
	1.6–6.4	Surfactant	Post, directed
Tree fruits			
Citrus	1.6–6.4	—	Established, directed
Deciduous[2]	3.0–3.2 or 2.0 + 2.0	—	Established; directed, split Application winter and spring
Deciduous[3]	1.6–4.0	—	Established, directed
Deciduous[4]	0.8–1.6	Terbacil	Established
Wheat, winter	0.8–1.6	—	Pre, post
	0.4–0.8	Bromoxynil	Post
Lawn, seed	1.6–3.2	—	See product label
Legumes,[5] seed	1.2–2.4	—	Dormant to semidormant
Shelterbelts	2.0–4.0	—	Directed under established trees

TABLE 14.2. (*Continued*)

Crop	Rate (lb/acre)	Combinations	Remarks
Flowers			
Bulbs[6]	3.2	—	Pre, field grown
Plumosus fern	2.4	—	Established, 3–5 days after mowing
Noncropland	4.0–16.0/ 16.0–48.0	± Surfactant	Pre, post; annuals/perennials

There are some restrictions on some of the above uses relative to geographic location, soil type, varieties, rainfall, and time of application and harvest; see product labels; all labels do not list all above uses.

[1] Blackberries, blueberries, boysenberries, caneberries, dewberries, gooseberries, loganberries, raspberries.
[2] Apples, pears.
[3] Peaches.
[4] Apples, peaches.
[5] Alfalfa, birdsfoot trefoil.
[6] Narcissus, bulbous iris, tulips.

Uses Linuron is used as a preplant, preemergence, or postemergence treatment to control many annual broadleaf and grass weeds in several crops and noncropland. Foliar applications are most effective when the weeds are young and succulent, temperatures are 70°F or higher, and humidity is high. Control of emerged weeds under drought stress is less effective. See Table A-0 for specific weeds controlled and Table 14-3 for detailed information on the crops.

Siduron

Siduron is the common name for *N*-(methylcyclohexyl)-*N*′-phenylurea. The trade name is Tupersan®. It is a white crystalline solid with a low water solubility of 18 ppm at 25°C. It is formulated as a wettable powder (50%). The acute oral LD_{50} is > 7.5 g/kg for rats.

Uses Siduron is a specialty herbicide used to control annual grass weeds in newly seeded or established grass lawns and turf. It has an unusual degree of selectivity among grass species; most turf grasses are tolerant even when germinating from seed. Most other herbicides used to control annual grass weeds in turf can be safely used only in established not newly seeded turf. Although most turf grasses are tolerant to siduron, bermudagrass and certain bentgrass strains (see product label) are subject to injury by siduron. It is particularly effective for the control of barnyardgrass, foxtails, and both smooth and hairy crabgrass. It does not control annual bluegrass, clovers, or most broadleaf weeds. In newly seeded turf, it is applied at 2 to 6 lb/acre immediately after seeding. In established

TABLE 14-3. Linuron Uses

Crop	Rate (lb/acre)	Combinations	Remarks
Asparagus	0.50–2.00	—	Pre, post; directed-seeded, crowns, established
Carrots	0.50–1.50	—	Pre
	0.75–1.50	—	Post, crop > 3 inches
Celery	0.75–1.50	—	Post; transplants < 8 inches
Cotton	0.50–0.75	Surfactant	Post; directed, crop > 8–15 inches
Corn			
Field, sweet	0.60–1.50	Surfactant	Post; directed, crop > 15 inches
Popcorn	0.33–1.50	Atrazine, alachlor, or propachlor	Pre, tank mix
Parsnips	0.75–1.50	—	Pre
Potatoes	0.50–2.00	± Surfactant	Pre
Sorghum	0.50–1.00	Surfactant	Post; directed, crop > 12–15 inches
	0.30–1.00	Propazine	Pre, tank mix
	0.33–1.50	Propachlor	Pre, tank mix
Soybeans	0.50–2.50	—	Pre
	0.33–1.50	Alachlor, chloramben, metolachlor, or pendimethalin	Pre, tank mix
	0.33–1.25	Oryzalin	Pre, tank mix
	0.16–1.00	Metribuzin + alachlor or metolachlor	Pre, tank mix

	0.45–0.90	Chlorimuron	Pre, package mix (Germini®)
	0.25–2.50	Trifluralin	PPI (trifluralin) followed by pre (linuron), sequential
	0.50–3.00	Paraquat	PP, pre; tank mix
	0.38–1.13	Oryzalin + glyphosate or paraquat	Pre, tank mix
	0.33–1.75	Alachlor or metolachlor + glyphosate or paraquat	PP, pre, tank mix
	0.50	2,4-DB	Post; directed, crop > 12 inches
	0.50–1.00	Surfactant	Post; directed, crop > 12 inches
	0.33–1.50	Propachlor	Pre, seed crop only
Noncropland	1.00–3.00	± Surfactant	Short control of annual weeds

There are some restrictions of some of the above uses relative to geographic location, soil type, temperature, and grazing and forage; see product labels; all labels do not list all above uses.

turf, it is applied at 8 to 12 lb/acre in the spring just before emergence of annual weedy grasses. Irrigate within 1 week after treatment if it has not rained.

Tebuthiuron

Tebuthiuron is the common name for *N*-[5-(1,1-dimethylethyl)-1,3,4-thiadiazol-2yl]-*N*,*N*′-dimethylurea. The trade name is Spike®. It is a colorless crystalline solid with a high water solubility of 2300 ppm. It is formulated as a dry flowable (85%) and pellets (20 and 40%). It has an acute oral LD_{50} of 644 mg/kg for rats.

Uses Tebuthiuron is relatively nonselective and is used to control weeds on noncropland and woody plants on rangeland. In general, lower rates are adequate for woody plant control but higher rates are used for total vegetation control. On rangeland, it is applied at 0.75 to 6 lb/acre depending on the species to be controlled either as a broadcast or individual plant treatment when desirable grass species are dormant. The treated area can be used for grazing when the application rate is not greater than 4 lb/acre. On noncropland, it is applied at 1 to 16 lb/acre depending on the species to be controlled, soil type, climatic conditions, and duration of control desired. It can be applied either as a preemergence or postemergence treatment although it is absorbed more readily by roots than by the foliage.

Soil Interaction of Urea Herbicides

This topic for each of these herbicides is summarized in the *Herbicide Handbook* (1989). As a class, urea-type herbicides are relatively persistent in soils. Under favorable moisture and temperature conditions with little or no leaching, most of them can be expected to remain phytotoxic for 6 months at the lower selective rates and 24 months or more at higher nonselective rates.

Fluometuron is the least persistent of this group with a half-life of about 30 days. Linuron and siduron have half-lives of 2 to 5 months and < 1 years, respectively. At selective rates, diuron has a half-life of < 1 year but at nonselective rates its half-life is > 1 year. Tebuthiuron is very persistent with a half-life of 12 to 15 months in areas receiving 40 to 60 inches annual rainfall and considerably greater in areas of low rainfall.

Principal factors affecting persistence of substituted ureas in the soil are microbial decomposition, leaching, adsorption on soil colloids, and photodecomposition. The latter is important only when the herbicide remains on the soil surface for an extended period of time. Researchers believe volatility and chemical decomposition are of minor importance in the persistence of the urea-type herbicides.

Decomposition by microorganisms is the most important factor in the loss of most of the urea-type herbicides from the soil. Although tebuthiuron has also been shown to be degraded by microorganisms, this does not appear to be the major cause of its loss from soils. Bacteria such as *Pseudomonas*, *Xanthomonas*,

TABLE 14-4. Water Solubility and Adsorption on Soil of Urea and Uracil Herbicides

Compound	Solubility in Water (ppm)	Adsorption on Keyport Silt Loam
Fenuron	3850	0.3
Tebuthiuron	2300	—
Bromacil	815	1.5
Terbacil	710	1.7
Monuron	230	2.6
Fluometuron	90	—
Linuron	75	5.5
Diuron	42	5.2
Siduron	18	2.5
Neburon	5	16.0

[1]Expressed as ppm (active ingredient) present on soil in equilibrium with 1 ppm in soil solution.

From Wolf et al. (1958).

Sarcina, and *Bacillus*, and fungi such as *Penicillium* and *Aspergillus* can use some substituted ureas as a direct source of energy (Hill and McGalen, 1955). Conditions such as moderate moisture and temperature with adequate aeration, favoring such organisms, would also favor decomposition. Therefore, under dry, cold, or very wet soil conditions (poor aeration), the chemicals normally persist for a long time.

The adsorptive forces between the chemical and the soil colloids directly affect the chemical's rate of leaching; its solubility is a less-important factor.

In a study using four urea-type herbicides, leachability was correlated with adsorption and water solubility (Wolf et al., 1958) (see Table 14-4). Fenuron was leached the most, followed in order by monuron, diuron, and neburon. The water solubilities and available adsorption values for Keyport silt loam of the other urea and uracil herbicides are given in Table 14-4. Some of these urea-type herbicides are no longer available in the United States but are included here because they help in understanding the soil interactions of the entire class.

Mode of Action of Urea Herbicides

Phytotoxic symptoms of urea-type herbicides can be seen largely in the leaves (Figure 14-1). They may show merely a slight chlorosis, which develops slowly with low-rate applications of the herbicide, or a water-soaked appearance, becoming necrotic in a few days at higher rates (Geissbühler et al., 1975; Ashton and Crafts, 1981).

Most of the urea herbicides are readily absorbed by roots and rapidly translocated to upper plant parts via the apoplastic system. Applications to the

Figure 14-1. Diuron-induced chlorosis in peaches, usually veinal chlorosis but sometimes interveinal. *Left to right*: Untreated to increasing rates. (C. L. Elmore, University of California, Davis.)

leaves are also translocated apoplastically with little if any translocated from the treated leaf. However, the actual amount absorbed and translocated from roots to shoot varies greatly with various compounds. Furthermore, differences in absorption and translocation between species have been sufficient with different urea herbicides to allow their selective use in certain crops (Geissbühler et al., 1975; Ashton and Crafts, 1981).

Inhibition of photosynthesis is generally acknowledged to be the primary action of the urea-type herbicides. This inhibition is caused by the binding of the urea herbicide to a specific proteinaceous electron acceptor in the electron transport chain between photosystem-I and photosystem-II. This prevents the formation of high-energy compounds (ATP and NADPH) that are required for carbon dioxide fixation and numerous other biochemical reactions. See Moreland and Hilton (1976) for a comprehensive discussion.

However, many workers do not believe that this explains the light-dependent phytotoxic symptoms of the urea herbicides. In other words, plants do not merely starve from lack of photosynthate; rather, it has been postulated that a secondary phytotoxic substance is formed. The nature of this secondary phytotoxic substance has not been determined (Ashton and Crafts, 1981).

URACILS

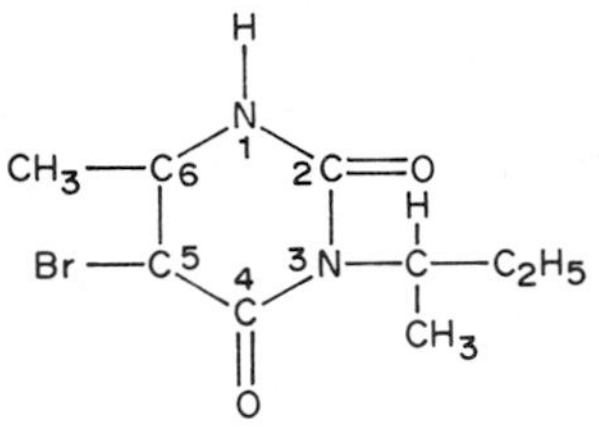

bromacil

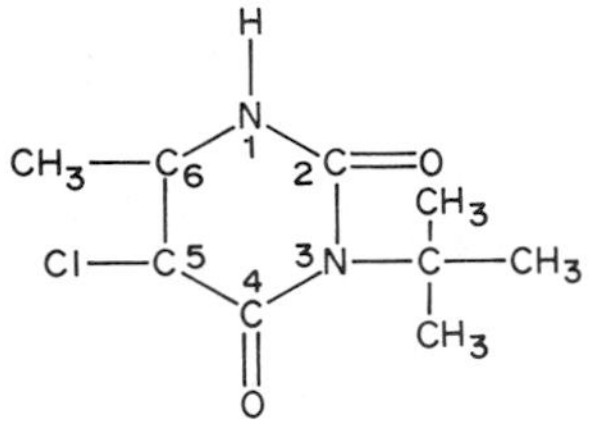

terbacil

Bromacil

Bromacil is the common name for 5-bromo-6-methyl-3-(1-methylpropyl)-2,4-(1*H*,3*H*)pyrimidinedione. The trade names are Hyvar® and Bromax®. Krovar® and Weed Blast® are trade names for a bromacil–diuron package mix. Bromacil is a white crystalline solid with a moderate water solubility of 815 ppm at 25°C. It is formulated as a wettable powder (80%), miscible liquid (2 and 4 lb/gal), and granule (4%). The acute oral LD_{50} is 5.2 g/kg for rats.

Uses Bromacil is used for selective weed control in citrus and pineapple crops and for general vegetation control on noncrop land. Many annual broadleaf and grass weeds are controlled at low rates. Some perennial weeds and brush species are controlled at moderate rates. Johnsongrass and other somewhat-resistant perennial weeds may require high rates. These higher rates are not selective to citrus or pineapple (Figure 14-2).

In established grapefruit, lemons, and oranges bromacil is applied to the orchard floor at 1.6 to 6.4 lb/acre to control annual weeds and certain perennial weeds, e.g., bermudagrass, johnsongrass, nutsedge, and pangolagrass. Lower rates are used for annual weed control and higher rates are required for perennial weed control. Lower rates are used on sand and sandy loam soils and higher rates are used on silt and clay loam soils.

In pineapple, bromacil is applied at 2 to 6 lb/acre before planting material begins to grow. This is followed by a 2 to 4 lb/acre application as a directed interline spray as needed or 2 lb/acre as a broadcast spray after plants are 8 months old, either application prior to difcrentiation. Do not apply more than 16 lb/acre total per plant crop. Other restrictions including geographic location, rainfall, trashmulch in field, and replant also apply; see product label. For ratoon crop, apply 2 to 4 lb/acre broadcast after harvest but before differentiation. Do not apply more than 4 lb/acre total per ratoon crop.

On noncropland areas the application rate of bromacil varies considerably for the various types of weeds. In general low rates (2.4 to 4.8 lb/acre) are used for annuals, intermediate rates (5.6 to 9.6 lb/acre) for easy-to-kill perennials, and high rates (12 to 24 lb/acre) for hard-to-control perennials. Certain woody species are

Figure 14-2. Chlorosis in walnuts induced by bromacil. (C. L. Elmore, University of California, Davis.)

also controlled at the higher rates. See product labels for the rate and special application methods for specific species. The addition of diuron broadens the spectrum of weed species controlled.

Terbacil

Terbacil is the common name for 5-chloro-3-(1,1-dimethylethyl)-6-methyl-2,4-(1*H*,3*H*)-pyrimidinedione. The trade name is Sinbar®. It is a white crystalline solid with a moderate water solubility of 710 ppm at 25°C. It is formulated as a wettable powder (80%). The acute oral LD_{50} is > 5.0 g/kg but < 7.5 g/kg for fasted rats.

Uses Terbacil is used to control many annual grass and broadleaf weeds and some perennial weeds in alfalfa, mints, sugarcane, and certain fruit and nut trees. Lower rates are used for annual weed control and higher rates are required for perennial weed control. Lower rates are used on sand and sandy loam soils and higher rates are used on silt and clay loam soils. It is preferably applied to the soil surface just before or during active weed growth. If dense, the emerged shoots should be removed prior to treatment. Control of perennial grasses may be improved by cultivation prior to treatment. See Table A-0 for specific weeds controlled and Table 14-5 for detailed information on the crops.

TABLE 14-5. Terbacil Uses

Crop	Rate (lb/acre)	Combination	Remarks
Alfalfa	0.4–1.2	—	Established, dormant, fall or spring
Mint[1]	0.8–1.6	—	Pre, spring after last cultivation
	0.8–1.2	—	Post, weeds 1–2 inches
Pecans	1.6–2.4	—	Established
	1.2–1.6	Diuron	Established
Small fruits[2]	1.6–3.2	—	Established
Strawberries	0.4–1.0	—	Established, dormant or after postharvest renovation
Sugarcane	0.8–2.0	—	Pre
Tree fruits			
Citrus	1.6–6.4	—	Established
Deciduous[3]	1.6–3.2	—	Established

There are some limitations on some of the above uses relative to geographic location, soil type, crop variety, stage of crop growth, time of application or harvest, and grazing or feeding forage; see product label.

[1] Peppermint, spearmint.
[2] Blackberries, blueberries, boysenberries, dewberries, loganberries, raspberries, youngberries.
[3] Apples, peaches.

Soil Interaction of Uracil Herbicides

Bromacil and terbacil are adsorbed less on soil colloids than the urea-type herbicides monuron, diuron, or neburon, but more tightly than fenuron (see Table 14-4). Therefore, they are leached more readily than monuron, diuron, or neburon, but less readily than fenuron.

Bromacil and terbacil have a half-life of about 5 to 6 months when applied at 4 lb/acre, but at sterilant rates they persist for more than one season (see Table 6-1). This loss is apparently a result of microbiological degradation. Soil diphtheroids, *Pseudomonas*, and *Pencicillium* species have been shown to be able to degrade bromacil (*Herbicide Handbook*, 1989).

Mode of Action of Uracil Herbicides

The uracil-type herbicides are similar to the urea-type herbicides in their mode of action (Gardiner, 1975; Ashton and Crafts, 1981). They are readily absorbed by roots and translocated apoplastically to the leaves, where they block photosynthesis. These herbicides cause chlorosis and necrosis in leaves, and bromacil has been shown to inhibit root growth. The structure of leaf chloroplasts is grossly altered and cell-wall development of roots is modified by bromacil.

OTHER ORGANIC HERBICIDES

This section covers those organic herbicides that do not have general chemical structures common to the chemical classes previously discussed. Each of these herbicides is in a class of its own and is referred to by its common name rather than a chemical class.

Amitrole

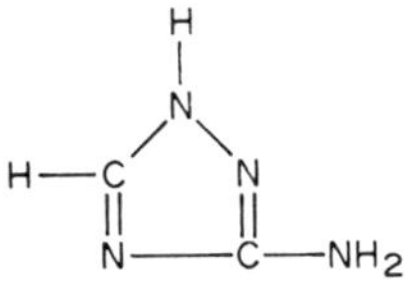

amitrole

Amitrole is the common name for 1*H*-1,3,4-triazol-3-amine. The trade name is Amazol®. Package mixes of amitrole plus ammonium thiocyanate (Amitrol T®) and amitrole plus simazine (Amizine®) are also available. Amitrole is an off-white crystalline solid with a very high water solubility of 280,000 ppm at 25°C. The acute oral LD_{50} of amitrole is 24.6 g/kg for rats. Two-year life time studies with rats indicate that amitrole is a goitrogen, causing enlargement of the thyroid gland.

Uses Amitrole is applied as a foliar spray to control essentially all emerged annual weeds, most emerged perennial weeds, and many woody plants. It is one of the few herbicides that is effective on poison ivy and poison oak. It is used on noncropland and in certain woody-plant nursery crops. Since amitrole is nonselective, use directed sprays in nurseries to avoid contact with stem and foliage of desirable plants. The application rate of amitrole is 1.8 to 9.0 lb/acre. In general, lower rates are used on annual weeds when they are 3 to 4 inches high and higher rates are used on perennial weeds when they are 6 to 10 inches high. Certain perennial weeds such as johnsongrass and bermudagrass may require treatment of regrowth for adequate control. The time of the year and the stage of growth when the initial application is made are critical with certain perennial weeds.

The combination of amitrole and ammonium thiocyanate is often used since the addition of ammonium thiocyanate increases the effectiveness of amitrole on many weedy species. The combination of amitrole and simazine is used since amitrole controls only emerged vegetation and the addition of simazine provides residual control.

Soil Influence Since amitrole is a foliar applied herbicide soil has no effect on its performance. However, it is rapidly degraded in soils and the average persistence is 2 to 4 weeks at recommended rates.

Mode of Action The most striking symptom of amitrole phytotoxicity is the albino appearance of developing leaves and shoots that are not exposed to the spray application. These organs that subsequently develop are white since all visible pigments are absent. Amitrole is absorbed relatively slowly but it is translocated very well in both the symplast and apoplast. It is one of the most readily translocated herbicides.

Amitrole reacts with the amino acid serine to form the conjugate β-(3-amino-1,2,4-triazol-1-yl)-α-alanine (Carter, 1975; Smith and Chang, 1973). This conjugate is considered to be nonphytotoxic and the terminal residue of amitrole metabolism in higher plants (Carter 1965, 1975). Amitrole has also been reported to form a complex with tannin (Kroller, 1966).

Amitrole blocks the development of chloroplast ribosomes (Bartels et al., 1967) and inhibits the chloroplast DNA formation (Bartels and Hyde, 1970). It also inhibits development of plastids and blocks the formation of grana. These effects explain the albinism observed following amitrole treatments. Amitrole also inhibits imidazoleglycerol phosphate dehydratase, an enzyme involved in histidine synthesis (Wiater et al., 1971a,b). However, this does not appear to be a primary site of action of amitrole in higher plants (Duke, 1985).

Bensulide

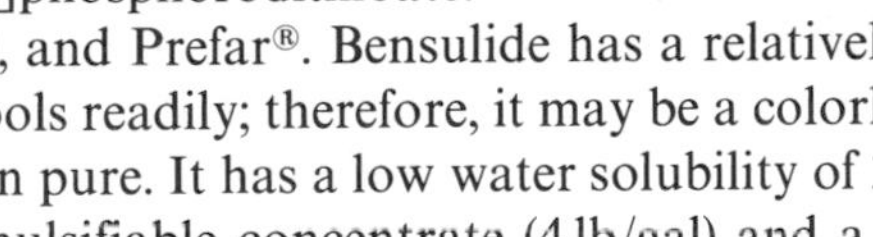

$$C_6H_5{-}SO_2NHCH_2CH_2S\overset{O}{\overset{\|}{P}}(OCH(CH_3)_2)_2$$

bensulide

Bensulide is the common name for *O,O*-bis(1-methylethyl) *S*-[2-[(phenylsulfonyl)amino]ethyl]phosphorodithioate. The trade names are Betamec®, Betasan®, and Prefar®. Bensulide has a relatively low melting point (34.4°C) and supercools readily; therefore, it may be a colorless liquid or a white crystalline solid when pure. It has a low water solubility of 25 ppm at 20°C. It is formulated as an emulsifiable concentrate (4 lb/gal) and a granule (3.6%). The acute oral LD_{50} is 770 mg/kg for rats.

Uses Bensulide is a selective soil-applied herbicide used to control many annual grass and certain broadleaf weeds in established grass and dichondra lawns and several crops. In general, preplant applications require soil incorporation and preemergence applications are used only on crops that are to be "irrigated up." There are exceptions to this generalization for established lawns and grass seed crops. See Table A-0 for the specific weeds controlled and Table 14-6 for detailed information on the crops.

TABLE 14-6. Bensulide Uses

Crop	Rate (lb/acre)	Remarks[1]
Carrots	5.0–6.0	PPI, pre
Cole crops[2]	5.0–6.0	PPI, pre
Cotton	2.0–6.0	PPI, pre
Cucurbits[3]	5.0–6.0	PPI, pre
Lettuce	5.0–6.0	PPI, pre
Onions[4]	5.0–6.0	PPI, pre; bulb
Peppers[5]	5.0–6.0	PPI, pre
Tomatoes[6]	4.0–5.0	PPI, pre
Grass seed[7]	8.0–10.0	Established
Lawns and turf[8]	7.5–10.0	Established

There are some restrictions on some of the above uses relative to geographic location, soil type, planting of nontolerant crops, and feeding of plant products.

[1] In general, preplant applications require soil incorporation and preemergence applications are used only on crops that are to be "irrigated up."
[2] Broccoli, brussels sprouts, cabbage, cauliflower.
[3] Cantaloupes, cucumbers, crenshaw melons, muskmelons, persian melons, pumpkins, squash (summer and winter), and watermelons. Bensulide can be used as a tank mix with naptalam in cantaloupes, cucumbers, muskmelons, and watermelons.
[4] Bulb.
[5] Seeded bell peppers, chili peppers.
[6] Direct seeded and transplants.
[7] Bentgrass and bluegrass only.
[8] All turfgrasses and dichondra.

Soil Influence Bensulide is tightly bound to organic matter in soil and is inactive in soils containing high amounts of organic matter. It is not subject to significant leaching in any type of soil. It is slowly degraded by soil microorganisms and at 70 to 80°F it has a half-life of about 4 months in moist loam soil and about 6 months in moist loamy sand soil. Treated areas should not be seeded to nonregistered crops within 18 months (12 months for soybeans) or turf species within 4 months.

Mode of Action

Bensulide inhibits the growth of roots and partially inhibits cell division (Cutter et al., 1968). It is adsorbed on root surfaces and a small amount is absorbed by the root. However, little, if any, is translocated upward to the leaves. It appears to be degraded by higher plants.

Bentazon

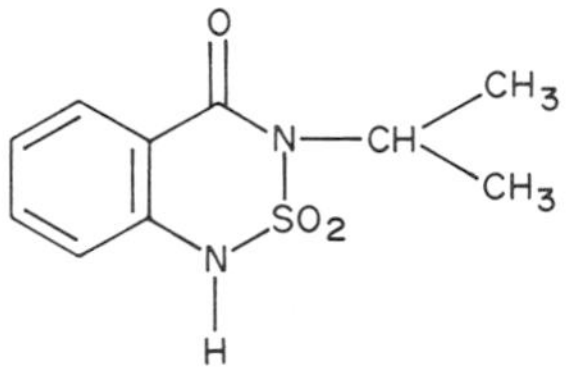

bentazon

Bentazon is the common name for 3-(1-methylethyl)-(1*H*)-2,1,3-benzothiadiazin-4-(3*H*)-one 2,2-dioxide. The trade name is Basagran®. It is a white crystalline solid with a moderate water solubility of 500 ppm at 20°C. It is formulated as an emulsifiable concentrate (4 lb/gal). The acute oral LD_{50} is 1.1 g/kg for rats.

Uses Bentazon is a selective postemergence herbicide used to control a number of annual and perennial broadleaf weeds and sedge weeds in most grass and many large-seeded legume crops. See Table A-0 for the specific weeds controlled and Table 14-7 for detailed information on the crops.

Soil Influence Bentazon is not adsorbed by soil particles but is rapidly incorporated into soil organic matter by microorganisms (Abernathy and Wax, 1973; *Herbicide Handbook*, 1989). Therefore, it does not leach below the plow layer. Persistence of bentazon varies with soil type but reaches undetectable levels within 6 weeks.

Mode of Action Bentazon is considered to be a contact herbicide but the foliar symptoms develop relatively slowly. The initial symptoms are a bronze cast to the leaves followed by necrosis. Modification of the differentiation of etioplasts into chloroplasts has also been reported (Meier and Lichtenthaler, 1981). It is absorbed moderately well by leaves but rain shortly after application may decrease its effectiveness (Doran and Andersen, 1975). Translocation from the leaves is minimal but following root absorption it is rapidly translocated to the shoots via the apoplast. (Mahoney and Penner, 1975; *Herbicide Handbook*, 1989).

The metabolism of bentazon has been studied by several investigators and this research is reviewed in the *Herbicide Handbook* (1989), Hatzios and Penner (1982), and Duke (1985). Bentazon is rapidly metabolized in tolerant species but slowly in susceptible species. This appears to be the basis of selectivity. It sequentially undergoes hydroxylation, conjugation, and incorporation into plant components.

Bentazon inhibits photosynthesis by interfering with photosynthetic electron transport.

TABLE 14-7. Bentazon Uses

Crop	Rate (lb/acre)	Combinations	Remarks[1]
Beans[2]	0.75–1.00	—	Post, early
Corn[3]	0.75–1.00	—	Post, early
	0.50–0.75	Atrazine	Post, early, tank mix
Mint[4]	1.00–2.00	—	Post, early
Peanuts	0.50–1.00	—	Post, early
	0.50–1.00	Acifluorfen	Post, early; tank mix
	0.75–1.00	2,4-DB	Post; tank mix
Peas	0.75–1.00	—	Post, early
Rice	0.75–1.00	—	Post
	0.75	Propanil	Post; tank mix
Sorghum	0.75–1.00	—	Post, early
	0.50–0.75	Atrazine	Post, early; tank mix
Soybeans	0.50–1.00	—	Post, early
	0.50–1.00	Acifluorfen	Post, early; tank mix
	0.75–1.00	2,4-DB	Post, early; tank mix
	0.50–0.75	Mefluidide	Post; tank mix
	0.50–1.00	Sethoxydim	Post; tank mix or sequential
	0.50–1.00	Acifluorfen plus sethoxydim	Post, early; tank mix
	0.50–1.00	Bentazon plus aclfluorfen (tank mix); sethoxydim (sequential)	Post
Lawns and turf	1.00–2.00	—	Established, yellow nutsedge control

There are certain limitations on some of the above uses relative to geographic location, stage of crop and/or weed growth, days to harvest, and grazing or use of forage; see product label.

[1] Oil concentrate or urea ammonium nitrate (UAN) can be added to the spray solution for the control of certain weeds in some of the above crops; see product label.
[2] Great northern, green, kidney, lima, navy, pink, pinto, red, snap, white.
[3] Field, silage, seed corn, sweet corn, popcorn.
[4] Peppermint, spearmint.

Clomazone

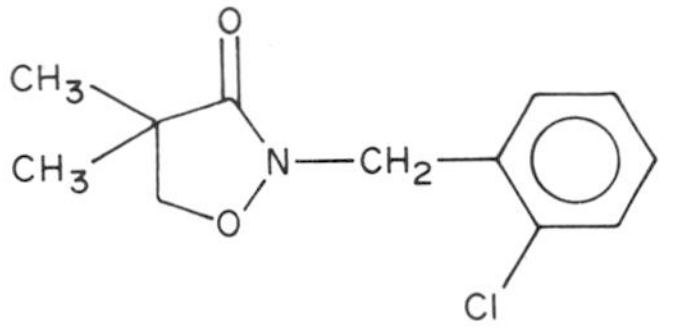

clomazone

Clomazone is the common name for 2-[(2-chlorophenyl)methyl]-4,4-dimethyl-3-isoxazolidinone. The trade name is Command®. It is formulated as an emulsifiable concentrate (4 lb/gal). Commence® is a package mix of clomazone plus trifluralin. Clomazone is a light brown viscous liquid with a high water solubility of 1100 ppm. The acute oral LD_{50} of technical clomazone is 2.1 and 1.4 g/kg for male and female rats, respectively.

Uses Clomazone is used as a preplant soil-incorporated (PPI) herbicide to control annual grass and broadleaf weeds in soybeans. See Table A-0 for specific weeds controlled. The use rate is 0.75 to 1.0 or 1.0 to 1.25 lb/acre in the northern or southern United States, respectively. These rates may vary with soil type and time of application relative to planting time. Clomazone can also be used as a tank mix with alachlor, ethalfluralin, imazaquin, metolachlor, metribuzin, pendimethalin, trifluralin, or metribuzin plus chlorimuron for a PPI treatment to broaden the spectrum of weeds controlled. Sequential treatments of clomazone and aciflurofen or imazaquin are also used. Commence®, the package mix of clomazone plus trifluralin, is registered as a PPI treatment with metribuzin or imazaquin or as a sequential treatment with aciflurofen or imazaquin for use in soybeans. It is also registered for use in pumpkins and peas for processing.

Soil Influence Clomazone has a low mobility in sandy loam, silt loam, and clay loam soils and an intermediate mobility in fine sand soils. It is subject to volatility losses and vapor drift when applied to the surface of moist or wet soils. Its degradation is more rapid in sandy loam soils than in silt loam and clay loam soils. Field studies indicated that its half-life ranges from 15 to 45 days depending on soil type. Degradation appears to proceed via binding to the soil matrix and mineralization to carbon dioxide.

Mode of Action Susceptible species emerge from clomazone-treated soil but are devoid of pigmentation and death occurs in a short period of time. It appears that clomazone inhibits the biosynthesis of both chlorophyll and carotenoids in susceptible species. Clomazone is absorbed by both roots and shoots. The translocation patterns indicate that movement in the plant is almost exclusively in the apoplast. Soybean tolerance appears to be primarily due to differential metabolism of clomazone relative to the susceptible species.

Dazomet

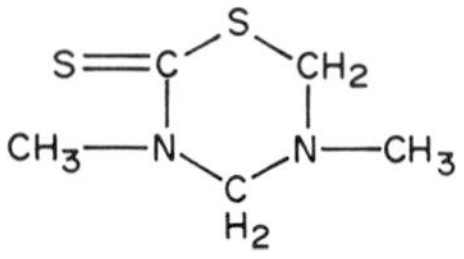

dazomet

Dazomet is the common name for tetrahydro-3,5-dimethyl-2*H*-1,3,5-thiadiazine-2-thione. The trade name is Basamid® Granular. It has also been marketed as Mylone® and Dazomet. It is a white crystalline solid with a high water solubility of 2000 ppm. The acute oral LD_{50} is 650 mg/kg for male albino mice.

Uses Dazomet is a preplant soil fumigant used to control annual and perennial weeds, nematodes, soil fungi, and certain soil insects. It is used on seedbeds of ornamentals, tobacco, and lawns and turf. Seedbeds should be well prepared and have adequate moisture for good plant growth. Plant of the crop should be delayed until dazomet and its toxic degradation products have disappeared, usually 10 to 30 days. The use rates range from 6.5 to 19 oz/100 ft^2 for ornamental seedbeds, 8 to 10 oz/100 ft^2 for lawn and turf seedbeds, and 13 oz/100 ft^2 for tobacco seedbeds.

Soil Influence Dazomet undergoes chemical degradation into methyliso-thiocyanate, formaldehyde, hydrogen sulfide, and monomethylamine in moist-warm soil. Soil moisture is essential for its biological activity. Approved respirator, gloves, and protective clothing are recommended during handling and application.

DCPA

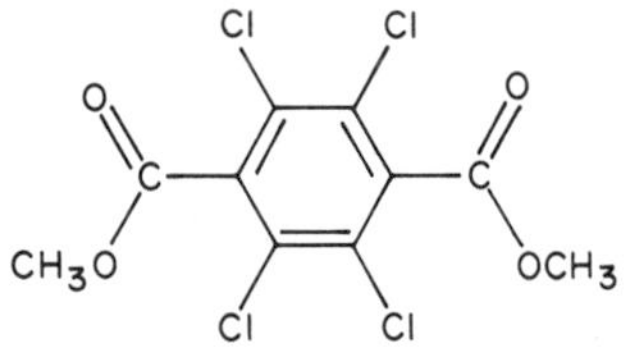

DCPA

DCPA is the common designation for dimethyl 2,3,5,6-tetrachloro-1,4-benzenedicarboxylate. The trade name is Dacthal®. It is a white crystalline solid with an extremely low water solubility of 0.5 ppm at 25°C. It is formulated as a wettable powder (75%). The acute oral LD_{50} of DCPA is > 3.0 g/kg for male albino rats.

Uses DCPA is applied as a preemergence or preplant soil-incorporated treatment to control most annual grasses and many broadleaf weeds in many crops. It is also used in grass turf, nursery stock, and established ornamentals. See Table A-0 for the specific weeds controlled and Table 14-8 for detailed information on the crops.

Soil Influence DCPA is adsorbed to organic matter in the soil and is not subject to significant leaching. Degradation by microorganisms is the primary factor involved in its disappearance from the soil. The average half-life is about 100 days and is dependent on microbial population, temperature, and soil moisture. This relatively long period of persistence limits the crops that can be planted in treated

TABLE 14-8. DCPA Uses[1]

Crop	Remarks
Beans	PPI, pre; dry, field, snap, mung, soybeans, blackeyed peas
Cole crops	Pre, post (transplants); broccoli, brussels sprout, cabbage, cauliflower
Cotton	Pre, post (at layby, not after first bloom)
Cucurbits	Post (4–5 true leaves); cucumbers, melons (cantaloupe, honeydew, watermelons), squash
Eggplant	Post (direct seeded, 4–6 inches; transplants, 4–6 weeks after transplanting)
Greens	PPI, pre; collards, kale, mustard (greens), turnip (greens, roots)
Horseradish	Pre
Onions, garlic	Pre (direct seeded), post (transplants and/or at layby)
Peppers	Post (direct seeded, 4–6 inches; transplants, 4–6 weeks after transplanting)
Potatoes	Pre (at planting after drug-off), post (at layby)
Radishes	Pre or post (up to 3-leaf stage)
Strawberries	PPI or at transplanting (new plantings); post (established plantings)
Sweet potatoes, yams	At transplanting up to 6 weeks after transplanting
Tomatoes	Post (direct seeded, 4–6 inches; transplants, 4–6 weeks after transplanting
Lawns and turf	Post (established)
	Post (ncwly sprigged or seeded), after greening of sprouted grass, 1–2 inches
Ornamentals	Immediately after lining out of stock or to established stock after cultivation

There are some limitations on some of the above uses relative to soil type, grazing and use of forage, days to harvest, and ornamental species; see product label.

[1] Rate of application 4.5–10.5 lb/acre for all crops except ornamentals, strawberries, and lawns and turf. Application rates: ornamentals (10.5–12.0 lb/acre), strawberries [new plantings (9.0 lb/acre), established plantings (6.0–9.0 lb/acre)], lawns and turf [established (10.5–15.0 lb/acre), newly sprigged or seeded (10.5 lb/acre)].

soil within 8 months following the application, including some of the crops listed on the label.

Mode of Action DCPA appears to act as a herbicide mainly by inhibiting the growth of germinating seeds. It interferes with cell division (Bingham, 1968; Nishimoto and Warren, 1971; Anderson and Shaybany, 1972; Chang and Smith, 1972). It also inhibits the growth of shoots and roots, rooting from nodes of bermudagrass, and abnormal swelling of hypocotyls; see Ashton and Crafts (1981) for references and a discussion of these effects.

DCPA is absorbed by roots, coleoptiles of grass seedlings, and hypocotyls of broadleaf species (Nishimoto and Warren, 1971). However, little translocation occurs. It is not considered to be metabolized by higher plants (*Herbicide Handbook*, 1989).

Endothall

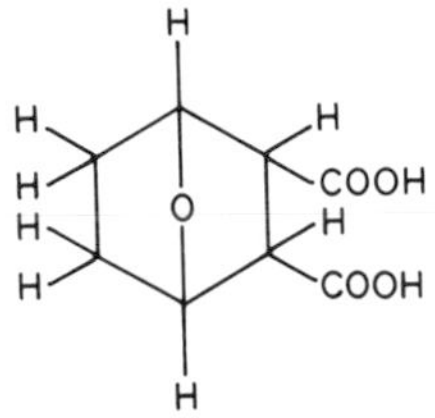

endothall

Endothall is the common name for 7-oxabicyclo[2.2.1]heptane-2,3-dicarboxylic acid. There are several trade names. It is a white crystalline solid with a water solubility of about 10%, formulated as water-soluble liquids and in granular forms. Various endothall salts are available, such as disodium, dipotassium, or amine. The acute oral LD_{50} of the technical acid ranges from 38 to 51 mg/kg for rats.

Uses Endothall is used to control certain annual grass and broadleaf weeds in sugar beets and established lawns and turf. It is also used to control aquatic weeds in still and moving water and in rice. In cotton, potatoes, and legumes for seed production, it is utilized as a preharvest crop desiccant or harvest aid.

In sugar beets, endothall is applied at 3.0 to 6.6 lb/acre to the soil as a preemergence or as a shallow (< 1.5 inches) preplant soil-incorporated treatment. Sugar beets can also be treated postemergence at 0.75 to 1.5 lb/acre. To broaden the spectrum of weeds controlled from a postemergence application of endothall, dalapon, desmedipham, phenmedipham, or chloridazon can be added to the spray solution. However, the rate of endothall in these tank mixes is 0.75 to 1.125 lb/acre for the first three herbicides and 0.5 lb/acre for the latter herbicide. The sugar beets should be in the four- to six-leaf stage at the time of application for these combination treatments.

In established lawns and turf, endothall is applied at 0.171 lb/acre when the temperature and soil moisture favors good growth. For aquatic weed control in rice, endothall is used at 2.5 lb/acre 25 to 60 days after sowing.

In still water (lakes, ponds), aquatic weeds are controlled with an application of 0.05 to 2.5 ppm of liquid or granular endothall. A combination of the dipotassium salt of endothall (0.74 ppm, endothalic acid) and mixed copper-ethanolamine complexes (0.11 ppm Cu) is also used in still water for aquatic weed control. In moving water (irrigation, drainage, bayous), aquatic weeds are controlled with a 1 to 5 ppm application of the dipotassium salt of endothall.

Mode of Action Endothall is absorbed readily by leaves and roots. It is translocated to a limited extent from roots to shoots of plants via the xylem, but it is not phloem mobile and is thus not translocated from leaves to other plant parts. Its action appears to be contact in nature, causing rapid desiccation to germinating seedlings, browning of the foliage, or both.

Endothall is subject to considerable leaching in soils. It is rapidly degraded in both soil and water.

Ethofumesate

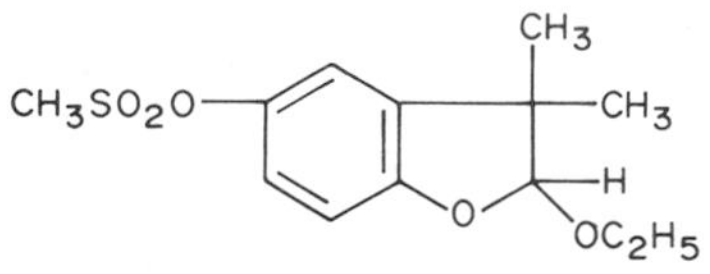

CH_3SO_2O — CH_3, CH_3, H, OC_2H_5, O

ethofumesate

Ethofumesate is the common name for (±)-2-ethoxy-2,3-dihydro-3,3-dimethyl-5-benzofuranyl methanesulfonate. The trade names are Nortron® and Prograss®. It is a white crystalline solid with a moderate water solubility of 110 ppm. It is formulated as an emulsifiable concentrate (1.5 lb/gal). The acute oral LD_{50} of technical ethofumesate is 6.4 g/kg.

Uses Ethofumesate is a selective herbicide used in sugar beets, established lawns and turf, and established grass seed crops to control annual grass and broadleaf weeds. It is primarily a soil-applied herbicide but appears to have some postemergence activity on young weeds. Postemergence activity may be increased when used in combination with certain other herbicides in sugar beets. See Table A-0 for specific weeds controlled and Table 14-9 for detailed information on the crops.

Soil Influence Ethofumesate appears to be adsorbed to organic matter in soils. Field studies have demonstrated that it is not readily leached below 6 inches and laboratory studies have shown that it does not leach in soils having an organic matter content greater than 1% (*Herbicide Handbook*, 1989). The effective

TABLE 14-9. Ethofumesate Uses

Crop	Rate (lb/acre)	Combinations	Remarks
Sugar beets	1.12–3.75	—	PPI, pre
	1.50–2.62	Chloridazon	PP, pre
	1.50–2.50	Diethalyl	PPI, pre
	1.50–2.00	Chloridazon plus diethalyl	Pre
	1.50–3.00	TCA	PP, pre
	3.00	EPTC	PPI, sequential; EPTC (fall), ethofumesate (spring)
	1.12–1.50	Desmedipham or phenmedipham	Post
Lawns and turf	1.00–2.00	—	Newly established or mature perennial ryegrass, dormant bermudagrass
Grass seed production	0.75–1.87	—	Established

There are some restrictions on some of the above uses relative to geographic location, soil type, and grazing and use of forage; see product labels.

application rate varies with soil type. Ethofumesate is biologically degraded in soils. The half-life ranges from less than 5 weeks under warm-moist conditions to more than 14 weeks under cold-dry conditions. The label states that crops other than sugar beets or ryegrass should not be planted within 12 months following application.

Mode of Action Ethofumesate is absorbed by the emerging shoots and roots of most plants (*Herbicide Handbook*, 1989). Foliar absorption is reduced as leaves mature and their cuticles develop. A preemergence application of this herbicide decreased epicuticular wax formation on sugar beet leaves (Duncan et al., 1982). A similar application of ethofumesate plus TCA increased ethofumesate foliar absorption. Ethofumesate is translocated to the foliage following emerging shoot or root absorption but is not translocated from treated leaves. This suggests that it is translocated in the apoplast. It is also translocated more in at least some susceptible species than in tolerant sugar beets (Duncan et al., 1982).

The tolerant species sugar beets and ryegrass metabolize ethofumesate into a major and a minor metabolite that subsequently forms conjugates (*Herbicide Handbook*, 1989). These data suggest that selectivity involves both translocation and metabolism of the herbicide. The crop protectant CGA-43089 (cyometrinil) has been reported to protect sorghum from ethofumesate injury (Leek and Penner, 1980).

Fluridone

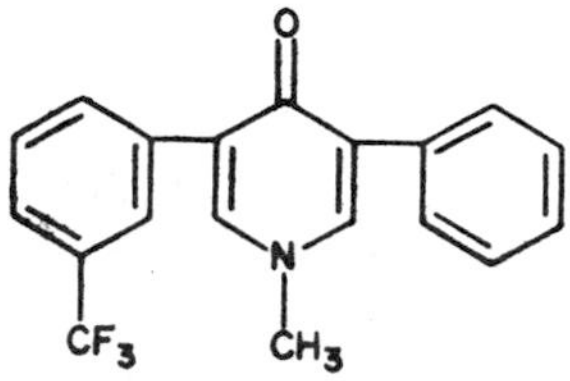

$$\text{fluridone}$$

Fluridone is the common name for 1-methyl-3-phenyl-5-[3-(trifluoromethyl)-phenyl-4(1*H*)-pyridinone. The trade name is Sonar®. It is a white crystalline solid with a water solubility of 12 ppm. It is available as an aqueous suspension (4 lb/gal) and two 5% pellet formulations. The acute oral LD_{50} of technical fluridone is > 10 g/kg for rats.

Uses Fluridone is used for the control of common duckweed, many emersed and submersed vascular aquatic weeds, and some shoreline grasses. It can be used in freshwater ponds, lakes, reservoirs, rivers, and drainage and irrigation canals. Common duckweed is controlled only with a surface application of aqueous suspension formulation. Most other aquatic weeds are controlled or partially controlled by any of the formulations; see product labels for species susceptibility. For best results, apply fluridone prior to initiation of weed growth or when weeds begin actively growing.

An optimum concentration of fluridone needs to be in contact with the weeds as long as possible. Therefore, rapid water movement or any condition that results in rapid dilution of fluridone will reduce its effectiveness. The application rate depends on the type of water (ponds, canals, etc.) treated and the water depth; see product labels for details. Irrigation with fluridone-treated water may result in injury to the irrigated vegetation. This risk is minimized by not using the treated water for irrigation for specified periods of time depending on type of water and type of plants irrigated. An environmental assessment providing information on the safety of this product to humans, wildlife, and the environment is available from Elanco.

Soil and Water Influence Fluridone is strongly adsorbed to organic matter in soil and leaches slowly (*Herbicide Handbook*, 1989). Microorganisms appear to be the major factor responsible for its slow degradation in soils. There is little relationship between rainfall, soil texture, or cultural practices and its persistence in terrestrial soils. It appears to persist over 12 months in terrestrial soils.

In contrast to the long persistence of fluridone in soils, it is degraded relatively rapidly in treated water. Under field conditions, its half-life in water ranges from 5 to 60 days with an average of 20 days.

Mode of Action Phytotoxic symptoms of fluridone include reduction in the rate of growth, a pinkish to albescent color of developing buds and leaves, and finally necrosis. In terrestrial species, it is absorbed by roots and emerging shoots of seedlings; however, there are quantitative differences in the rate of absorption among species (Albritton and Parka, 1978; Rafii and Ashton, 1979). In aquatic species, foliar uptake also appears to be important. It is mainly translocated in the apoplast, but the rate of transport also varies among species (Bernard et al., 1978). The tolerance of cotton to fluridone has been reported to be related to site of uptake and reduced translocation and effect on metabolic processes (Albritton and Parka, 1978; Bernard et al., 1978; Rafii and Ashton, 1979).

Inhibition of carotenoid synthesis appears to be a major biochemical response of fluridone action (Bartels and Watson, 1978). However, it has also been reported to inhibit photosynthesis and RNA and protein synthesis (Rafii and Ashton, 1979). No significant metabolism of fluridone has been detected in higher plants.

Methazole

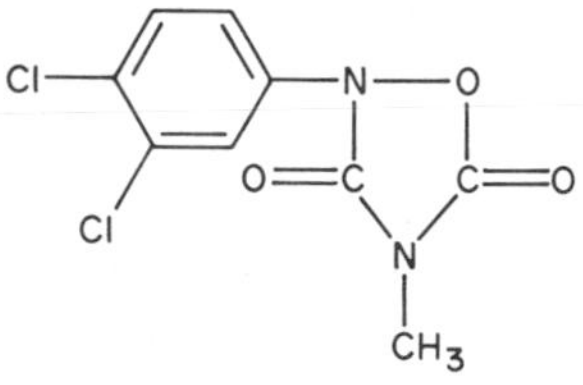

methazole

Methazole is the common name for 2-(3,4-dichlorophenyl)-4-methyl-1,2,4-oxadiazolidine-3,5-dione. The trade name is Probe®. It is a tan solid with a water solubility of 1.5 ppm at 25°C. It is formulated as a wettable powder (75%). The acute oral LD_{50} of technical methazole is 2.5 g/kg for albino rats.

Uses Methazole is used as preemergence or directed postemergence application in cotton to control certain annual broadleaf weeds; see Table A-0 for the specific weeds controlled. Preemergence application rates range from 1.0 to 2.5 lb/acre depending on soil type. It is not recommended for use in soils containing more than 4% organic matter. Shallow incorporation can be used if rainfall is not anticipated. Directed postemergence applications may be made when the cotton is 3 inches or greater in height, up to layby. The use rates are 0.50 to 1.5 lb/acre, but not greater than 1.0 lb/acre on cotton less than 6 inches in height. Repeat postemergence applications may be made, but do not exceed a total of 3.0 lb/acre per cropping season.

Methazole can also be used as a tank mix in combination with DSMA, MSMA, or cyanazine as a directed postemergence application. It can also be used as a sequential soil treatment after the soil has been previously treated with one or more of the following herbicides: diuron, fluometuron, norflurazon, pendimethalin, prometryn, trifluralin. See product labels for detailed application and limitation information on methazole alone and on all of these combinations.

Soil Influence Methazole is adsorbed on soil colloids, especially organic matter. Adsorption is much greater on organic soils than on medium loam or sandy soil. Leaching is minimal in most soils. It is subject to degradation by soil microorganisms and has a residual half-life of less than 30 days.

Mode of Action Methazole delays emergence of seedlings, inhibits shoot and root growth, and induces foliar chlorosis (Keeley et al., 1972). It is readily absorbed by leaves and roots. It is rapidly translocated to mature leaves following root uptake but translocation from treated leaves is minimal (Jones and Foy, 1982; Wills, 1976). This translocation pattern suggests apoplastic transport.

Methazole is metabolized into 1-(3,4-dichlorophenyl)-3-methylurea (DCPMU) and this compound is subsequently metabolized into 1-(3,4-dichlorophenyl)urea (DCPU) (Jones and Foy, 1972; Dorough et al., 1973; Dorough, 1974). Although other metabolites and conjugates are formed, they will not be considered here, see Ashton and Crafts (1981). These reactions are considered to be critical to the biochemical mechanism of action methazole and the tolerance of cotton to this herbicide. DCPMU is a strong inhibitor of the Hill reaction (photosynthesis) (Good, 1961), whereas methazole and DCPU have little effect on this process. It is probable that DCPMU is responsible for the herbicidal activity of methazole and cotton is tolerant because DCPMU is rapidly converted to DCPU in this species (Ashton and Crafts, 1981; Corbett et al., 1984).

Oxadiazon

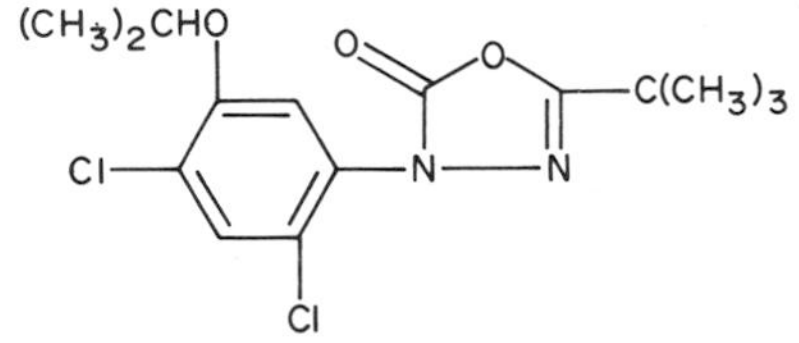

oxadiazon

Oxadiazon is the common name for 3-[2,4-dichloro-5-(1-methylethoxy)phenyl]-5-(1,1-dimethylethyl)-1,3,4-oxadiazol-2-(3*H*)-one. The trade name is Ronstar®. It is a white crystalline solid with a moderate water solubility of about 700 ppm at 20°C. It is formulated as a wettable powder (50%) and a granule (5%). The acute oral $LD5_{50}$ of technical oxadiazon is > 8.0 g/kg for rats.

Uses Oxadiazon a preemergence herbicide used to control certain annual grass and broadleaf weeds in lawns, turf, and ornamental nurseries. In lawns and turf, the wettable powder formulation can be applied to dormant-established bermudagrass, St. Augustinegrass, and zoysia grass at 2.0 to 3.0 lb/acre. The granular formulation can be applied to established bermudagrass, perennial bluegrass, perennial ryegrass, St. Augustinegrass, tall fescue, and zoysia grass at 2.0 to 4.0 lb/acre. In ornamental nurseries, the wettable powder formulation is used in field grown culture as a directed spray or an over-the-top spray for certain species, see label, at 2.0 or 4.0 lb/acre. Do not apply this formulation during or within 4 weeks after bud break. The granular formulation is used at these same rates to dormant or actively growing plants in newly transplanted or established container culture or field grown ornamentals. The rate used in ornamentals depends on the weed species expected; see label.

Soil Influence Oxadiazon is strongly adsorbed on soil colloids and humus and very little leaching occurs. There appears to little information its degradation and half-life in soil.

Mode of Action Oxadiazon appears to be adsorbed by young seedlings as they emerge through the soil and to some extent by the leaves of young seedlings. However, it has been reported that it is not actively absorbed by foliage (*Herbicide Handbook*, 1989). In very susceptible species, it may be translocated to the roots when applied postemergence (*Herbicide Handbook*, 1989). It appears to function by contact action on young shoots as they emerge through the soil and in the leaves from postemergence treatments (*Herbicide Handbook*, 1989).

Pyridate

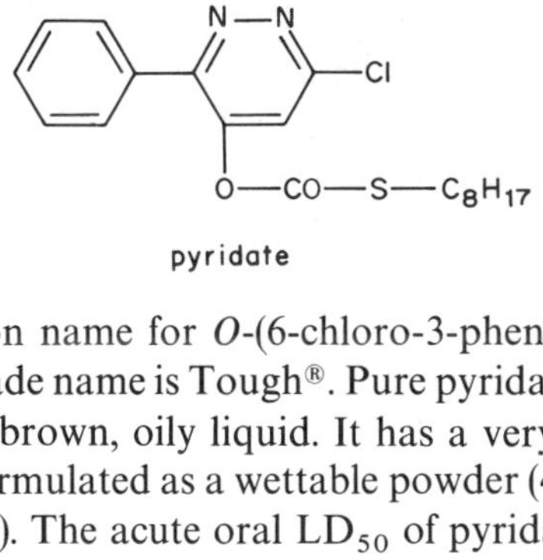

pyridate

Pyridate is the common name for *O*-(6-chloro-3-phenyl-4-pyridazinyl)-*S*-octyl carbonothioate. The trade name is Tough®. Pure pyridate is a colorless solid and technical pyridate is a brown, oily liquid. It has a very low water solubility of 1.5 ppm at 20°C. It is formulated as a wettable powder (45%) and an emulsifiable concentrate (3.75 lb/gal). The acute oral LD_{50} of pyridate has been reported as two values: 4.6 g/kg for rat, and 2.0 and 2.4 g/kg for male and female rats, respectively.

Uses Registration of pyridate is still pending, but is expected soon. Experimental Use Permits have been issued for its use as a postemergence herbicide for the control of certain annual weeds in peanuts, corn, and winter wheat. Other tolerant crops appear to be alfalfa, other cereals, cabbage, rice, poppy, onion, leek, chickpeas, oil seed rape, asparagus, and forest nurseries. The rates proposed range from 0.9 to 4.0 lb/acre depending on the crop and weeds. There appears to be little risk of carryover problems from soil residues.

Mode of Action Plants effected by pyridate show leaf scorch and ultimately desiccation. It is rapidly absorbed by leaves and rainfall shortly after application has no adverse effect on its performance. Translocation appears to be limited since metabolic activity is restricted to the treated areas of the plant; there is no systemic action. Pyridate is rapidly converted into a major metabolite in higher plants and this metabolite is further decomposed rapidly. Depending on climatic conditions, pyridate persistence in plants ranges from a few hours to a few days. Its biochemical mechanism of action involves an irreversible blocking of photosynthesis.

Tridiphane

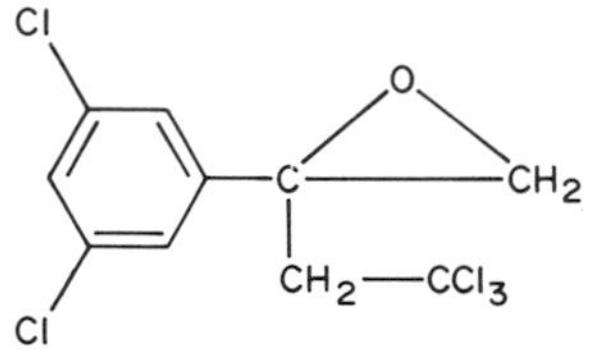

Cl
O
C
CH_2
CH_2—CCl_3
Cl

tridiphane

Tridiphane is the common name for 2-(3,5-dichlorophenyl)-2-(2,2,2-trichloroethyl)oxirane. The trade name is Tandem®. Tridiphane is a white crystalline solid with a low water solubility of 1.8 ppm at 21°C. It is formulated as a liquid (4 lb/gal). The acute oral LD_{50} is > 2.0 g/kg for rats.

Uses Tridiphane is a selective postemergence herbicide applied as a tank mix in combination with atrazine or cyanazine for the control of annual grass and broadleaf weeds in field corn. It is not recommended that it be used without one of these two triazine herbicides. Emulsifiable crop oil concentrate should be added to the spray solution to improve wetting of the weed foliage when atrazine is used but not when cyanazine is used. For optimum control, apply after the first flush of annual grass has emerged and is in the one- to three-leaf stage. The use rate of tridiphane is 0.50 to 0.75 lb/acre depending on soil type. In addition to the foliar contact action of these combinations of herbicides, residual preemergence activity on subsequently emerging weeds is also normally obtained.

Soil Influence Tridiphane's preemergence performance in influenced by soil type but its postemergence performance is not. In laboratory studies, it's half-life in soil under aerobic conditions ranged from 7 to 54 days with an average of 28 days. Under anaerobic conditions, a half-life of only 3 days was reported. It's half-life in sterilized soil was 109 to 148 days suggesting that soil microorganisms play a major role in its loss from soil. Under field conditions, it does not appear to leach through soil or accumulate as a result of annual applications.

Mode of Action The first symptoms of tridiphane–triazine combinations on the treated weeds are burning of leaf tips and margins about 2 to 3 days after treatment. The leaf surface may also appear limp and almost transparent suggesting membrane damage. As the symptoms progress, the entire leaf surface turns brown and will eventually disintegrate. Tridiphane acts as a synergist for the triazine herbicides.

Petroleum Oils Petroleum oils have been used to control weeds for many years. They were extensively used from the 1940s through 1973. However, mainly two events have greatly decreased their use in recent years. There was a dramatic increase in the price of all petroleum products in 1973 and many new, more effective herbicides have been introduced in recent years. Since petroleum oils currently have limited use our coverage of these herbicides is greatly reduced from what was presented in earlier editions of this book. Detailed information on these products can be found in the first and second editions (Klingman and Ashton, 1975, 1982).

Uses Basically there are two types of petroleum oils used as herbicides, selective and nonselective.

Selective weed oils are highly purified petroleum distillates similar to drycleaning fluids or paint thinners. They are clear and smell like refined naphtha. They are referred to as selective weed oil, petroleum spirits, mineral spirits, Stoddard solvent, or Varsol®. They are used selectively in carrots, parsnips, celery, parsley, and conifer-tree nurseries.

Nonselective weed oils are less refined than selective weed oils and similar to low-grade fuel oils or diesel oil. However, in some products produced specifically for weed control distillate fractions high in unsaturated and ring-structure compounds are added to these two products to increase their phytotoxicity.

Mode of Action Petroleum oils are hydrocarbons derived from petroleum. Size and structure of the oil molecule influence toxicity of these oils to plants. Range in size and structure is partially reflected by the number of atoms and the degree of saturation of the molecule. In general, products containing the most unsaturated and/or ring molecules are the most phytotoxic. Boiling point, viscosity, and specific gravity also influence the phytotoxicity of these oils.

Leaves of susceptible plants sprayed with oil rapidly become translucent taking on a water soaked appearance. Soon there is loss of turgidity and drooping of stems and leaves. Ultimately the plant becomes desiccated. Once oil reaches the inside of the leaf, it solubilizes the lipoids of the cell membranes. This makes the semipermeable membranes more permeable, cell sap leaks into the intercellular spaces, and the cells collapse (Van Overbeek and Blondeau, 1954). The membranes of some plants, e.g., carrots, resist this effect from selective weed oil; however, the mechanism is unknown.

INORGANIC HERBICIDES

Inorganic herbicides are those weed-control chemicals that contain no carbon atoms in their molecules. The principal ones are ammonium sulfamate (AMS), sodium chlorate, and various sodium borate salts. Many other inorganic salts are also toxic to plant tissues if applied in high concentrations. Certain of these have been used as contact-foliar sprays for weed control and found to be selective in certain species that retain a minimal amount of the spray solution on their leaves or by using directed sprays, e.g., ammonium nitrate or potassium chloride. The addition of a wetting agent to the spray solution may increase their contact but may also reduce their selectivity. However, all inorganic herbicides are generally considered to be relatively nonselective.

Although inorganic herbicides were used to a considerable extent in the past, they have been gradually and largely replaced by more active organic herbicides. This trend began in the mid-1940s when the introduction of 2,4-D heralded the modern era of weed science and has continued with the subsequent development of the many organic herbicides available today.

AMS still has considerable usage as a weed control material. Package mixes of sodium chlorate, sodium tetraborate, and various triazines are often used for total vegetation control on noncrop land. Sodium chlorate plus a fire retardant is widely used as a harvest aid to desiccate weeds and the mature crop and/or reduce the moisture content in crop seeds and seed heads of certain crops.

AMS

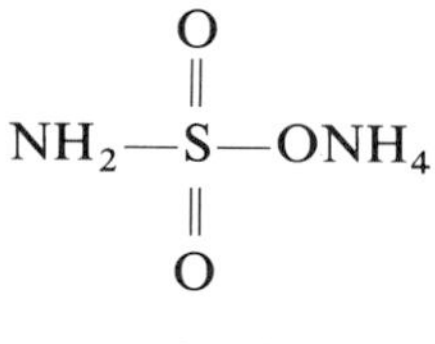

$$NH_2\text{—}\underset{\underset{O}{\|}}{\overset{\overset{O}{\|}}{S}}\text{—}ONH_4$$

AMS

AMS is the common designation for ammonium sulfamate. The trade name is Ammate®. In pure form, AMS is a colorless crystalline solid with a very high water solubility of 684,000 ppm at 25°C. The commercial form is available as yellow crystals (95%). The acute oral LD_{50} of AMS is 3.9 g/kg.

Uses AMS is particularly useful for the control of undesirable vegetation adjacent to areas that contain desirable species that could be injured by auxin type herbicides, e.g., 2,4-D. These areas include industrial sites, roadsides, fence rows, and drainage ditches (see Figure 14-3). It can also be used on rangelands and permanent pastures adjacent to reservoirs, lakes, ponds, and streams. Care should be taken to minimize drift onto desirable species and contamination of water. It is very effective for the control of many hardwood and coniferous woody

Figure 14-3. Ammonium sulfamate used as a foliar spray to control brush on the left side of the road. The tobacco on the right side of the road was not injured. (E. I. du Pont de Nemours and Company.)

species but also controls many annual and perennial herbaceous species, both grass and broadleaf weeds. It is applied as a foliar spray to brush any time after full leaf until the foliage begins to discolor. The rates of application for rangelands and permanent pastures are 57 lb/acre in 100 gallon of water for brush and 95 lb/acre in 100 gallons of water for herbaceous species. The latter rate is also used for all types of vegetation on noncrop land. The addition of a spreader-sticker or surfactant improves wetting of the foliage and performance. It has also been used as crystals or concentrated solution applied to cut surfaces (frills, notches, or cups cut in bark, or freshly cut stumps) for woody plant control.

In solution the chemical corrodes some metals, especially brass and copper; it also affects steel surfaces exposed to air such as the pump, the outside of tanks, and truck or tractor parts. Stainless steel, aluminum, and bronze are resistant. Metal parts covered by the spray solution inside the tank, pulp, and lines corrode much more slowly. Exposed areas should be coated with protective paints or covered with oil. Rinsing exposed surfaces with water after each day's use is important. Thorough cleaning inside and out at the end of the season, followed by a coating of oil, will preserve the equipment. Corrosive effects are considered negligible on fences, guy wires, and telephone wires.

Soil Influence AMS does not appear to be adsorbed by soils and is subject to leaching (Crafts, 1945). It is degraded by soil microorganisms and certain ones have been shown to use it as a source of nitrogen (*Herbicide Handbook*, 1989). It is relatively nonpersistent; at normal use rates its phytotoxicity disappears in 6 to 8 weeks.

Mode of Action AMS is rapidly absorbed by the foliage and green stems and translocated in woody plants (Carvell, 1955) and herbaceous species (Cupery and Gordon, 1942). Little is known about the biochemical basis for its herbicidal action.

Borates

Boron, the phytotoxic element of the borate herbicides, is an essential minor element for plant growth. In areas that are deficient in boron, small amounts of boron must be applied for optimum growth. However, in large quantities boron is toxic to plants.

Borax is the common name for sodium tetraborate ($Na_2B_4O_7$) and its hydrated forms, sodium tetraborate pentahydrate ($Na_2B_4O_7 \cdot 5H_2O$) and sodium decahydrate ($Na_2B_4O_7 \cdot 1OH_2O$). Their water solubilities are very high; 25,600 ppm (20°C), 38,200 ppm (20°C), and 59,300 ppm (25°C), respectively. The acute oral LD_{50} of these compounds ranges from 2.0 to 5.6 g/kg for rats.

Uses The various sodium borate salts are seldom used alone for weed control today. In the past, they were extensively used alone or in combination with sodium chlorate for total vegetation control on noncropland. These included railroads, farms, and industry and the paving industry to prevent weeds from growing through asphalt. They are currently used in package mix combinations with sodium chlorate as a harvest aid and certain triazine herbicides for total vegetation control on noncrop land. Atratol® 8P contains atrazine, borax, and sodium chlorate; Pramitol® 5PS contains prometon, simazine, borax, and sodium chlorate. In these combinations, the borates not only act as herbicides but also function as a fire retardant for sodium chlorate and an inhibitor of microorganism growth. The latter delays the degradation of the triazine herbicides by soil microorganisms. In the recent past, package mixes of borax plus bromacil (Borocil®) and borax plus monuron (Ureabor®) were also available. Borax is also used as a fire retardant for sodium chlorate when it is used as a harvest aid (see Sodium Chlorate section of this chapter).

Soil Influence Borax is moderately adsorbed by inorganic components of the soil and subject to slow leaching. This is in contrast to sodium chlorate, which is readily leached. This difference plus the fact that borax acts as a fire retardant for sodium chlorate explains why borax and sodium chlorate are often used together for weed control. The overall advantage of a borate–chlorate combination is that they both are nonselective and borax controls shallow-rooted weeds and sodium chlorate controls deep-rooted weeds for total vegetation control on noncropland.

The sodium borate salts are relatively persistent usually lasting one or more years. The period of persistence depends on soil type and rainfall. They are less persistent in acid soils and areas of high rainfall.

Mode of Action Herbicidal rates of borates cause plant desiccation beginning with necrosis of leaf margins, which progressively continues throughout the leaves. They are principally absorbed by roots and translocated through the xylem to all parts of the plant, accumulating in the leaves. The herbicide is most effective on young and tender plants. Therefore, treatment should be applied early enough to allow the material to be leached into the root-absorption zone by the time weed growth is just beginning. Virtually nothing has been published on the mechanism of action of borate as a herbicide (Brian, 1976). However, Crafts (1964) suggested that boron compounds apparently "tie up" calcium in the plant since the injury symptoms from excessive boron resemble those of calcium deficiency.

Sodium Chlorate

Sodium chlorate is the common name for $NaClO_3$. It is a white crystalline solid when pure and commercial forms are white to pale yellow. It has a very high water solubility of 2,300,000 ppm. The acute oral LD_{50} of sodium chlorate is 5.0 g/kg for rats.

Sodium chlorate has a salty taste. "Salt-hungry" animals may eat enough of treated plants to be poisoned. One pound of this chemical per 1000 pounds of animal is considered lethal. An additional hazard is the fact that certain poisonous plants that are ordinarily avoided by livestock become palatable when treated with sodium chlorate.

Sodium chlorate is 30 to 50 times more toxic to higher plants than common table salt, NaCl.

Fire Danger Sodium chlorate is a strong oxident; contact with combustible materials such as clothing, leather, wood, and plants may cause a fire. They are *dangerously flammable* and have been ignited by the sun's rays, clothing friction, or shoes scraping a rock. Numerous precautions related to this in regard to application, spillage, storage, and container disposal are given on the label and in the *Herbicide Handbook* (1989). Some of these include (1) wear *rubber* boots and apron, (2) remove contaminated clothing promptly and immediately wash with water, and (3) apply only in dry form to dry vegetation for weed control.

Uses Sodium chlorate is seldom used alone for weed control today. In the past, it was used alone as a foliar contact spray or as crystals for a soil treatment even when emerged weeds were present. The foliar contact treatment has been largely discontinued mainly because of the fire hazard presented when the foliage dried. When dry crystals are applied to dry vegetation there is usually no fire hazard because most of the material falls to the ground. However, sodium chlorate-sodium borate combinations as a soil treatment usually gave superior total vegetation control than either product alone for the reasons presented previously; see Borates section.

Currently, sodium chlorate is used in package mixes of borates plus various triazine herbicides for total vegetation control on noncropland; see Borates section of this chapter for details. Sodium chlorate plus a fire retardant in widely used as a harvest aid to desiccate weeds and the mature crop and/or reduce the moisture content of seeds and seed heads of many crops. These crops include cotton, corn (field, sweet, and popcorn), flax, gaur beans, peppers (chili, processing), rice, safflower, sorghum, soybeans, and sunflower.

Soil Influence Sodium chlorate is not adsorbed to soils to a significant degree and is subject to leaching (Seely et al., 1948). Soil microorganisms degrade sodium chlorate to sodium chloride. This occurs most rapidly in moist soils at temperatures above 70°C. Its phytotoxicity may persist 5 years or longer in areas of low rainfall and low microbial activity in the soil. In areas of high rainfall and high microbial activity in the soil, its phytotoxicity may be lost in 12 months in heavy soils and 6 months in sandy soils. Heavy rains or irrigation soon after a sodium chlorate application may leach the chemical from the upper 2 to 3 inches of the soil rendering this zone free of the herbicide. This would allow weed seeds to germinate and shallow-rooted weeds to continue growth. See Borates section of this chapter for the basis of the advantages of a sodium chlorate–sodium borate combination for total vegetation control related to their differential behavior in soils.

Mode of Action Sodium chlorate desiccates foliage of plants quickly and is injurious to the roots and other organs and living tissues that it contacts. It is rapidly absorbed by both leaves and roots. The stomates need not be open for the chemical to enter the leaves (Meadly, 1933). It penetrates the cuticle and comes in direct contact with the living cells (Loomis et al., 1933). Since it kills living cells rapidly, translocation from the leaves to the rest of the plant via the living phloem is minimal. However, it is rapidly translocated from the roots to the shoots through the nonliving xylem.

In addition to the contact action of sodium chlorate, which probably is related to an alteration of cell membranes, it has been reported to have an effect on certain metabolic processes. Plants treated with sodium chlorate are rapidly depleted of their food reserves (Bakke et al., 1939; Crafts, 1935; Latshaw and Zahnlcy, 1927); mainly carbohydrates, apparently by an increased rate of respiration (Wort, 1964). However, Gorenflot (1947) found that sodium chlorate inhibited respiration and photosynthesis as well as protoplasmic streaming of leaf cells of *Elodea canadensis* but the effect was reversible if treatment was not too prolonged. Sodium chlorate has also been reported to decrease catalase activity (Neeler, 1931), which could induce an increase in hydrogen peroxide which is toxic to plants. Presumably many other effects on normal plant metabolism could be demonstrated from a sodium chlorate treatment considering its high chemical reactivity.

SUGGESTED ADDITIONAL READING

Abernathy, J. R., and L. M. Wax, 1973, *Weed Sci.* **21**, 224.

Albritton, R., and S. J. Parka, 1978, *Proc. South Weed Sci. Soc.* **31**, 253.

Anderson, J. L., and B. Shaybany, 1972, *Weed Sci.* **20**, 434.

Ashton, F. M., and A. S. Crafts, 1981, *Mode of Action of Herbicides*, Wiley, New York.

Bakke, S. L., W. G. Gaesser, and W. E. Loomis, 1939, *Iowa Agric. Expt. Sta. Bull.* No. 254.

Bartels, P. G., and A. Hyde, 1970, *Plant Physiol.* **46**, 825.

Bartles, P. G., G. K. Matsuda, A. Siegel, and T. E. Weier, 1967, *Plant Physiol.* **42**, 736.

Bartels, P. G., and C. W. Watson, 1978, *Weed Sci.* **26**, 198.

Bernard, D. F., D. P. Rainey, and C. C. Lin, 1978, *Weed Sci.* **26**, 252.

Bingham, S. W., 1968, *Weed Sci.* **16**, 449.

Brian, R. C., 1976, The history and classification of herbicides, in L. J. Audus, Ed., *Herbicides*, p. 41, Academic Press, New York.

Bucha, H. C., and C. W. Todd, 1951, *Science* **114**, 493.

Carter, M. C., 1965, *Plant Physiol.* **18**, 1054.

Carter, M. C., 1975, Amitrole, in P. C. Kearney and D. D. Kaufman, Eds., *Herbicides*, Vol. 1, pp. 377–398, Dekker, New York.

Carvell, K. L., 1955, *Forest Sci.* **1**, 41.

Chang, C. T., and D. Smith, 1972, *Weed Sci.* **20**, 220.

Corbett, J. R., K. Wright, and A. C. Baillie, 1984, *The Biochemical Mode of Action of Pesticides*, Academic Press, New York.

Crafts, A. S., 1935, *Plant Physiol.* **10**, 699.

Crafts, A. S., 1945, *Hilgardia* **16**, 483.

Crafts, A. S., 1964, Herbicide behavior in the plant, in L. J. Audus, Ed., *The Physiology and Biochemistry of Herbicides*, p. 93, Academic Press, New York.

Crop Protection Chemical Reference, 1989, Wiley, New York (revised annually).

Cupery, M. E., and W. E. Gordon, 1942, *Ind. Eng. Chem.* **34**, 792.

Cutter, E. G., F. M. Ashton, and D. Huffstutter, 1968, *Weed Res.* **8**, 346.

Doran, D. L., and R. N. Anderson, 1975, *Weed Sci.* **23**, 105.

Dorough, H. W., 1974, *Bull. Environ. Toxicol.* **12**, 473.

Dorough, H. W., D. M. Whitacre, and R. A. Cardona, 1973, *J. Agric. Food Chem.* **21**, 797.

Duke, S. O., Ed., 1985, *Weed Physiology*, Vol II, *Herbicide Physiology*, CRC Press, Boca Raton, FL.

Duncan, D. N., W. F. Meggett, and D. Penner, 1982, *Weed Sci.* **30**, 191, 195.

Farm Chemicals Handbook, 1987, Meister, Willough, OH (revised annually).

Fedtke, C., 1982, *Biochemistry and Physiology of Herbicide Action*, Springer-Verlag, New York.

Gardiner, J. A., 1975, Substituted uracil herbicides, in P. C. Kearney and D. D. Kaufman, Eds., *Herbicides*, Vol. 1, pp. 293–321, Dekker, New York.

Geissbühler, H., H. Martin, and G. Voss, 1975, The substituted ureas, in P. C. Kearney and D. D. Kaufman, Eds., *Herbicides*, Vol. 1, pp. 209–291, Dekker, New York.

Good, N. E., 1961, *Plant Physiol.* **36**, 788.

Gorenflot, R., 1947, *Rev. gen. Bot.* **54**, 153.

Hatzios, K. K., and D. Penner, 1982, *Metabolism of Herbicides in Higher Plants*, Burgess, Minneapolis.

Herbicide Handbook, 1989, Weed Science Society of America, Champaign, IL.

Hess, F. D., 1985, Herbicide absorption and translocation and their relationship to plant tolerances and susceptibility, in S. O. Duke, Ed., *Weed Physiology*, Vol. II, *Herbicide Physiology*, pp. 191–214, CRC Press, Boca Raton, FL.

Hill, G. D., and J. W. McGalen, 1955, *Proc. 8th Southern Weed Conf.*, p. 284.

Jones, D. W., and C. L. Foy, 1972, *Weed Sci.* **20**, 8, 116.

Keeley, P. G., C. H. Carter, and J. H. Miller, 1972, *Weed Sci.* **20**, 71.

Klingman, G. C., and F. M. Ashton, 1975 and 1982, *Weed Science: Principles and Practices*, Wiley, New York.

Kroller, E., 1966, *Residue Rev.* **12**, 162.

Latshaw, W. L., and J. W. Zahnley, 1927, *J. Agric. Res.* **35**, 757.

Leek, G., and D. Penner, 1980, *Abstr. Weed Sci. Soc. Am.*, No. 182.

Loomis, W. E., V. E. Smith, R. Bissey, and L. E. Arnold, 1933, *J. Am. Soc. Agron.* **25**, 724.

Mahoney, M. D., and D. Penner, 1975, *Weed Sci.* **23**, 265.

Meadly, G. R. W., 1933, *J. Dept. Agric. W. Aust.* **10**, 481.

Meier, D., and H. K. Lichtenthaler, 1981, *Protoplasma* **107**, 195.

Moreland, G. D., and J. L. Hilton, 1976, Action on photosynthetic systems, in L. J. Audus, Ed., *Herbicides*, pp. 493–523, Academic Press, New York.

Neeler, J. R., 1931, *J. Agric. Res.* **43**, 183.

Nishimoto, R. K., and G. F. Warren, 1971, *Weed Sci.* **19**, 152, 256, 343.

Rafii, Z. E., and F. M. Ashton, 1979, *Weed Sci.* **27**, 321, 422.

Seely, C. E., K. H. Klages, and E. G. Schafer, 1948, *Washington Agric. Expt. Sta. Bull.* No. 505.

Smith, L. W., and F-Y. Chang, 1973, *Weed Res.* **13**, 339.

Van Overbeek, J., and R. Blondeau, 1954, *Weeds* **3, 55.**

Weed Control Manual and Herbicide Guide, 1988, Meister, Willough, OH (revised annually).

Wiater, A., T. Klopotonski, and G. Bagdasarian, 1971a, *Acta Biochim. Pol.* **18**, 309.

Wiater, A., K. Krajewska-Grynkiewicz, and T. Klopotonski, 1971b, *Acta Biochim. Pol.* **18**, 299.

Wills, G. D., 1976, *Weed Sci.* **24**, 370.

Wolf, D. E., R. S. Johnson, G. D. Hill, and R. W. Varner, 1958, *Proc. 15th North Central Weed Conf.*, p. 7.

Wort, D. J., 1964, Effects of herbicides on plant composition and metabolism, in L. J. Audus, Ed., *The Physiology and Biochemistry of Herbicides*, p. 312, Academic Press, New York.

For herbicide use, see the manufacturer's or supplier's label and follow these directions. Also see the Preface.

PART 3
Practices

15 Small Grains and Flax

Small grains discussed here include wheat, oats, barley, rye, and rice. Flax is included because its cultural pratices and weed problems are similar to those of small grains.

Winter varieties are planted in the fall, live through winter, and are harvested the following summer. Spring varieties are planted in early spring and harvested in mid to late summer. As an average, tolerance to cold is in this order: rye, wheat, barley, oats, and rice. Therefore, winter rye is grown in far northern areas. Rice in the United States is normally planted in the spring.

In general, winter varieties are most often infested with winter-annual weeds and to a lesser extent by summer annuals that germinate in early sring. Spring varieties are primarily infested by summer annuals that germinate in the early spring. Perennial weeds are also troublesome in certain areas.

EFFECT OF WEEDS ON YIELD

Weeds compete with small grain and flax crops for light, carbon dioxide, and soil moisture and nutrients. Certain weeds have also been shown to reduce yields by allelopathic effects (Chapters 1 and 2).

Chandler et al. (1984) reported on crop losses due to weeds in small grains and flax seed in Canada and the United States (Table 15-1). The mean values in this table indicate that yield reductions due to weeds are greater in rice (18.8%) than in other grain crops (11.6%). Although there are differences in the mean values among the other grain crops, an examination of the range values suggests that they may not be significantly different. The full report shows great differences in crops losses due to weeds for a given crop in the various geographic areas, e.g., 6 to 25% in oat. Yield reductions for a flax seed crop are also considerable, about 21.3%. In general, yield reductions caused by weeds in these crops are similar to those reported earlier by Friesen (1957).

Weed competition early in the season reduces yields more than late-season competition. Although yields are not greatly reduced by late-season weeds, they may cause difficulty in harvesting. Weeds also lower crop quality and may reduce the protein content of grain.

TABLE 15-1. Estimated Average Annual Yield Reduction due to Weeds in Small Grains and Flax in the United States[1,2]

	Yield Reduction	
Crop	Mean (%)	Range (%)
Wheat	12.5	9.0–20.0
Oat	14.7	6.0–25.0
Barley	9.3	6.0–18.0
Rye	10.0	6.0–18.0
Rice	18.8	12.0–34.0
Flax, seed	21.3	20.0–24.0

[1] Mean and range values are from the major production areas and based on 5 years of data, 1975–1979.
[2] From Chandler et al. (1984).

WEED CONTROL METHODS

Weed control is small grains and flax includes "clean seed," crop rotation, good seedbed preparation, crop competition, and application of herbicides.

Clean Seed

The use of crop seed free of weed seeds is commonly referred to as "clean seed." The importance of using clean seed in these crops can hardly be overemphasized. Most commercial crop seed currently available is relatively free of weed seed, but this aspect should be confirmed by checking the label. This is also discussed in Chapter 3.

Crop Rotation

When small grains or flax are continuously grown in the same field year after year, weeds that grow during the same season and are favored by crop-management programs flourish. Their populations increase with time due to an increasing seed bank and/or vegetative propagules.

Experience taught farmers that small grains and flax could usually be grown for several years on "new land" before weeds became a serious problem. Then they were forced to use more effective methods of weed control. Crop rotation proved to be one of the first effective methods.

Crop rotation is a strong link in the chain of weed control practices. Small grains and flax can be rotated with a number of crops that provide variations in cultural practices that break the monoculture cycle favoring certain weed species.

In some areas, the rotation of cotton, corn, and small grain with lespedeza illustrates this point. The variations in time of seedbed preparation, time of cultivation, and periods of growth of these four crops are detrimental to certain weed species. Herbicides can be combined with crop rotation to broaden the spectrum of weeds controlled. These crops are tolerant to different herbicides and the different herbicides control different weed species. Consequently, a more effective weed control program can be developed by utilizing both crop rotation and herbicide rotation in a concerned action rather than using either type of rotation alone.

Seedbed Preparation

Weed control is one of the principal purposes of seedbed preparation. Since small grains cannot be effectively cultivated after sowing, the importance of controlling weeds before sowing is obvious. A presowing cultivation can eliminate most annual weed seedlings that appear before sowing and often reduce early competition from perennial weeds. Summer fallow can reduce weed–seed production and the growth of perennial weeds.

Crop Competition

Dense stands of fast-growing small grains provide considerable competition to weeds (Figure 15-1). Maximal crop competition can be obtained by using high seeding rates, well-adapted varieties, and proper planting date. Adequate soil moisture and fertility may also increase crop competition. Under less favorable conditions, weeds can flourish and not only reduce yields but also produce seeds and vegetative propagules that increase the weed problems for subsequent years.

Flax is only partially effective in competing with weeds because its initial growth is slow and the leaf surface area is small. Nevertheless, thick stands of flax are somewhat helpful for weed control.

Chemical Control

Weeds can often be a serious problem in small grains and flax, even with the use of clean seed, crop rotation, seedbed preparation, and crop competition. Some weeds thrive under the same management practices of these crops.

Crop competition is often increased with the use of herbicides. The herbicide may only stunt weed growth with little or no reduction in the growth of the crop plant. This gives the crop plant a competitive advantage over the weeds.

The performance of herbicides often varies under different climatic and edaphic conditions. Therefore, their use may vary or be restricted in specific geographic areas; see labels. It is also advisable to consult with local public authorities, agricultural consultants, and/or company field representatives about their use.

Figure 15-1. Competition from the crop crowds out many weeds. (North Carolina State University.)

Barley, Oats, and Wheat Many herbicides are registered for use in these crops, more in barley and wheat than in oats (Table 15-2). Each of these herbicides has specific uses in regard to time and method of application and weeds controlled. For example, paraquat can be applied preplant or preemergence to control most emerged weeds, triallate can be applied preplant or preemergence as a soil-incorporation treatment for wild oat control, 2,4-D can be applied postemergence to control many broadleaf weeds, and difenzoquat can be applied postemergence to control wild oats in barley and wheat. Most postemergence herbicides used in these crops must be applied at specific stages of crop and weed growth to obtain optimal selectivity. Combinations of certain herbicides can also be used to broaden the spectrum of weeds controlled. A detailed discussion of all of these herbicides, the combinations, and their uses is not possible because of

Table 15-2. Small Grains and Flax Herbicides[1,2]

Herbicide	Barley	Oats	Rye	Wheat	Rice	Flax
Acifluorfen					×	
Atrazine				×		
Bentazon					×	
Bromoxynil	×	×	×	×		×
Chlorsulfuron	×	×		×		
Clopyralid	×			×		
2,4-D	×	×	×	×	×	
Dicamba	×	×		×		
Diclofop	×			×		×
Difenzoquat	×			×		
Diuron	×	×		×		
Endothall					×	
EPTC						×
Glyphosate	×	×	×	×	×	
Imazamethabenz	×			×		
MCPA	×	×	×	×	×	×
Metribuzin	×			×		
Metsulfuron	×			×		
Molinate					×	
Paraquat	×			×		
Picloram	×	×		×		
Pendimethalin					×	
Propachlor						×
Propanil	×			×	×	
Sethoxydim						×
Thiobencarb					×	
Triallate	×			×		
Thiameturon	×			×		
Trifluralin	×			×		×

[1] Refer to herbicide section for information on rates, herbicide combinations, and restrictions (locate through index); see label for complete information.

[2] Refer to pages 440–442 for weeds controlled and pages 443–445 for trade names.

space limitations. However, 2,4-D and related compounds are covered because of their wide use and importance.

2,4-D. This herbicide is widely used in small grains because it is relatively inexpensive and controls a broad spectrum of broadleaf weeds. In 1976, 2,4-D was applied to 71% of the wheat acreage in the United States. It has been said that more acres of small grain are treated with 2,4-D than any other crop with any other herbicide.

Good weed control without crop injury usually depends on proper timing in the application of herbicides. This is particularly true for 2,4-D in small grains.

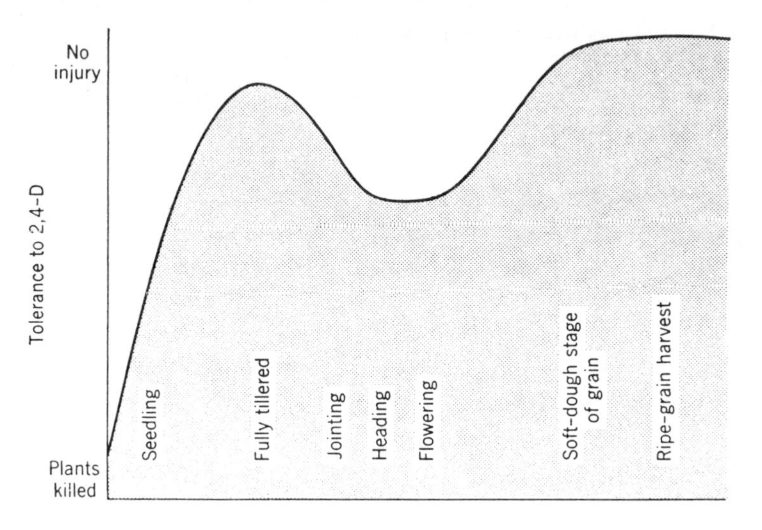

Figure 15-2. Stages of small-grain growth and the degree of tolerance to 2,4-D. (North Carolina State University.)

Periods of *greatest susceptibility* of small grains to 2,4-D are during periods of rapid growth. These are the zero- to four-leaf stage and boot stage through flowering (Figure 15-2). Applications during the zero- to four-leaf stage will usually cause many malformations of the head and "onion" leaves, general stunting of the plant, and reduced yields (Figure 15-3). Applications from the beginning of the boot stage through flowering will usually reduce yields seriously. During the boot stage the internodes elongate rapidly. This stage is also known as the jointing stage. Injury during these susceptible periods are associated with periods of high meristematic activity that provides very active sinks for increased symplastic translocation of the herbicide to these sites.

Periods of *least susceptibility* of small grains to 2,4-D are the four-leaf to boot stage and soft-dough-grain stage to maturity (Figure 15-2). The four-leaf stage to just before the boot stage, including the fully tillered stage, is considered to be the *most desirable time* to apply 2,4-D. At this time, the grain is relatively tolerant, the weeds are usually small and easily controlled, the weeds have not caused serious competitive damage, and ground-spray equipment causes only slight mechanical damage to the grain. Although grain is very tolerant from the soft-dough-grain stage to maturity, treatment at this time is not usually recommended because weed competition has already had its major effect, ground-spray equipment will damage the grain, and there are possible residues of 2,4-D in the harvested grain. In general, wheat varieties are most tolerant to 2,4-D, with barley being intermediate and oats the least tolerant.

MCPA and *dicamba* are two other auxin-type herbicides used as a postemergence treatment to control broadleaf weeds in small grains. MCPA is similar to 2,4-D in both chemical structure and action. It is less injurious to small

Figure 15-3. Wheat injured by a premature treatment with 2,4-D. The wheat was about 3 inches tall when treated and 60 to 70% of the heads were abnormal. (J. B. Harrington, University of Saskatchewan.)

grains, especially oats, and most broadleaf weeds. However, it is more effective on a few weeds, e.g., hempnettle and Canada thistle. It can also be applied somewhat earlier than 2,4-D without injury to small grain. MCPA usually costs more than 2,4-D in the United States. In contrast, dicamba is more injurious than 2,4-D to small grain but controls some broadleaf weeds better than 2,4-D, e.g., Canada thistle, chickweeds, field bindweed, and wild buckwheat.

Wild Oats. Wild oat is a serious problem weed in spring wheat, spring barley, spring oats, and flax. It is less of a problem in winter cereals. Yield losses in North America have been estimated as high as $ 1 billion per year. Crop yield losses are due to both allelopathy and competition. More than 50 different strains of wild oat give the weed a wide area of adaptation. Many of the seeds are dormant, making eradication impossible. Most seeds germinate with soil temperatures of 50°–60°F.

Crop rotation and summer fallow provide only limited control. Delayed spring seeding, with effective spring cultivation just before seeding, is reasonably effective. However, seeding delayed later than the recommended date of seeding may result in a serious drop in crop yield.

Several herbicides can be used to control wild oats in barley and wheat. These include diclofop, difenzoquat, glyphosate, imazamethabenz, triallate, and trifluralin. Glyphosate can also be used in oat crops. Glyphosate is applied as a preplant or preemergence treatment to emerged wild oats. Triallate is applied to the soil before or after seeding the crop and requires soil incorporation.

Trifluralin is applied to the soil preplant and crop seed must be planted below the depth of soil incorporation. Diclofop can be used preplant soil-incorporated, preemergence, or postemergence. Difenzoquat and imazamethabenz are used postemergence to both grain and wild oats. Specific stages of growth of both crop and wild oats are essential for all postemergence treatments to obtain good wild oat control without crop injury. See labels for detailed information on rates, time and methods of application, and restrictions.

Tillage Substitutes. Although small grains are grown in many areas of the United States, the major production area is the Great Plains. This is particularly true for wheat. The region is subject to high winds and severe thunderstorms, which cause considerable soil erosion if sufficient plant cover is not present (Wicks, 1986). In addition, rainfall for maximum crop production is frequently limited. These factors require soil and moisture conservation practices for a sustainable agriculture. These practices include stubble mulch, chemical fallow, ecofarming, and lo-till or no-till (Wicks, 1986). Each system has been adapted to specific areas in the Great Plains.

Stuble mulch uses shallow tillage with V sweeps and a rod weeder to control weeds during the fallow year between wheat crops, leaving much of the previous wheat residue on the soil surface. *Chemical fallow* uses herbicides to control weeds during the fallow period. *Ecofarming* is a system of controlling weeds and managing crop residues throughout a crop rotation with herbicides and minimum tillage, e.g., winter wheat-sorghum or corn-fallow. *Lo-till* or *no-till* practices are usually used where rainfall is sufficient for continuous winter wheat, no fallow year. Herbicides are applied as needed with minimal cultivation for planting and crop emergence. The objective of these practices is to reduce soil erosion by wind and rainfall and conserve soil moisture. A comprehensive coverage of this topic is available in *No-Tillage and Surface-Tillage Agriculture*, edited by Sprague and Triplett (1986).

Rice Rice culture utilizes many of the same weed control practices common to wheat, barley, and oats. However, some of these are discussed further because of the unique cultural practices of rice. Rice is grown under "lowland" (paddy, flooded with water) or "upland" (dryland) conditions. The cultural practices and the weed species present and control methods are somewhat different under these two growing conditions. Water management is a very important aspect of weed control in water-sown rice (Bayer et al., 1985). A continuous water depth of 3 to 6 inches improves the effectiveness of rice herbicides and reduces weed competition. Maintaining a water depth of 6 to 8 inches for 21 to 28 days after planting can provide partial control of barnyardgrass. In areas of low rainfall, deep plowing (8 to 14 inches) to expose underground stems of cattails, tubers of river bulrush, and winter buds of American pondweed can reduce their populations, if sufficient drying of reproductive organs is attained in the spring.

Sixty percent of the herbicides used in rice are not used in the other small grains (Table 15-2). Most rice herbicides are applied postemergence to both crop and weed. The time of application relative to the growth stage of crop and weed and water management for all rice herbicides is critical for crop safety and good weed control. Each herbicide has its own specific requirements; see labels. However, some brief comments should be informative. MCPA is less phytotoxic to rice than 2,4-D and therefore safer to use. With foliar-applied herbicides such as MCPA, bentazon, and propanil, the weed foliage must be exposed to the herbicide spray. This may require lowering the water level. To allow time for herbicide absorption, the water level should not be raised over the weeds for 24 to 48 hours. Small submerged weeds not sprayed with the herbicide will not be controlled. Molinate can be applied as a preplant or preflood soil incorporation treatment, or either preemergence or postemergence postflood. Pendimethalin is used only in combination with propanil, propanil controls certain emerged weeds, and pendimethalin provides residual control of certain weeds. Pesticide combinations or sequential applications may cause rice damage, e.g., propanil with a carbamate or organophosphate insecticide. Since glyphosate is nonselective, it must not contact the crop and rice fields and levees must not be treated when the fields contain water. There are many other limitations and precautions for the use of rice herbicides; see labels.

Flax Nonchemical methods of weed control in flax are essentially the same as barley, oats, and wheat. The herbicides used in flax are listed in Table 15-2. Bromoxynil and MCPA are postemergence herbicides used to control broadleaf weeds. Diclofop and sethoxydim are postemergence herbicides used to control wild oats and other annual grasses. Sethoxydim also controls certain perennial grasses. Propachlor is a preemergence herbicide used to control annual broadleaf and grass weeds, but not wild oats. EPTC and trifluralin are preplant soil-incorporated herbicides that control many grass and broadleaf weeds. EPTC is applied in the late fall befor the ground freezes and trifluralin is applied just before planting. EPTC controls wild oats but trifluralin does not control this weed. Appropriate combinations of some of these herbicides are used to broaden the spectrum of weeds controlled. Limitations and precautions for the use of these flax herbicides are given on the labels.

SUGGESTED ADDITIONAL READING

Bayer, D. E., J. E. Hill, and D. E. Seaman, 1985, Rice, in *Principles of Weed Control in California*, Thomson Publications, Fresno, CA.

Chandler, J. M., A. S. Hamell, and A. G. Thomas, 1984, *Crop Losses due to Weeds in Canada and the United States*, Weed Science Society of America, Champaign, IL.

Crop Protection Chemical Reference, 1989, Wiley, New York.

Farm Chemicals Handbook, 1987, Meister, Willough, OH.

Friesen, G., 1957, *North Central Weed Control Conf. Proc.* **14**, 40.

Herbicide Handbook, 1989, Weed Science Society of America, Champaign, IL.

Sprague, M. A., and G. B. Triplett, Eds., 1986, *No-Tillage and Surface-Tillage Agriculture*, Wiley, New York.

Weed Control Manual and Herbicide Guide, 1988, Meister, Willough, OH.

Wicks, G. A., 1986, Substitutes for tillage on the Great Plains, in M. A. Sprague and G. B. Triplett, Eds., *No-Tillage and Surface-Tillage Agriculture*, pp. 183–196, Wiley, New York.

For herbicide use, see the manufacturer's or supplier's label and follow these directions. Also see the Preface.

16 Small-Seeded Legumes

Several small-seeded legumes are used for hay, pasture, and soil improvement. Alfalfa, white clover, and ladino clover are perennial crops usually grown for hay and pasture. Red clover, alsike clover, crimson clover, sweet clover, and lespedeza are hay and soil-improvement crops. They can be interplanted in small grain. Large-seeded legumes planted in rows, such as soybeans and peanuts, are discussed in Chapter 17.

Weeds in small-seeded legumes can reduce yields, lower quality, cause premature loss of stand, and present harvesting problems. The estimated losses due to weeds in certain small-seeded legume seed crops are pesented in Table 16-1. The overall average loss was 16.5% (Chandler et al., 1984). Weeds can also increase disease and insect problems in these crops. Certain weeds in hay or pastures can adversely effect animals and/or their products. Some weeds, such as spurge, fiddleneck, and common ground, are toxic to livestock when consumed in large amounts (Figure 1-4). Spiny weeds, such as thistles and downy brome, are irritating to livestock. Some weeds, such as wild onion and wild garlic, cause off-flavors in milk. When these crops are grown for certified seed production they must be essentially free of weeds.

Weed control in these crops during the seedling stage and established stage present distinctly different problems and are discussed separately. However, conditions that increase the vigor of the crop make it more competitive at any stage of growth. These include favorable environmental conditions (temperature, soil moisture, fertilization), appropriate variety, disease and insect control, and other management practices inductive to establishment and maintenance of the stand.

SEEDLING STAGE

Since small-seeded legume seedlings do not grow as vigorously as many common weeds, it is essential that control measures be taken to establish a strong stand. These control measures include (1) clean seed, (2) weed control before seeding, (3) proper date of seeding, and (4) chemicals. Mowing weeds that "outgrow" legume seedlings can also be an effective method of control.

TABLE 16-1. Estimated Average Annual Yield Reduction due to Weeds in Certain Small-Legume seed Crops in the Continental United States[1,2]

	Reduction	
Crop	Average (%)	Range (%)
Alfalfa	20.0	17–25
Crimson clover	13.0	13
Red clover	17.2	10–22
Lespedeza	15.6	8–19

[1] Mean and range values are form the major production areas and based on 5 years of data, 1975–1979.
[2] Adapted from Chandler et al. (1984).

Clean Seed

The use of *certified* seed is the best method to avoid sowing weed seeds along with the crop. This type of crop seed is certified by the state as to its high purity in regard to the presence of weed seeds. Less expensive crop seed is frequently contaminated with weed seeds. Once introduced, these weed seeds may develop into serious weed problems that persist for many years and are difficult and expensive to control. This is especially true for dodder, whose seeds are about the same size as most small-seeded legume seeds.

The seeds of many serious weeds are very similar to seeds of small-seeded legumes. Once the legume seeds are contaminated with these weed seeds, they can only partially be cleaned by present methods. Many of the mechanical seed cleaners are quite ingenious and take advantage of small physical differences between the seeds to be separated. These differences include size, shape, weight, seed coat, hairiness, and appendages. One example is the cleaner used to remove dodder seed from small-seeded legume seed. It takes advantage of the fact that the legume seed is smooth and waxy and the dodder seed is rough and pitted. Consequently, dodder seed adhere to felt cloth and legume seed do not. Adjoining felt-covered rollers rotate in opposite directions and seeds pass over them. The rollers are slanted so that the legume seed flow out the lower end; the dodder seed is removed from the rollers.

Weed Control before Seeding

Small-seeded legume *seedlings* are not very competitive to aggressive weeds because they do not grow vigorously. Therefore, it is desirable to eliminate seeds and propagules of both annual and perennial weeds before seeding these crops. Control methods must be appropriate for the weed species present.

Annual weeds are often controlled by *two methods*. The *first method* is crop rotation before seeding. For example, a rotation of row crops and small grain with good weed control for 2 or more years usually reduces weed–seed populations in the soil. The *second method* involves killing one or more "crops" of weeds by cultural or chemical methods before, during, or after seed bed preparation, but before seeding the crop. Cultivations to eliminate these small weeds should be shallow to prevent bringing deeply buried weed seed near the soil surface where they may germinate. Deeply buried weed seed usually remain dormant and present no problem if undisturbed.

Perennial weeds should be controlled by cultural or chemical methods *before* planting small-seeded legumes. Once these crops have been planted, the options for perennial weed control without crop injury is limited.

Date of Seeding

Date of seeding may determine weediness of the crop (Klingman, 1970). Most small-seeded legume crops can be either fall-planted or spring-planted. Weediness will depend on whether winter- or summer-annual weeds are more serious. Summer-annual weeds are often more serious. Thus alfalfa is usually fall planted. With fall planting, the legume crop is well established by spring and competes well with summer-annual weeds. However, there are the hazards of fall drought and winter injury in some areas and excessive rainfall in other areas.

Spring planting may be preferred if winter-annual weeds are the major problem. Spring planting avoids the flush of winter-annual-weed growth. However, with spring planting, summer-annual weeds may crowd out legume seedlings before they become established. More-effective herbicides may make it possible to take advantage of more-ideal seeding conditions usually found in the spring.

Chemicals

Herbicide practices for small-seed legumes at the seedling stage are usually different from those used in the established crop. In general, the seedling stage is more likely to be injured by a given herbicide at a specific rate than the established stage. Small-seeded legumes very in their tolerance to various herbicides. This tolerance and susceptibility of the weeds are most important in choosing an appropriate herbicide treatment.

Herbicides are applied preplant, preplant soil-incorporated, preemergence, or postemergence to the crop for the establishment of a stand from seed. Table 16-2 lists the herbicides that can be used at both the seedling stage and established crop. Refer to herbicide section for information on stage of growth, rates, herbicide combinations, and restrictions (locate through index); see label for complete information. Refer to Table A-0 for weeds controlled and Table A-1 for trade names.

TABLE 16-2. Herbicides Used in Small-Seeded Legumes[1,2,3]

Herbicide	Alfalfa	Birdsfoot Trefoil	Clovers[4]
Benefin	×	×	×
Chlorpropham	×	×	×
2,4-DB	×	×	×
Diuron	×	×	×
EPTC	×	×	×
Glyphosate	×		×
Hexazinone	×		
Metribuzin	×		
Paraquat	×		
Pronamide	×	×	×
Propham	×	×	×
Sethoxydim	×		
Terbacil	×		
Trifluralin	×		

[1] Refer to herbicide section for information on rates, herbicide combinations, and restrictions (locate through index); see label for complete information.
[2] Refer to pages 440–442 for weeds controlled and pages 443–445 for trade names.
[3] Each herbicide may be used at the preplant, seedling, and/or established stage, but often not at all stages.
[4] Some labels list the type of clovers, others merely state clovers.

ESTABLISHED STAGE

Weeds are often a serious problem in established stands of small-seeded legumes. The weed problem may increase with time as the crop becomes less vigorous due to any one of a number of environmental conditions or cultural practices. Weeds can reduce the number of crop plants in an established stand by competition. This results in an increasing loss in yield and quality with time. Weed control methods in established stands include (1) crop competition, (2) mowing, (3) flaming, (4) cultivation, and (5) chemicals.

Crop Competition

Although seedlings of small-seeded legumes often do not compete favorably with weeds, a vigorous established stand is very competitive to many weeds. Therefore, maintaining a thick established stand is very important in weed suppression in these crops. Proper adapted varieties, fertilization, drainage, moisture conservation, irrigation, mowing time, and disease, insect, and weed control help maintain a thick stand of legumes. These practices also ensure rapid regrowth of the crop after mowing, which increases its competitive ability.

Mowing

Many erect annual weeds may be killed and the vigor of erect perennial weeds reduced by mowing. However, prostrate weeds are not controlled by mowing. Some erect weeds may develop prostrate growth patterns with repeated mowing. In fact, at least one erect weed, yellow foxtail, has developed a genetically stable biotype in some areas that is prostrate under mowing regimes. If the crop is consistently mowed too frequently or when it is immature, its vigor will be reduced and weed problems may increase. Mowing controls many weeds especially well when a thick legume stand is maintained.

Flaming

Propane or diesel burners have been used to control weeds in established alfalfa. Winter-annual broadleaf weeds can be controlled with flaming just before the crop resumes growth in the spring. This treatment also suppresses alfalfa weevil larval populations. Flaming just after cutting has controlled established dodder plants. However, it usually results in a few day's suppression of alfalfa growth. Flaming has been generally replaced by other methods of weed control because of the current high price of petroleum.

Cultivation

Although tillage of alfalfa fields to control annual weeds has been recommended and practiced, there is little experimental evidence to support this practice. The early research of Kiesselback and Anderson (1927) showed that cultivation of alfalfa neither increased nor decreased yields. However, in a good stand of alfalfa, one cultivation may control some annual weeds. The use of the spring-tooth harrow is preferred since it kills many weeds without injury to alfalfa crowns. The spike-tooth harrow is effective only on very small weeds and the disc harrow may cause considerable damage to alfalfa by cutting the crowns.

When these crops are grown for seed production they are often planted in rows. The areas between the rows can be cultivated as in any row crop.

Chemical

Small-seeded legumes vary in their tolerance to various herbicides. This tolerance and susceptibility of the weeds are most important in choosing an appropriate herbicide treatment. Most herbicides applied to these established crops are applied in the fall immediately after the last cutting, winter, and/or early spring, preemergence or postemergence to the weeds. The crop is usually dormant or semidormant at these times. Table 16-2 lists the herbicides that can be used at both the seedling stage and established crop. Refer to herbicide section for information on stage of growth, rates, herbicide combinations, and restrictions (locate through index); see label for complete information. Refer to Table A-0 for weeds controlled and Table A-1 for trade names.

SUGGESTED ADDITIONAL READING

Chandler, J. M., A. S. Hamill, and A. G. Thomas, 1984, *Crop Losses due to Weeds in Canada and the United States*, Weed Science Society of America, Champaign, IL.

Crop Protection Chemical Reference, 1989, Wiley, New York (revised annually).

Kiesselback, T. A., and A. Anderson, 1927, *Nebraska Expt. Sta. Bull.*, 222.

Klingman, D. L., 1970, Brush and weed control on forage and grazing lands, *FAO Internat. Conf. on Weed Control*, pp. 401–424, Weed Science Society of America, Champaign, IL.

Mitich, L. W., 1985, Alfalfa (*Medicago sativa*), in *Principles of Weed Control in California*, pp. 232–237, Thompson Publications, Frenso, CA.

Weed Control Manual and Herbicide Guide, 1988, Meister, Willough, OH (revised annually).

For chemical use, see the manufacturer's or suppliers's label and follow these directions. Also see the Preface.

17 Field Crops Grown in Rows

Several field crops are grown in rows to facilitate their culture. This is in contrast to other field crops such as small grains, flax, and small-seeded legumes that are not grown in rows. The nature of the individual crop plant usually determines whether or not it is grown in rows. In general, crops grown in rows have larger seeds, develop into larger and more vigorous plants, and require more space per plant. These characteristics permitted earlier farmers to utilize "horse-hoeing" in the development of the row crop concept. However, with the demise of the horse and advances in agriculture, the rows have gradually become closer and yields have increased. Some of these advances include cultural implements, new varieties, and weed control practices, especially herbicides. Losses due to weeds in row crops are given in Table 17-1.

Weed control in most annual crops can be divided into early-season and late-season phases. Early-season weeds usually have a greater effect on crop yields than late-season weeds. Late-season weeds may make harvesting difficult, reduce the quality of the crop (e.g., grass in cotton), and reinfest the soil with weed seeds.

WEED CONTROL METHODS

Weed control in field crops grown in rows involves an integrated approach utilizing the methods discussed in Chapter 3. These include mechanical, competition, crop rotation, and chemical methods. Biological control through predators and diseases has not been perfected for weed control in cultivated crops.

Mechanical

Mechanical weed control in row crops primarily involves cultivation. Cultivations are often used primarily for weed control. However, some crops may benefit from one early cultivation to loosen the soil if it becomes hard and packed when dry. On other soils, cultivation is of little value if the weeds are controlled by other means.

Cultivation has some disadvantages. Some crops grow slowly and weeds may get ahead of the crop before they can be cultivated. Cultivation frequently fails to control weeds in the crop row, may injure roots, and may result in slower crop growth and ultimately yield reductions. Heavy weed growth may develop after the last cultivation. Repeated cultivation, especially in wet soils, injures the

TABLE 17-1. Estimated Average Annual Yield Reduction due to Weeds in Row Crops in the Continental United States[1,2]

	Yield Reduction	
Crop	Mean (%)	Range (%)
Corn	11.8	8–18
Cotton	10.9	8–16
Milo	13.8	11–18
Peanuts	15.0	10–25
Soybeans	16.6	13–27
Sugarbeets	10.2	5–20
Sugarcane	20.0	10–26

[1] Mean and range values are from the major production areas and based on 5 years of data, 1975–1979.
[2] Adapted from Chandler et al. (1984).

physical condition of the soil. Cultivation also consumes costly fuel and large tractors are an expensive investment.

Competition

Providing conditions favoring crop competition is paramount for good weed control in field crops grown in row. These conditions include all cultural and environmental factors that provide rapid seed germination, vigorous seedling growth, and subsequent maximum crop growth. For most field crops grown in rows, the crop canopy covers the row middles within 7 to 8 weeks providing major competition to late emerging weeds. A herbicide that keeps the crop weed free until the crop closes-in provides an integrated weed control approach. In the absence of good crop competition, other integrated methods of weed control often fail to reach their optimum effectiveness. Dense vigorous crop stands provide maximum competition to weeds (Figure 15-1).

Crop Rotation

Crop rotation is a very useful component of an integrated weed control program. A row crop can be rotated with a number of crops that provide variations in cultural practices that break the monoculture cycle favoring certain weed species. The variations in time of seedbed preparation, time of cultivation, and periods of growth of various crops are detrimental to certain weed species. Herbicides can be combined with crop rotation to broaden the specrum of weeds controlled. Crops are tolerant to different herbicides and the different herbicides control different weed species. Consequently, a more effective weed control program can be developed by utilizing both crop rotation and herbicide rotation in a concerted action rather than using either type of rotation alone.

Chemical Weed Control

Chemical weed control in field crops grown in rows should be integrated with other weed control methods for optimum effectiveness. Some examples of this integration have been presented earlier in this chapter and others will be given later in the discussion of specific crops.

The following chemical weed control methods have proven effective in various parts of the United States and Canada, but not necessarily in all areas. Herbicides are applied at four different times in row crops: (1) *preplant*—before planting, (2) *preemergence*—after planting but before crop emergence, (3) *postemergence*—after emergence of the crop, and (4) *minimum-till* or *no-till*—small amount of tillage or no tillage.

Preplant Preplant treatments can be applied to emerged weeds, the soil surface, or soil incorporated. Preplant treatments to emerged weeds with nonselective, nonpersistent herbicides, e.g., glyphosate or paraquat, can be used in most field crops grown in rows. Soil-surface applications must be followed by rain or sprinkle irrigation to be effective. Overhead water and soil incorporation places the herbicide into the area of the soil where the weed seeds germinate. In general, soil incorporation is more effective than surface application with overhead water. Soil incorporation is also required for certain volatile herbicides such as the carbamothioates, e.g., EPTC. Crop seeds must be tolerant to the herbicide or with some herbicides the seeds can be placed below the herbicide-treated layer, e.g., cotton-trifluralin (Figure 6-1).

With soil incorporation, the herbicide is applied to the soil surface and incorporated to a depth of 2 to 3 inches with appropriate equipment. Incorporation equipment includes power-driven rotary hoes with L-shaped knives, ordinary rotary hoes, discs, sweeps, drag-harrows, and other similar soil-tillage implements (Chapter 7). The equipment must break up large clods and mix the herbicide evenly throughout the treated soil profile. Overhead water is not needed to "activate" the herbicide with soil incorporation; therefore, the herbicide seldom fails to be effective. Other preplant herbicide treatments are discussed under the section on minimum- or no-till.

Preemergence Preemergence treatments are applied to the soil surface after planting but before the crop emerges. Rain or sprinkle irrigation is required to leach the herbicide into the soil. Comments related to surface applications of herbicides and overhead water made in the preplant section also apply to preemergence treatments.

Postemergence Postemergence treatments are made any time after the crop emerges. Highly selective herbicides may be applied as a broadcast spray over the top of the crop and weeds. The crop is not injured and the susceptible weeds are controlled. Moderately selective or nonselective herbicides can also be applied postemergence to the crop utilizing methods that apply the chemical to the weeds but minimize or essentially avoid application of the chemical to the crop. Moderately selective herbicides can be applied as a directed spray that minimizes crop exposure, e.g., 2,4-D to small broadleaf weeds in the taller corn crop. Drop nozzles are often used for this type of application. Nonselective herbicides can be applied with a recirculating sprayer or rope-wick applicator that essentially avoids crop exposure, e.g., glyphosate to johnsongrass that is taller than cotton. These selective methods of application are discussed in Chapter 7.

Minimum-Till and No-Till The concept of minimum-till or no-till evolved after the development of herbicides that could replace certain cultivation practices. Many of the sequential tillage operations were primarily for weed control. Minimum-till and no-till methods result in reduced cultivation for weed control and the retention of plant residues as a mulch cover on the soil surface. This reduces soil erosion from wind and rain, increases water conservation, improves mineral nutrition (Sprague and Triplett, 1986). Other advantages of these methods include more efficient use of labor, fuel, land, fertilizers, pest control, and water.

CORN

Modern weed control in corn utilizes an integrated program that includes field selection, crop rotation, crop competition, cultivation, and herbicides. Yields are usually reduced when weeds are not controlled before the corn is 6 to 8 inches high.

In the selection of cultivation methods, consideration needs to be given to the growth pattern of the crop, e.g., root distribution (Figure 17-1). Since corn roots are relatively shallow, corn should be cultivated only deep enough to remove or cover the weeds to minimize root pruning. Rotary weeders, finger weeders, and harrows are used for cultivation when the corn is small. When it is 3 to 4 inches tall, shovel or sweep cultivation can be started to remove weeds from the interrow area and throw soil into the crop row to bury small weeds. Cultivation is repeated as often as needed until the corn is 20 to 24 inches tall. These cultivations are usually made three or four times when cultivation is the only method of weed control used. The last cultivation is often referred to as a "layby" cultivation.

Much of the root system will be pruned off by a cultivation within 6 inches of the stem to a depth of 6 inches (Figure 17-1). Serious wilt may follow a deep cultivation in dry weather. Deep, late-season cultivation will nearly always reduce corn yields. Therefore, shallow sweeps should be used if a late cultivation is necessary. Roots on the cultivator shank indicate that the cultivation is too deep.

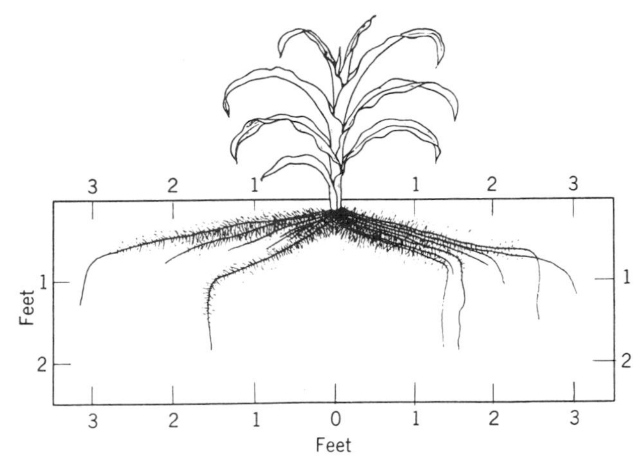

Figure 17-1. Corn root distribution. Cultivation should be shallow to avoid root damage, especially after the plant is 15 inches tall. (Adapted from *Nebraska Research Bulletin* No. 161, 1949.)

The use of herbicides reduces the number of postplant cultivations required. A program of reduced cultivation with the use of herbicides can provide very effective, season-long weed control in corn.

Chemical Weed Control in Corn

Since 2,4-D was introduced in the mid-1940s to control broadleaf weeds in corn, many other herbicides have been developed to control most weeds in corn (Table 17-2). In fact, more herbicides have been developed for corn and soybeans than for any other crop. Most corn herbicides are used to control annual weeds; however, some also suppress the growth or control perennial weeds. Combinations of two or more herbicides are often used as a tank mix or sequential treatments to extend the period of control and/or broaden the spectrum of weeds controlled. Herbicides should be selected primarily on the basis of weed species present, stage of crop growth, and succeeding crop rotation. Certain limitations relative to soil type, geographic location, and other factors must be considered for the use of many herbicides; see labels. Depending on the specific herbicide, they can be applied preplant, preemergence, or postemergence to the crop. Many, but not all, of the herbicides listed in Table 17-2 can also be used in nonfield corn crops, e.g., silage, sweet, pop, or seed.

Corn has susceptible and tolerant development stages relative to 2,4-D. It is applied as an overall postemergence spray from the three-leaf stage until the corn is less than 10 inches tall. Thereafter until tassle initiation directed sprays with drop nozzles are used. Corn is most susceptible to 2,4-D injury during or after tassle to the dough stage; do not apply during this period.

TABLE 17-2. Field Corn Herbicides[1,2]

Preplant Incorporated	Preemergence to Crop		Postemergence to Crop[3]	
Alachlor	Alachlor	Linuron	Alachlor	Dicamba
Atrazine	Atrazine	Metolachlor	Ametryn	Diuron
Butylate[4]	Chloramben	Metribuzin	Atrazine	Linuron
Cyanazine	Cyanazine	Paraquat	Bentazon	Oxyfluorfen
EPTC[4]	2,4-D	Pendimethalin	Bromoxynil	Paraquat
Metolachlor	Dicamba	Propachlor	Cyanazine	Pendimethalin
Simazine	Diuron	Simazine	2,4-D	Tridiphane
	Glyphosate		DCPA	Trifluralin

[1] Refer to herbicide section for information on rates, herbicide combinations, and restrictions (locate through index); see label for complete information.

[2] Refer to pages 440–442 for weeds controlled and pages 443–445 for trade names.

[3] Some of those herbicides require special application methods, e.g., directed sprays.

[4] Special formulation with antidote.

Figure 17-2. Atrazine applied in a band over the corn row at planting time. Usually, the weeds between the herbicide bands are removed by cultivation when the weeds are much smaller. The rows to the left were not treated. (Ciba-Geigy Corporation.)

MILO

Nonchemical weed control in milo (grain sorghum) is similar to corn. This includes an integrated program of field selection, crop rotation, and cultivation. These methods are further integrated with chemical methods.

Chemical Weed Control in Milo

In general, milo is less tolerant to most herbicides than corn. The development of herbicides to give season-long weed control with minimum tillage has allowed the

TABLE 17-3. Milo Herbicides[1,2]

Preplant Incorporated	Preemergence to Crop		Postemergence to Crop[3]	
Atrazine	Atrazine	Metolachlor	Atrazine	Glyphosate
Metolachlor	Cyanazine	Paraquat	Bentazon	Linuron
	Glyphosate	Propachlor	Bromoxynil	Paraquat
	Linuron		2,4-D	Pendimethalin
			Dicamba	Trifluralin
			Diuron	

[1] Refer to herbicide section for information on rates, herbicide combinations, and restrictions (locate through index); see label for complete information.
[2] Refer to pages 440–442 for weeds controlled and pages 443–445 for trade names.
[3] Some of these herbicides require special application methods, e.g., directed sprays.

rows to be planted closer together increasing their competitiveness to weeds. The herbicides use in milo are given in Table 17-3.

As with corn, milo has susceptible and tolerant development stages relative to 2,4-D (Figure 17-3). Milo is most tolerant from the 4- to 12-inch stage and this is the best time to treat with this herbicide. Earlier at the seedling to four-leaf stage and later at the 12-inch stage (head starts to develop rapidly) to soft-dough stage milo is susceptible to 2,4-D injury. When the grain is half-formed, milo again becomes tolerant; however, by this time weeds have already produced most of their adverse effects.

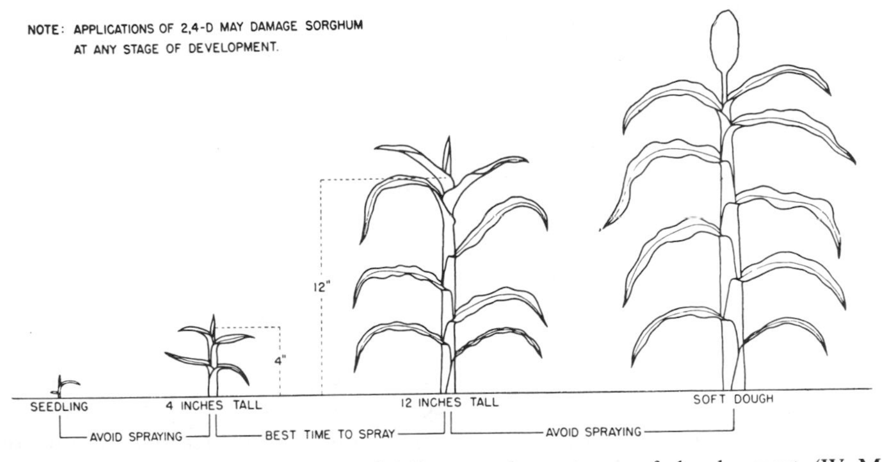

Figure 17-3. Sorghum tolerance to 2,4-D at various stages of development (W. M. Phillips, Kansas Agricultural Experiment Station and USDA.)

Figure 17-4. Control of green foxtail in soybeans by rotary hoeing. *Left*: Before rotary hoeing. *Right*: After rotary hoeing: note the lack of both weeds and soybean injury. (Iowa State University, Ames, IA.)

SOYBEANS

Since soybeans are not planted until the soil is relatively warm, early germinating and emerged weed seedling can be controlled by cultivation before planting. Good management practices such as seedbed preparation, selection of high-quality well-adapted-variety seed, and favorable soil moisture contribute to a vigorous crop that will compete with weeds.

The rotary hoe is an effective and economical method of weed control in soybeans (Figure 17-4). For best results, it should be used when the soil surface is dry and slightly crusted and when the weeds are just emerging, not more than 1/4 inch high. The rotary hoe is also effective on these small weeds when the soil surface is moist.

Chemical Weed Control in Soybeans

Many herbicides have been developed for the control of weeds in soybeans. A chemical weed-control program in soybeans usually begins with either a preplant soil-incorporated herbicide treatment or a surface-applied preemergence herbicide after planting. If necessary, this may be followed by a rotary hoeing and one or two cultivations. The use of postemergence treatments usually depends on the number of weeds that have survived earlier treatments. The herbicides used in soybeans are listed in Table 17-4.

PEANUTS

The peanut root grows very fast after planting and germination, even though leaves emerge slowly. Under favorable conditions, the root may be 2, 6, and 15 inches long in 4, 6, and 12 days, respectively. In contrast, the first leaves usually do not emerge until 7 to 12 days after planting. This unique growth pattern allows an adventitious use of both mechanical and chemical weed control methods.

TABLE 17-4. Soybean Herbicides[1,2]

Preplant Incorporated	Preemergence to Crop	Postemergence to Crop[3]	
Alachlor	Alachlor	Acifluorfen	Glyphosate
Chloramben	Chloramben	Alachlor	Imazaquin
Chlorpropham	Chlorpropham	Bentazon	Lactofen
Clomazone	DCPA	Chlorimuron	Linuron
DCPA	Glyphosate	2,4-DB	Metribuzin
Ethalfluralin	Imazaquin	Dichlofop	Naptalam
Imazaquin	Linuron	Fenoxaprop	Paraquat
Metolachlor	Metolachlor	Fluazifop-P	Quizalofop
Metribuzin	Metribuzin	Fomesafen	Sethoxydim
Norflurazon	Naptalam		
Pendimethalin	Norflurazon		
Trifluralin	Paraquat		
Vernolate	Propachlor		

[1] Refer to herbicide section for information on rates, herbicide combinations, and restrictions (locate through index); see label for complete information.
[2] Refer to pages 440–442 for weeds controlled and pages 443–445 for trade names.
[3] Some of these herbicides require special application methods, e.g., directed sprays.

The rotary weeder, flexible-spike weeder, and cultivator are effective tools for weed control in peanuts. The rotary weeder can be used just before and again just after peanuts emerge to control small weed seedlings. Before wide array of herbicides available for use in soybeans, the timely use of the rotary weeder was considered to reduce cultivating time by 25% and hand-hoeing by 50%. Cultivating two to four times has been common, but this has been reduced by the use of herbicides.

Chemical Weed Control in Peanuts

Herbicides are used in peanuts as preplant soil-incorporated, preemergence, and postemergence treatments with herbicide combinations and sequential applications common. The herbicides used in peanuts are listed in Table 17-5.

Preplant soil-incorporated treatments are usually applied as part of the final seedbed preparation and planting operation. In one pass through the field, the herbicide is applied to the soil surface and incorporated into the soil, the seed planted, and the seedbeds shaped. Preemergence herbicides are applied after planting but before weed emergence. This may be immediately after planting, a delayed application, or at cracking time. The delayed application is applied about 6 days after planting to allow time for the root tip to grow deep into the soil and reduce the risk of injury from certain herbicides. A cracking-time treatment is also a delayed preemergence application applied when the emerging peanut shoot cracks the soil surface but has not emerged. This treatment usually is 6 to 10 days

TABLE 17-5. Peanut Herbicides[1,2]

Preplant	At Planting or Delayed Planting	At Cracking	Postemergence
Alachlor	Acifluorfen	Acifluorfen	Acifluorfen
Benefin	Alachlor	Alachlor	Bentazon
Ethalfluralin	Chloramben	Chloramben	2,4-DB
Metolachlor	Metolachlor		Metolachlor
Pendimethalin	Naptalam		Sethoxydim
Trifluralin	Trifluralin		
Vernolate	Vernolate		

[1] Refer to herbicide section for information on rates, herbicide combinations, and restrictions (locate through index); see label for complete information.
[2] Refer to pages 440–442 for weeds controlled and pages 443–445 for trade names.

after planting. However, both of these delayed applications may reduce the herbicide's effectiveness if the weed seedlings are beyond their optimal control period.

COTTON

Cotton was once planted thick and thinned to stand during the first hand-hoeing. This required considerable hand labor and cultivation. Even so, cotton was often still weedy at harvest time. With mechanical harvesting, weed control became increasingly important because weedy trash stained and contaminated cotton lint. Herbicides have replaced most hand labor and reduced cultivation. Now cotton is usually planted at final stand spacing, usually about 45,000 plants/acre.

Good production methods are needed that provide for the best possible growth and yields of cotton. Among these are well-prepared seedbeds, high-quality seed, well-adapted variety, adequately fertility, and proper control of diseases, insects, and weeds. Planting should be delayed until the soil temperature has reached at least 60°F which usually provides the desired stand. A vigorous crop of cotton helps control weeds, especially late-season weeds, through competition.

Flaming

The use of fire by means of a directed flame toward the base of the cotton stem can control many small weed seedlings. The flat-type burner usually uses volatile gases such as butane. The specialized equipment required is called a flame cultivator. It may be used when cotton stems are 1/4 inch in diameter. At this stage of growth, the cotton stem is tolerant to the small amount of heat required

TABLE 17-6. Cotton Herbicides[1,2]

Preplant Incorporated	Preemergence to Crop	Postemergence to Crop[3]	
Alachlor	Alachlor	Cyanazine	Linuron
Bensulide	Bensulide	DCPA	Methazole
Cyanazine	Cyanazine	Diuron	Oils, selective
DCAA	DCAA	DSMA	MSMA
Fluometuron	Diphenamid	EPTC	Oxyfluorfen
Metolachlor	Diuron	Fluazifop-P	Prometryn
Norflurazon	Fluometuron	Fluometuron	Sethoxydim
Pendimethalin	Metolachlor	Glyphosate	Trifluralin
Prometryn	Norflurazon		
Trifluralin	Pendimethalin		

[1] Refer to herbicide section for information on rates, herbicide combinations, and restrictions (locate through index); see label for complete information.
[2] Refer to pages 440–442 for weeds controlled and pages 443–445 for trade names.
[3] Some of these herbicides require special application methods, e.g., directed sprays.

to control small weed seedlings. However, the current high cost of fuel has greatly reduced the use of this practice.

Chemical Weed Control in Cotton

Herbicides used in cotton can be applied preplant, preplant soil-incorporated, preemergence, or postemergence. Preplant treatments to emerged weeds with nonselective, nonpersistent herbicides, e.g., glyphosate or paraquat, can be used in cotton. The other herbicides applied in cotton are listed in Table 17-6. Most preplant soil-incorporated or preemergence herbicides are applied just before or just after planting, often at the same time of planting with the application and planting equipment attached to the same tractor. Certain relatively persistent preplant soil-incorporated herbicides, e.g., trifluralin, can be applied in the fall or at any time up to the date of planting.

Selective herbicidal oils such as Varsol or Stoddard are effective postemergence sprays on cotton. They are applied when the cotton and weeds are small. When used without other herbicides, repeat applications are required to control weeds that emerge later. If these oils are used after cracks appear in the bark of the cotton stem injury may occur. The use of this practice is no longer widely used because of the high cost of petroleum products.

SUGAR BEETS

Sugar beet "seed" is actually a fruit and may contain more than one seed. However, monogerm "seed," *often* containing only one seed, is usually planted. If

Figure 17-5. Diuron applied in a band over the cotton row at planting time. Areas between the rows have been cultivated. This treatment usually controls annual weeds for at least 6 weeks. (E. I. du Pont de Nemours and Company.)

more than one plant develops from these "seed," they should be thinned to one plant for maximum production. The "seed" is usually planted at a high rate and thinned to the desired stand mechanically or by hand. Some farmers use precision planting and thereby achieve lower seeding rates. The weeds that escape the usual previous herbicide treatment can also be removed during the thinning operation.

The major weed problem in sugar beets is annual weeds, both grasses and broadleaf weeds. They are particularly troublesome at emergence through thinning time and after layby. Favorable production methods, discussed in the previous crops, induce a vigorous crop that helps control the weeds through competition.

Chemical Weed Control in Sugar Beets

Preplant and preemergence herbicides used in sugar beets are degraded fairly rapidly in the soil and control weeds for only about 4 to 6 weeks. Therefore, an additional herbicide application is required later to give season-long weed control. The herbicides used in sugar beets are listed in Table 17-7.

TABLE 17-7. Sugar Beet Herbicides[1,2]

Preplant Incorporated	Preemergence to Crop	Postemergence to Crop
Cycloate	Diethatyl	Dalapon
Diethatyl	Endothall	Desmedipham
Ethofumesate	Ethofumesate	Endothall
EPTC	Paraquat	EPTC
Pebulate	Pyrazon	Ethofumesate
Pyrazon		Phenmedipham
		Pyrazon
		Sethoxydim
		Trifluralin

[1] Refer to herbicide section for information on rates, herbicide combinations, and restrictions (locate through index); see label for complete information.
[2] Refer to pages 440–442 for weeds controlled and pages 443–445 for trade names.

SUGARCANE

Sugarcane is a perennial crop with harvests at 12 to 24 month intervals. Following the initial planting, two to four sequential harvests are made before productivity declines and replanting is necessary. The crop is reproduced vegetatively by planting 12- to 24-inch pieces of the stem containing two to four buds, one bud at each node. These vegetative propagules are referred to as "seed pieces." The initial planting is referred to as *plant cane* and subsequent crops as *ratoon* or *stubble* cane.

Sugarcane requires a tropical or semitropical climate with a continuous or long-growing season, high soil fertility, and abundant rainfall or irrigation. Under these conditions, all types of weeds flourish—annuals, biennials, and perennials. Weeds need to be controlled from planting until the row middles are covered by sugarcane foliage. Depending on environmental conditions, this usually takes 4 to 7 months for plant cane and 2 to 5 months for a ratoon crop. Weeds developing after "close-in" are usually controlled by competition of the sugarcane plant.

Chemical Weed Control in Sugarcane

A chemical weed control program for sugarcane usually requires at least two herbicide applications, one immediately after planting but before crop emergence and another postemergence to the crop but before "close-in." Glyphosate is used preplant or as a spot treatment within the crop, usually for perennial weed control. However, it should not be applied to vegetation in or around ditches, canals, or ponds containing water to be used for irrigation. Glyphosate can also

TABLE 17-8. Sugarcane Herbicides[1,2,3,4]

Ametryn	Dicamba	Matribuzin
Asulam	Diuron	Paraquat
Atrazine	Glyphosate	Simazine
2,4-D	Hexazinone	Terbacil
		Trifluralin

[1] Refer to herbicide section for information on rates, herbicide combinations, and restrictions (locate through index); see label for complete information.

[2] Refer to pages 440–442 for weeds controlled and pages 443–445 for trade names.

[3] All of these herbicides except asulam and glyphosate can be applied preemergence and/or postemergence to the crop; asulam is only preemergence and glyphosate only preplant or spot.

[4] Some of these herbicides require special application methods, e.g., directed sprays.

be used to control undesirable sugarcane plants. The herbicides used in sugarcane are listed in Table 17-8.

TOBACCO

Tobacco has very small seed and initially the emerging seedlings grow very slowly. Therefore, the plants are started from seed in plant beds and later transplanted to the field. Different weed control programs are required in each of these situations.

Seed Beds

Soil-borne diseases, insects, and nematodes, as well as weeds, may create a problem in tobacco seed beds. A soil fumigant that controls all of these plant pests is usually used, such as methyl bromide, metham, or dazomet. The seedbed should be well prepared without clods and porous so that the fumigant thoroughly permeates the soil. Adequate soil moisture is also required to convert metham and dazomet into their toxic degradation products. Adequate soil moisture also initiates seed germination, which makes all three of these fumigants more effective. Dormant seeds and seeds with impermeable seed coats (hard seed) may not be killed by these fumigants.

Methyl bromide is usually injected into the soil and the soil covered with a gas-proof plastic tarp in a single operation. After at least 48 hours the tarp can be removed and planting must be delayed for at least an additional 72 hours.

The label for metham recommends that it be applied in the fall if possible. It can be applied using either a tarp or drench method. In the tarp method, metham is diluted with water (40 gal/100 yd^2) and applied to the soil surface and immediately covered with a plastic tarp for no less than 1 day but not more than 2 days. Seven days later the soil should be "loosened" to a depth of 2 inches and planting delayed until at least 21 days after the metham application. In the drench method, metham is diluted with water (150 to 200 gal/100 yd^2). Application may be made with sprinklers, sprayers with nozzles, or any suitable equipment. Additional manipulations are the same as for the tarp method.

Dazomet is applied as a preplant soil-incorporation treatment using conventional equipment. The application is made 3 weeks before seeding in the fall or summer and 4 weeks before seeding in the early spring.

Field Transplants

The influence of cultivation on tobacco transplants in the field was studied in North Carolina. Cultivation did not increase yields on sandy soils if the weeds were otherwise controlled. However, on loam or clay-loam soils increased yields were observed with one or two cultivations; additional cultivations did not further increase yields.

Herbicides are commonly used in field-transplanted tobacco. Most of these are applied as a pretransplant soil-incorporation treatment. These include benefin, diphenamid, isopropalin, napropamide, pebulate, and pendimethalin. Diphenamid can also be applied posttransplant over the top of the transplants and at layby. Napropamide and pendimethalin can also be applied at layby.

SUGGESTED ADDITIONAL READING

Chandler, J. M., A. S. Hamill, and A. G. Thomas, 1984, *Crop Losses due to Weeds in Canada and the United States*, Weed Science Society of America, Champaign, IL.

Crop Protection Chemical Reference, 1989, Wiley, New York (revised annually).

Sprague, M. A., and G. B. Triplett, Eds., 1986, *No-Tillage and Surface-Tillage Agriculture*, Wiley, New York.

Weed Control Manual and Herbicide Guide, 1988, Meister, Willough, OH (revised annually).

For chemical use, see the manufacturer's or suppliers's label and follow these directions. Also see the Preface.

18 Vegetable Crops

Losses caused by weeds in vegetable crops often mean the difference between profit and losses. Monaco et al. (1981) reported severe losses in direct seeded tomatoes with full season competition of cocklebur, tall morningglory, redroot pigweed, and large crabgrass. Mendt (1979) found that 4 weeks of competition from cocklebur or a mixed stand of weeds in transplanted pepper reduced yields. Friesen (1978, 1979) found similar results with cucumbers and transplanted tomatoes. Most of this work indicates that weed competition in the first 4 weeks of crop growth can result in severe crop reduction. A review on weed crop competition by Zimdahl (1980) gives many other examples of serious weed losses in vegetable crops.

Vegetable crops are generally more vulnerable to weed competition than agronomic crops because many of them are short season crops and they are usually weak competitors with weeds. Weeds can reduce efficiency of disease and insect pest protection and thereby lower quality and marketability and can cause crop losses by interfering with mechanical and hand harvesting.

Past methods of weed control in vegetable crops have centered around the cultivator and hoe, but hoeing is expensive and labor is in short supply. With increased labor costs, more efficient methods of weed control are necessary if the grower is to make a profit. Tillage (cultivation) is discussed in Chapter 3. Experienced growers usually follow these practices. Growers can control weeds far more efficiently with less human effort by using selective herbicides. Savings in labor are important to the grower as well as to the consumer; both reap the benefits. Comprehensive listings of herbicides registered for vegetable crops and other crops are published annually. See the suggested additional reading for the listing.

Vegetable crop production is a very specialized business with crops grown intensively under varying conditions of soil and climate. A herbicide may produce excellent results under one set of conditions and either injure the crop or fail to control weeds under other conditions. The herbicide label should be checked for specific instructions and precautions.

Anyone not experienced in using herbicides on a given crop should proceed cautiously. With experience the farmer can expand the control program.

MINOR-CROP HERBICIDES

Development of an agricultural pesticide is expensive, currently \$10–\$50 million per product. Costs of synthesis, screening, toxicology, environmental studies, development, registration, and service to the product must be paid ultimately through sales of the product. See Chapter 4 for additional information on registration requirements.

Average of any one vegetable crop is small when compared with that of small grain, corn, soybeans, or cotton. Therefore, vegetables are considered minor crops with reference to pesticide use. It is generally not economically feasible for a chemical company to develop a herbicide for a minor crop if it is not suitable for a major crop as well. As a result, most herbicides used in vegetable crops were previously developed for a major crop. Also, most minor crops are expensive high-cost crops. The liability or risk of lawsuits further reduces a company's interest in adding such crops to the label.

An additional problem with developing herbicides for minor crops is that of obtaining federal registration for a particular use. Although a considerable amount of data used for registration of the herbicide for use on a major crop also applies to its use on minor crops (e.g., toxicology of the compound, environmental impact, see Chapter 4), a considerable amount of specific information about the chemical's effect on the minor crop must also be developed. The cost of this research may exceed potential profit to the chemical company. A federal government program in cooperation with state experiment stations, Inter-regional Project No. 4 (IR-4), was established in the 1960s to alleviate this problem. This project coordinates and supports federal and state research directed toward the development of the information required for registration of a pesticide for use in a minor crop by the Environmental Protection Agency (EPA).

ARTICHOKES

Artichokes are perennials, and once established they are not replanted for many years. In addition to cultivation, there are several herbicides that can be used in this crop.

Napropamide, diuron, or simazine are applied to the soil as a directed spray to control annual weeds as they germinate. Diuron and simazine are relatively persistent in soils and rotational restrictions for sensitive crops must be observed. Additionally, for crop safety simazine should not be applied to sandy soils.

Oxyfluorfen can be applied as a directed spray for postemergence and preemergence control of annual weeds. Spray contact with the crop will cause injury, therefore care must be exercised with application of oxyfluorfen. Glyphosate can be used for postemergence control of weeds prior to crop emergence. In California only, sethoxydim can be used in artichokes to selectively control emerged annual grasses and perennial bermudagrass, johnsongrass, and ryegrass.

ASPARAGUS

Asparagus is a perennial crop that may remain in production 10–20 years before replanting or rotating to another crop; therefore, year-round weed control must

TABLE 18-1. Herbicides Used in Asparagus[1,2]

Herbicide	Seeded Asparagus (Crown Production)	Crown Planting	
		New	Established
Methyl bromide	SF	SF	—
Linuron	Pre, post-1	Post-1	Pre, post-1
Chloramben	Pre	—	—
Paraquat	SB	SB	Post-2
Glyphosate	SB	SB	Post-2
Fluazifop	Post-1	Post-1	Post-1
Sethoxydim	Post-1	Post-1	—
Napropamide	—	—	Pre
Simazine	—	—	Pre
Diuron	—	—	Pre
Terbacil	—	—	Pre
Metribuzin	—	—	Pre
2,4-D	—	—	Post-2
Dalapon	—	—	Post-2
Dicamba	—	—	Post-2

[1] There are limitations on some of the above uses relative to geographic location, formulation, soil type, and stage of crop growth. See product labels for specifics.
[2] Pre, preemergence to weeds and crop; Post-1, postemergence to weeds and crop; Post-2, either directed to base of crop or applied to weeds when crop is unemerged; SB, stale bed application made after weed emergence, but prior to planting or after planting prior to crop emergence; SF, soil fumigation.

be provided without injuring the crop. There is a wide selection of herbicides for weed control in asparagus (Table 18-1). Herbicide choice(s) depends on many factors such as crop age, crop growth stage, geographic location, target weed species, and soil type.

Asparagus is established by transplanting either crowns that are produced from seed or seedlings produced in the field or in greenhouses.

Crown Production

For crown production from direct seeding in the field, methyl bromide fumigation of soil can destroy most weed seeds and many harmful soil organisms prior to planting. If fumigation is not used there are several herbicides that can aid in reducing weed competition. Paraquat and glyphosate can be applied either prior to seeding or after seeding but prior to asparagus emergence to control emerged weeds. This is called a *stale bed application.* The stale bed technique consists of preparing the soil for seeding, waiting 7 to 10 days for weed germination, applying paraquat or glyphosate, and then following with seeding. Herbicide application must be followed by minimum soil disturbance to prevent bringing up additional weed seed. Additionaliy, this technique would not be feasible on soil types that crust severely thereby interfering with crop seeding.

Since asparagus seeds are slow to germinate, 2 to 3 weeks after planting, paraquat and glyphosate can be applied to emerged weeds during this period of time. This treatment is preemergence to the crop and postemergence to the weeds and is a variation of the stale bed technique.

Preemergence herbicides that can be applied subsequent to seeding include chloramben and linuron. Linuron can also be applied postemergence to weeds when the asparagus fern is 6 to 18 inches tall. Since selectivity of linuron is based on limited spray solution retention by the asparagus fern, no surfactant or crop oil is used in the spray mixture.

Newly Planted Crowns

There are only a limited number of herbicides for first year crown plantings, therefore it is essential that fields for these plantings be free of noxious and difficult-to-control weeds such as field bindweed, Canada thistle, nutsedges, bermudagrass, and johnsongrass. Methyl bromide fumigation can be utilized to control problem weeds prior to planting but this is generally too expensive for crown plantings.

Linuron applied postemergence to new crown plantings will control emerged and germinating annual broadleaf weeds. Fluazifop or sethoxydim applied postemergence will control emerged annual grasses and some perennial grasses such as johnsongrass and bermudagrass. No preemergence herbicides are available for first year plantings so weed control is accomplished by cultivation and the above postemergence herbicides.

Established Plantings

There is a vast array of herbicides for established plantings. Because asparagus is dormant in the winter and crowns are planted relatively deep (8 to 12 inches), growers can control weeds by discing the soil over the crop rows during this period. Herbicides are effectively used during the rest of the year.

Napropamide can be applied early in the season prior to weed and crop emergence. Diuron, simazine, terbacil, and metribuzin are preemergence herbicides that can be applied prior to crop emergence and/or immediately after last harvest. Choice of material and time of application will depend on weeds to be controlled, soil type, economics, and geographic location.

Herbicides that can be applied postemergence to control emerged annual and/or perennial weeds include linuron, 2,4-D, dicamba, paraquat, glyphosate, dalapon, fluazifop, and sethoxydim. Paraquat, glyphosate, or 2,4-D can be applied to emerged weeds in the spring prior to spear emergence or immediately after last harvest. Paraquat would be the choice for control of annual broadleaf and grass weeds, while glyphosate should be used for perennial weeds and 2,4-D for certain annual and perennial broadleaf weeds. Dicamba or dicamba plus 2,4-D also controls annual and perennial broadleaf weeds in asparagus; only one application per season is permitted.

Paraquat, glyphosate, 2,4-D, and dicamba will cause crop damage if they contact emerged spears and/or ferns.

Linuron can be used as a directed spray to the base of the ferns to control annual broadleaf weeds. Dalapon can be used to control quackgrass and bermudagrass in asparagus with applications made during the cutting season (immediately after harvest) and later directed under fern growth. To avoid injury with dalapon, contact with crop must not occur.

Fluazifop can be applied to any age planting for control of emerged annual grasses. Contact with crop does not have to be avoided. Sethoxydim can be used only on nonbearing plantings, that is, it cannot be applied within 1 year of the first harvest.

BEANS

Beans tolerate weed competition more successfully than many other vegetable crops because of their rapid emergence and early growth (Zimdahl, 1980). Most harmful are (1) annual weeds emerging soon after planting and not removed, and (2) tall weeds that compete for light and other factors. Nutsedges and quackgrass are also serious competitors. Preirrigation and planting into moisture is used in arid areas of the country to minimize early-annual-weed competition.

There are several herbicides and methods of application that can be used in bean culture. The several types of beans grown (see Table 18-2) commercially vary considerably in their tolerance to various herbicides; herbicide labels must be checked for specific use instructions.

Choice of a herbicide(s) for inclusion in a weed management program depends on weed species to be controlled, geographic limitations, type of bean grown, and soil type.

Paraquat or glyphosate can be used in the stale bed technique to non-selectively control emerged annual weeds and, in the case of glyphosate, certain perennial weeds. Dalapon can be applied preplant to the crop to control emerged quackgrass. This can be applied either in the fall or in the spring but at least 4 to 5 weeks before planting. Trifluralin, pendimethalin, or ethalfluralin is applied only preplant soil-incorporated, whereas chloropropham, alachlor, and metolachlor can be applied preplant soil-incorporated or preemergence to the soil surface immediately following planting. EPTC can be applied preplant soil-incorporated or at time of last cultivation referred to as crop layby.

Chloramben and DCPA are strictly preemergence soil-surface applied while bentazon and sethoxydim are postemergence herbicides. The latter two require addition of crop oil concentrate at 1% (v/v) to the spray solution for optimal activity. Timing of postemergence treatments is guided by crop growth stage, target weed, growth stage, and preharvest interval limitations.

TABLE 18-2. Herbicides Used in Beans[1,2]

Herbicide	Application[3]
Paraquat	SB
Glyphosate	SB
Dalapon	PP
Trifluralin	PPI
Pendimethalin	PPI
EPTC	PPI, LBY
Ethalfluralin	PPI
Chloropropham	PPI, pre
Alachlor	PPI, pre
Metolachlor	PPI, pre
Chloramben	Pre
DCPA	Pre
Bentazon	Post
Sethoxydim	Post

[1] Every herbicide listed in not suitable for use on all types of beans (dry, lima, snap, pole); *check labels.* Additionally, geographic restrictions may apply.
[2] Specific combinations of these herbicides are used for beans; *check labels.*
[3] LBY, layby at time of last cultivation; PP, preplant to crop; PPI, preplant soil incorporated; pre, preemergence to weeds and crop; post, postemergence to weeds and crop; SB, stale bed application, applied prior to or after seeding, but before crop emergence.

BEETS, RED

Red or table beets and sugarbeets have similar tolerances to herbicides and the same application methods are used. However, all sugarbeet herbicides are not registered for use in red beets.

Stale bed application of glyphosate to emerged weeds is useful for eliminating most annual weeds and certain perennial weeds. Cycloate is applied preplant soil incorporated while diethatyl can be applied preplant incorporated or surface applied after planting. Pyrazon can be applied to the soil surface immediately after planting or postemergence when beets are at the two-leaf stage and before weeds have four leaves. Phenmedipham is applied postemergence to beets past the two-leaf stage and when weeds are in the cotyledon to four-leaf stage. Restrictions concerning soil type, growth stage, and environmental conditions exist for these herbicides.

CARROT FAMILY (CARROTS, CELERY, DILL, PARSLEY, PARSNIPS)

Several herbicides are registered for weed control in crops in the carrot family. Only one of these products (petroleum solvents) can be used for all members of this crop group. Highly refined petroleum solvents represent one of the oldest selective weed-control methods in vegetables. These solvents are like paint solvent or drycleaning fluid. They are sold under several names including Stoddard solvent, Varsol, 350° thinner, selective weed oil, and carrot oil. Kerosene or No. 1 fuel oil has been used in the past, but they may cause a kerosene-like flavor and should be avoided. Even the highly refined petroleum solvents must be applied before the carrot root reaches 1/4 inch in diameter to avoid off-flavors. The solvent used should be fresh from the refinery. These solvents control most emerged annual weeds and are applied anytime after the carrot has two true leaves and before the root is 1/4 inch in diameter.

If needed, petroleum oils can be applied two to three times early in the season. Weeds are easiest to kill when less than 2 inches tall. Wetting sprays are applied.

High temperatures, but not over 85°F, usually provide best results. Carrots are more likely to be injured if the temperature is high, humidity is high, or plants are wet when they are sprayed. Increased plant toxicity when plants are damp can be used to the grower's advantage. These conditions permit use of less oil, or provide more effective weed control. Therefore, some producers have a standard practice of spraying petroleum solvents between dusk and dawn.

TABLE 18-3. Herbicides Used in the Carrot Family[1]

		Crop				
Herbicide	Application[2]	Carrot	Celery	Dill	Parsley	Parsnips
Allyl alcohol	SF	NR	R	NR	NR	NR
Metham	SF	R	R	NR	NR	R
Glyphosate	SB	R	R	NR	NR	NR
Paraquat	SB	R	NR	NR	NR	NR
Bensulide	PPI	R	NR	NR	NR	NR
Trifluralin	PPI	NR	R	NR	NR	NR
Prometryn	Pre, post	NR	R	NR	NR	NR
Metribuzin	Post	R	R	NR	NR	NR
Fluazifop	Post	R	NR	NR	NR	NR
Linuron	Post	R	R	NR	NR	R
Petroleum solvents	Post	R	R	R	R	R

[1] Check labels for precautions, restrictions, and geographic limitations.

[2] SF, soil fumigation for seedbeds in production of transplants; SB, stale bed applications after weed emergence, but prior to planting and/or crop emergence; PPI, preplant soil incorporated; post, postemergence to crop and weeds; pre, preemergence to crop and weeds.

[3] R, registered; NR, not registered.

Nine of the 11 products listed in Table 18-3 are registered for use in carrots, eight for celery, three for parsnips, and only one for dill and parsley.

COLE CROPS (BROCCOLI, BRUSSELS SPROUTS, CABBAGE, CAULIFLOWER)

All cole crops appear to respond similarly to herbicides; however, one must check the herbicide label for the registered use for each before using.

Cole crops may be directly seeded in the field or transplanted. In general, transplated cole crops are more tolerant of herbicides than direct-seeded crops. However, of the nine herbicides registered for the various cole crops (see Table 18-4) only oxyfluorfen is restricted to transplants. Selectivity of the herbicides in direct-seeded cole crops is achieved by application timing and, in some cases, rate adjustments.

Bensulide, trifluralin (Figure 18-1), DCPA, napropamide, and sethoxydim are herbicides that are used in direct-seeded and transplanted broccoli, brussels sprouts, cabbage, and cauliflower. Consult manufacturer's labels for rates, geographic restrictions, and precautions.

Oxyfluorfen is applied to the soil surface prior to transplanting broccoli, cabbage, or cauliflower. Minimal soil disturbance during the transplanting process is essential for this treatment to be effective.

TABLE 18-4. Herbicides Registered in Cole Crops[1]

		Registered in[3]			
Herbicide	Application[2]	Broccoli	Brussel Sprouts	Cabbage	Cauliflower
Methyl bromide	SF	S, T	—	—	S, T
Glyphosate	SB	S, T	—	S, T	S, T
Paraquat	SB	S, T	—	S, T	S, T
Bensulide	PPI, pre	S, T	S, T	S, T	S, T
Trifluralin	PPI	S, T	S, T	S, T	S, T
Oxyfluorfen	PT	T	—	T	T
DCPA	Pre	S, T	S, T	S, T	S, T
Napropamide	Pre	S, T	S, T	S, T	S, T
Sethoxydim	Post	S, T	S, T	S, T	S, T

[1] Consult labels for precautions, restrictions, and geographic limitations.

[2] SF, soil fumigation; SB, stale bed application after weed emergence but prior to planting and/or crop emergence; PPI, preplant soil incorporated; pre, preemergence to crop and weeds; PT, pretransplant soil surface; post, postemergence to crop and weeds.

[3] S, registered for use in direct-seeded crops; T, registered for use in transplanted crops; M, not registered.

Figure 18-1. Annual weed control in broccoli with a preplant soil-incorporation application of trifluralin. *Center*: Treated band. *Left and right*: Untreated.

Glyphosate or paraquat is applied to emerged weeds prior to seeding or transplanting broccoli, cabbage, and cauliflower. The usefulness of these applications is also dependent on minimal soil disturbance during the planting process.

CORN, SWEET

Many of the weed-control methods described for field corn in Chapter 17 apply to sweet corn. Some sweet corn varieties may be more sensitive than field corn to certain herbicides, therefore, herbicide labels must be carefully reviewed for precautions.

CUCURBIT FAMILY (CUCUMBERS, MELONS, PUMPKINS, AND SQUASH)

Members of the cucurbit family are primarily warm-weather crops; thus their major weed problems are summer-annual weeds, both grasses and broadleafs such as barnyardgrass, crabgrass, pigweed, and lambsquarters.

Herbicides registered for the various cucurbit crops are listed in Table 18-5. Although there are 13 herbicides registered for weed control in cucurbit crops, only a few can be used on all four types of cucurbit crops and even fewer are used in all the various growing regions. Restrictions with respect to cucurbit type are generally based on differences in herbicide selectivity between the varieties. In the case of a material such as methyl bromide it is merely the lack of sufficient information for registration.

TABLE 18-5. Herbicides Used in Curcurbits[1,2]

Herbicide	Cucumber	Melon	Pumpkin	Squash
Methyl bromide	SF	SF	—	—
Paraquat	SB	SB[3]	SB	SB
Glyphosate	SB	SB[4]	SB	SB
Bensulide	PPI	PPI[5]	PPI	PPI
Clomazone	—	—	PPI	—
Napropamide	—	PPI, pre[6]	—	—
Chloramben	Pre[7]	Pre[7]	PPI, pre	PPI, pre
Ethalfluralin	Pre	Pre[8]	—	—
Propachlor	—	—	Pre	—
Naptalam[9]	Pre, post-1	Pre, post-1[10]	—	—
DCPA	Post-2	Post-2[11]	—	Post-1
Trifluralin	Post-3	Post-3[8]	—	—
Sethoxydim	Post-4	Post-4	Post-4	Post-4

[1] There are limitations on the above uses relative to geographic location, formulation, crop, and stage of growth. See product label for specifics.
[2] SF, soil fumigation; SB, stale bed, application prior to seeding and transplanting or after seeding, but before crop emergence; PPI, preplant soil incorporated; pre, pre emergence to weeds and crop; post-1, immediately after transplanting or just before vining; post-2, 4–5 true leaf stage of crop, weeds unemerged; post-3, directed spray at 3–4 true leaf stage soil incorporated; post-4, postemergence to crop and weed.
[3] Cantaloupe/muskmelon, crenshaw, watermelon.
[4] Cantaloupe, watermelon.
[5] Cantaloupe, muskmelon, persian melon, crenshaw, watermelon.
[6] Crenshaw, persian melon, watermelon.
[7] Activated charcoal–vermiculite mix placed in seed furrow prior to herbicide application. See label for specifics.
[8] Cantaloupe, watermelon.
[9] A tank mix combination of bensulide and naptalam can be used PPI or pre; cantaloupe, cucumber, muskmelon, watermelon.
[10] Cantaloupe, muskmelon, watermelon.
[11] Cantaloupe, honeydew, watermelon.

LETTUCE AND GREENS

Unlike cucurbit crops, lettuce and greens and cool-weather plants. Their annual-weed complex may includc annual bluegrass, chickweed, henbit, common groundsel, and many others. Herbicides to control these weeds in commercial leaf crops are given in Table 18-6.

Lettuce

Although lettuce is usually a cool-weather crop, it may be subjected to high temperatures. In Arizona and southern California, lettuce is planted in late August and September when air and soil daytime temperatures may exceed 100°F. Lettuce will not germinate above 90°F. However, the beds are kept wet, so

TABLE 18-6. Herbicides Used in Lettuce and Greens[1,2]

Lettuce		Collards, Kale, Mustard Greens, Turnip Greens		Spinach	
Methyl bromide	SF	Paraquat	SB	Glyphosate	SB
Glyphosate	SB	Glyphosate	SB	Chloropropham	Pre
Paraquat	SB	Trifluralin	PPI	Diethatyl	Pre
Benefin	PPI	DCPA	Pre	Phenmedipham	Post
Bensulide	PPI			Sethoxydim	Post
Pronamide	PPI, pre			Fluazifop	Post
	Post				
Sethoxydim	Post				

[1]SF, soil fumigation; SB, stale bed, prior to or after seeding, but before crop emergence; PPI, preplant soil incorporated; pre, preemergence; post, postemergence.
[2]Consult product labels for specifics relative to limitations.

they are cooled by the constant evaporation of water from the bed surface in this arid climate. This cooling allows shallow-planted lettuce seeds to germinate. Under these conditions, a summer annual-weed complex is more common than those weeds named above.

In specialized situations, such as lettuce on plastic mulch, methyl bromide is utilized preplant as a soil fumigant to eliminate annual and perennial weeds. Another option for weed management in lettuce is the use of glyphosate or paraquat to control emerged weeds prior to planting or after seeding but before crop emergence. Obviously, emerged crop plants will be killed.

Preplant soil-incorporated applications of benefin (see Figure 18-2), bensulide, or pronamide control annual weeds in lettuce. Pronamide can also be applied preemergence or postemergence. Sethoxydim is used as a postemergence treatment for control of emerged annual and some perennial grasses.

Figure 18-2. Annual weed control in lettuce with a preplant soil-incorporation application of benefin. *Center*: Treated band. *Leaft and right*: Untreated.

Greens

Greens included here are collards, kale, mustard greens, turnip greens, and spinach. Herbicide tolerances of these first four crops are quite similar; however, spinach reacts differently to herbicides (see Table 18-6). Collards, kale, mustard, and turnips are closely related members of the mustard family while spinach is from a different plant family.

ONIONS

Although the home gardener often grows onions from bulbs (sets) or transplants, most commercial growers use seed. Weeds are particularly serious in onions produced from seeds because onions germinate and emerge slowly, plus their cylindrical-upright leaves do not shade the soil to suppress weed growth. However, these same leaves are an advantage when using certain contact herbicides whose spray droplets bounce off the onion leaves, but remain on leaves of many broadleaf weeds.

Several herbicides used as early postemergence treatments for annual-weed control in onions must be applied only at certain stages of growth to avoid injury to the crop. These stages are classified as loop (crook), flag, one-true-leaf, and two-true-leaf (Figure 18-3). The following discussion applies primarily to *direct-seeded onions.*

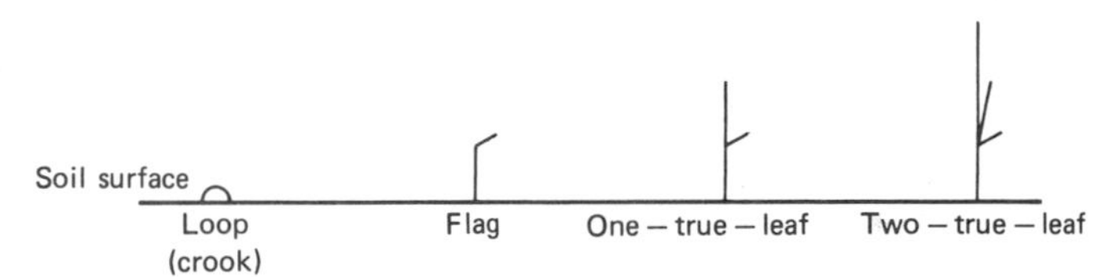

Figure 18-3. Stages of onion growth.

Figure 18-4. Barnyardgrass control in onions with a preemergence application of DCPA.

Glyphosate or paraquat can be used to control annual weeds that emerge before the onions. Neither herbicide has any residual soil activity and hence a second flush of weeds may appear, particularly if the soil has been disturbed through cultivation. Soil applied and postemergence herbicides are available to aid in suppressing weeds in onion plantings. However, these herbicides are specific for the type of onion grown. Green bunching onions are generally short season, nonbulbing, and the entire shoot is eaten. Dry bulb onions require a longer growing season and only the bulb is consumed.

Herbicides that can be used for green and dry bulb onions include glyphosate, paraquat, DCPA preemergence (see Figure 18-4), and chloropropham applied either preemergence or postemergence.

For dry bulb onions only, methyl bromide as a preplant soil fumigant will control many annual and perennial weeds. Due to the expense involved its use is warranted in only very specialized farming operations. Bensulide can be applied preplant soil incorporated to onions only in Texas and New Mexico. Oxyfluorfen controls many annual broadleaf weeds when applied postemergence after the onions have at least two fully developed true leaves. Fluazifop can be used to control emerged annual grasses in onions.

Injury to onions from some of the above mentioned herbicides can occur if the products are used incorrectly or under adverse or poor growing conditions. Consult labels for specifics.

PEAS

English (Green, Garden)

Because this type of pea is a cool-weather crop, the primary weed problem is annual cool-weather weeds such as chickweed, henbit, shepherdspurse, wild oat, and annual bluegrass. The perennial weed Canada thistle is also a problem in some areas.

Several herbicides have been developed to control annual and, in some cases, perennial weeds in English peas (see Table 18-7). Diallate, EPTC, triallate, trifluralin, clomazone, propham, or metolachlor is used as preplant soil-incorporation treatments. Oryzalin, propachlor, alachlor, and metolachlor can be used preemergence. Postemergence applications for peas include bentazon, dalapon, diclofop, MCPA, MCPB, and sethoxydim.

Although the list of herbicides for weed control in English peas is lengthy there are geographic limitations and special application methods for use of certain herbicides. Additionally, peas show varietal differences in their tolerance to some of the herbicides listed in Table 18-7. The label must be checked before use.

Southern (Blackeyed, Cowpea, Crowder)

Unlike English peas this type of pea is a warm-weather crop, hence weed problems in this crop consist of warm-season annuals such as crabgrass,

TABLE 18-7. Herbicides Used in Peas[1,2]

Herbicide	English Peas (Green, Garden)	Southern Peas (Cowpea, Blackeyed, Crowder)
Glyphosate	SB	SB
Paraquat	SB	SB
EPTC	PPI	PPI
Propham	PPI	—
Trifluralin	PPI	PPI
Clomazone	PPI	—
Metolachlor	PPI, pre	PPI, Pre
Oryzalin	Pre	—
Propachlor	Pre	—
DCPA	—	Pre
Chlorpropham	—	Pre
Diallate	PPI	—
Triallate	PPI	—
Alachlor	Pre	—
Bentazon	Post	Post
Dalapon	Post	—
Diclofop	Post	—
MCPA	Post	—
MCPB	Post	Post
Sethoxydim	Post	Post

[1]Consult product labels for specifics. Some uses for English peas are restricted to type of planting, i.e., processing, dry, fresh market, etc. Also geographic limitations exist for many of the products.
[2]SB, stale bed, prior to or after seeding, but before crop emergence; PPI, preplant soil incorporated, pre, preemergence to crop and weeds; post, postemergence.

goosegrass, pigweeds, cocklebur, and perennials such as bermudagrass and johnsongrass. Although English and Southern peas are both legumes, they are dissimilar enough that not all herbicides listed in Table 18-7 apply to both types.

Glyphosate and paraquat can be used in the stale bed technique while preplant soil-incorporation treatments include EPTC, trifluralin, and metolachlor. Chloropropham, DCPA, and metolachlor can be used preemergence while bentazon, MCPB, and sethoxydim can be used postemergence.

POTATOES

White (Irish)

The potato is the leading vegetable in the world and ranks with the major cereals as one of the leading food crops. The major weed problem in potatoes is annual broadleaf and grass weeds; in some areas perennial weeds such as nutsedges, quackgrass, or johnsongrass can also be serious.

Dalapon controls quackgrass as a preemergence treatment, applied in the fall or spring. Dalapon or glyphosate can be used as preplant treatments to emerged quackgrass. Preplant treatments are usually plowed before planting potatoes. EPTC is incorporated into the soil preplant and/or layby to suppress quackgrass and nutsedge growth. Postemergence applications of sethoxydim are used to control quackgrass and johnsongrass.

Glyphosate or paraquat applied preplant or postplant control many annual weeds. Postplant applications must be made to emerged weeds prior to crop emergence to prevent crop injury.

Many herbicides can be applied to the soil to control annual weeds in potatoes (see Table 18-8). Preplant soil-incorporated treatments include EPTC and trifluralin while preemergence herbicides are DCPA, pendimethalin, oryzalin, metolachlor, linuron, and metribuzin.

Many of the preemergence herbicides are used as tank mix combinations. The majority of these tank mixes include metribuzin which controls many annual broadleaf weeds coupled with herbicides such as pendimethalin, oryzalin, or metolachlor to provide grass control. Linuron, also good for annual broadleaf control, is frequently tank mixed with these grass control herbicides.

TABLE 18-8. Herbicides Used in White Potatoes[1,2]

Herbicide	Application
Glyphosate	SB
Paraquat	SB
Dalapon	PP
EPTC	PPI, LBY
Trifluralin	PPI
DCPA	Pre, LBY
Pendimethalin	Pre
Oryzalin	Pre
Metolachlor	Pre
Linuron	Pre
Metribuzin	Pre, post
Sethoxydim	Post

[1] Some of these treatments have specific limits such as geographic areas, soil types, application time relative to harvest and potato variety.

[2] Certain herbicide combinations are also used. Consult product labels for specifics.

[3] SB, stale bed, prior to or after planting but prior to emergence; PP, preplant treatment to quakgrass; PPI, preplant soil incorporated; pre, preemergence to crop and weed; post, postemergence to crop and weeds; LBY, at crop layby.

EPTC or DCPA is used at crop layby to provide extended weed control in the crop cycle. Metribuzin or sethoxydim is used postemergence to control annual broadleaf weeds or annual grasses, respectively. Injury to potato can occur with postemergence applications of metribuzin if several days of cloudy weather precede applications. Also, potato varieties differ in metribuzin tolerance. Consult label for specifics.

Ametryn, paraquat, or diquat can be used for vine killing in potatoes.

Sweet

Worldwide, sweetpotato is an important vegetable and ranks seventh in production, based on weight, among the food crops of the world. The sweetpotato is used for human consumption, animal feed, and industrial purposes. It is a dicotyledonous plant in the morningglory (Ipomoea) family and is not related to the *yam* which is a monocotyledonous plant in the family Dioscoreaceae.

The sweetpotato is a warm-weather crop having a predominately prostrate vine growth habit that expands very rapidly and develops a relatively dense canopy. As with many crops, weed control is critical during the period from planting to canopy closure. Although the sweetpotato is an extremely important crop worldwide, relatively few herbicides are registered for use in this crop.

The crop is generally planted as stem pieces referred to as cuttings, slips, sprouts, or transplants. The transplants are produced by bedding roots in the soil and subsequently pulling the sprouts as they emerge. Chloramben can be applied preemergence to sweetpotato plant beds to control annual weeds.

Glyphosate can be applied to emerged weeds prior to transplanting in the field.

EPTC is used preplant soil-incorporated in certain regions for suppression of nutsedge and control of annual weeds. DCPA, chloramben, or a tank mix of DCPA plus chloramben can be applied preemergence. Generally, preemergence herbicides are applied as banded treatments over the row and cultivation is relied on for weed control in the row middles. Fluazifop can be applied postemergence for the control of emerged annual grasses, bermudagrass, and johnsongrass. Two applications are generally used for the latter two weeds.

FRUITING VEGETABLES

Eggplant, pepper, and tomato are similar in tolerance to herbicides. All three crops can be grown commercially from transplants or direct-seeded in the field. Although annual weeds are the major problem by either planting practice, they are particularly serious in direct-seedings. Weeds can be removed from direct-seeded plantings of these crops during thinning, but by this time competition may have already delayed the growth of young seedlings. Furthermore, labor costs for thinning may be doubled or tripled by heavy annual-weed infestations. Lack of early season weed control delays crop maturity and, in the case of tomatoes,

TABLE 18-9. Herbicides Used in Eggplant, Pepper, and Tomato[1,2]

Herbicide	Eggplant	Pepper	Tomato[3]
Methyl bromide	SF	SF	SF
Glyphosate	SB	SB	—
Paraquat	SB	SB, post-1	SB, post-1
Pebulate	—	—	PPI
Trifluralin	—	PPI	PPI
Bensulide	—	PPI, pre	PPI, pre
Napropamide	PPI, pre	PPI, pre	PPI, pre
Metolachlor	—	Pre	—
Metribuzin	—	—	PPI, post-2
Chloramben	—	Pre	Pre
DCPA	Pre	Pre	—
EPTC	—	—	LBY
Sethoxydim	—	Post-3	Post-3

[1]Some of the above herbicides can be used on both seeded and transplanted crops, whereas others can be used only on transplanted crops. See product labels for specifics and limitations.
[2]SF, soil fumigation; SB, stale bed, prior to or after seeding but prior to crop emergence, or transplanting; PPI, preplant soil incorporated; pre, preemergence to crop and weeds; post-1, directed postemergence between rows; post-2, directed or overtop postemergence; post-3, postemergence overtop crop.
[3]Specific combinations of these herbicides are used for tomatoes; *check labels.*

greatly reduces the efficiency of machine harvesting. Some common annual weeds of these crops are barnyardgrass, crabgrass, foxtails, lambsquarters, purslane, mustard, smartweed, and nightshades. Nightshades are particularly difficult to control because they are in the same family as these crops, the Solanaceae family.

In general, transplants are more tolerant to herbicides than the direct-seeded crop. However, when the direct-seeded crop is at a transplant size the plants are generally tolerant to herbicides. The size of the plant, method of application, and other critical factors vary for each herbicide; therefore, detailed information from the herbicide label must be followed.

When these crops are grown on plastic mulch, methyl bromide fumigation immediately prior to laying the plastic is generally used to control annual and perennial weeds as well as many soil borne diseases and insects. Bare ground between the plastic can be treated with herbicides listed for the respective crops in Table 18-9.

Eggplant

Relatively few herbicides can be used in eggplant, a minor vegetable generally handled as a transplanted crop. Prior to transplanting glyphosate or paraquat can be used to eliminate emerged weeds. Napropamide is used preplant soil-incorporated or preemergence after planting the crop. DCPA is also used preemergence.

Peppers

Pepper is an important vegetable and the two types grown are referred to as bell and nonbell. The bell type is produced primarily for the fresh market while the nonbell types are generally processed (dried, pickled) prior to consumption.

In direct-seeded peppers, glyphosate or paraquat is applied to control emerged weeds prior to planting or after seeding but before crop emergence. Both materials can also be applied prior to transplanting.

Preemergence herbicides used in direct-seeded peppers are DCPA, bensulide, and napropamide. The latter two can be used either as preplant soil-incorporation treatments or preemergence. Sethoxydim can be applied postemergence to direct-seeded and transplanted pepper plantings for control of annual grasses as well as quackgrass, bermudagrass, and johnsongrass.

In transplants, preplant soil-incorporation treatments include trifluralin as well as bensulide and napropamide. Chloramben and DCPA are applied 4 to 6 weeks after transplanting while napropamide can be applied immediately following transplanting. Metolachlor can be used preemergence on nonbell chili peppers in New Mexico.

Tomatoes

Paraquat can control emerged weeds when applied prior to seeding or after seeding but before emergence, or as a directed shielded application between rows of established tomato plants. Bensulide and napropamide can be applied preemergence to control annual weeds in direct-seeded tomatoes. In areas of limited rainfall or to ensure activity, these herbicides are usually incorporated into the soil before planting. In addition to the above, preplant soil-incorporated herbicides for transplanted tomatoes include metribuzin, pebulate, trifluralin, and various tank mixes of these products. EPTC can be used in certain California counties. Chloramben is another preemergence herbicide that can be used in transplanted tomatoes.

Weed competition in staked tomato plantings can be a problem throughout the life of the planting because the crop canopy never fully shades the soil. This necessitates a season-long weed management program involving preplant, preemergence, and postemergence treatments.

PLANT BEDS FOR TRANSPLANT PRODUCTION

Many vegetable transplants are produced in beds both in enclosed structures (greenhouses or cold frames) and in the field. Methyl bromide applied as a soil fumigant can be used in plant beds for the production of any vegetable transplant. Furthermore, any herbicide registered for direct-seeded plantings of the various vegetable crops can also be used in plant beds utilized for transplant production of that vegetable.

VEGETABLE PRODUCTION ON PLASTIC MULCH

There is increased emphasis on the use of plastic culture in the production of vegetables. Weed management in these types of plantings differs from production in bareground settings.

When methyl bromide fumigation is used under the plastic there is no need for a herbicide under the plastic. Also if black plastic is utilized without methyl bromide fumigation, a herbicide is generally not necessary under the plastic. An exception is fields containing high populations of nutsedge. If clear plastic is used without soil fumigation, a herbicide will be required under the mulch. Herbicides applied to the soil surface prior to laying plastic should be activated by rainfall or overhead irrigation before the plastic mulch is placed over the soil. Herbicides that are safe to use under field conditions for various vegetable crops may be injurious when covered by plastic. Higher temperatures under the plastic enhance herbicide volatility and this can cause increased crop injury. Consult labels for precautions and specifies.

If weeds are present between plastic strips before planting, and where glyphosate or paraquat is registered for the crop to be planted under the stale bed provision, these herbicides can be used to control all existing vegetation. If glyphosate is used as a broadcast application, overhead irrigation or rainfall is required prior to planting to wash glyphosate off the plastic. Do not use glyphosate as a broadcast spray if holes for the crop have been punched in the plastic.

Only herbicides registered for a specific crop grown can be used for controlling weeds between the plastic strips. Crop roots generally extend beyond the plastic mulch into row middles and use of a nonregistered herbicide can result in crop damage and/or illegal herbicide residues in the crop. Preemergence herbicides can be applied prior to or after crop planting. However, in either case the surface of the plastic should not receive any herbicide. The herbicide can be concentrated by washing off the plastic into the holes where the crop plants are and cause crop damage.

Postemergence herbicides such as paraquat, sethoxydim, and metribuzin can be applied as directed sprays to row middles where permitted by label. Where paraquat is utilized, physical shielding is recommended to prevent contact with the crop to avoid injury.

SUGGESTED ADDITIONAL READING

Crop Protection Chemicals Reference, 1989, Wiley, New York, (updated annually).

Dallyn, S., and R. Sweet, 1970, *FAO International Conference on Weed Control*, WSSA, Champaign, IL, p. 210.

Danielson, L. L., 1970, *FAO International Conference on Weed Control*, WSSA, Champaign, Il, p. 245.

Friesen, G. H., 1978, *Weed Sci.* **26**, 626.

Friesen, G. H., 1979, *Weed Sci.* **27**, 11.

Meister Publishing Company, 1989, *Weed Control Manual and Herbicide Guide*, Willoughby, OH (updated annually).

Mendt, R. D., 1979, M.S. dissertation, North Carolina State University, Raleigh.

Menges, R. M., and T. D. Longbrake, 1970, *FAO International Conference on Weed Control*, WSSA, Champaign, IL, p. 229.

Monaco, T. J., D. C. Sanders, and A. Grayson, 1981, *Weed Sci.* **29**, 394.

Orsenigo, J. R., 1970, *FAO International Conference on Weed Control*, WSSA, Champaign, IL, p. 198.

Romanowski, R. R., 1970, *FAO International Conference on Weed Control*, WSSA, Champaign, IL, p. 184.

USDA, SEA, 1980, *Suggested Guidelines for Weed Control*, U.S. Government Printing Office, Washington, D.C. (621-220/SEA 3619).

USDA, SEA, 1981, *Compilation of Registered Uses of Herbicides.*

Zimdahl, R. L., 1980, *Weed-Crop Competition*, International Plant Protection Center, Oregon State University, Corvallis.

For herbicide use, see the manufacturer's or supplier's label and follow these directions. Also see Preface.

19 Fruit and Nut Crops

Weeds can damage fruit and nut crops seriously and in many ways. In newly planted crops, weeds compete directly with young trees for soil moisture, soil nutrients, carbon dioxide, and perhaps light. A recent review on orchard floor management (Skroch and Shribbs, 1986) indicated that groundcovers can vary in their competitive ability. Commonly used ground covers such as orchardgrass, tall fescue, and Kentucky bluegrass were reported to be more competitive with young apple trees than a number of other species. Allelopathy has also been implicated as a detrimental effect of groundcovers on growth of orchard crops. In older plantings, weed competition may be less serious, but still may reduce yields noticeably.

In addition, weeds may harbor plant diseases, insects, and rodents such as field mice or pine voles that girdle trees. Weeds such as poison ivy may interfere with harvest. When nut crops are harvested from the ground, weeds may seriously interfere with harvesting. The wasteful use of water by weeds is always important, especially in arid regions.

Crops discussed in this chapter are perennials. They are propagated in nurseries and transplanted to fields. The discussion is limited to field-weed control. Although nurseries have serious weed problems, they are not covered here because crop tolerance to most herbicides varies due to several factors including plant age, soil type, climatic conditions, and geographic location. The herbicide label should be consulted for more specific instructions.

Troublesome perennial weeds, such as quackgrass, johnsongrass, bermudagrass, nutsedge, field bindweed, or Canada thistle should be brought under control before setting a new orchard or making a small fruit planting. Perennial weeds can be controlled much easier and at less cost before planting than afterward.

If perennial weeds are absent or controlled before planting the crop, the primary weed problem is annual weeds. However, perennial weeds may invade the area later. This is especially true if perennial weeds are tolerant to the herbicides used. Tolerant perennial and annual weeds flourish when competition from susceptible weeds has been removed by a herbicide.

A total weed-control system in fruit and nut crops may combine several methods including *cultivation, mowing, mulching*, and *herbicides.*

Weed control in deciduous fruit and nut crops was reviewed by Lange (1970). Jordan and Day (1970) reviewed weed control in citrus crops.

CULTIVATION

When considered as a single weeding operation in fruit and nut crops, cultivation is just as effective and economical as with other tilled crops. The advantages of cultivation are well known by growers. As for the disadvantages, shallow feeder roots are damaged, soil structure is changed, soil erosion increases (especially on hillsides), weeds under trees are difficult to control by mechanized methods, and cultivation brings new weed seeds to the surface, where they may germinate and grow. For one or more of these reasons, growers are depending less and less on cultivation by itself.

MOWING

Mowing is popular in borders and areas between trees of many fruit and nut crops, especially where soil erosion is serious. These areas can be maintained as short turf, effectively controlling erosion while keeping weed competition to a minimum. Mowed turf gives a clean and neat appearance. The area under trees and vines can be kept weed free by appropriate use of herbicides or mulches.

MULCHING

Mulching usually controls weeds by depriving them of light. Most mulches also conserve soil moisture. Decomposable organic matter, such as straw, is used in many horticultural crops including strawberries and young fruit-tree plantings. In recent years, both black plastic sheets and paper sheets impregnated with asphalt have been used. Plants are planted through small holes made in these sheets.

HERBICIDES

Orchards

Groundcover management in orchards can be achieved with cultivation or noncultivation methods. Noncultivation methods are usually mowing or the use of herbicides. The specific method used depends somewhat on the crop, but perhaps more important is the slope of the land and soil type. Some type of ground cover is desirable on hillsides to prevent erosion; here mowing and herbicides are used. On relatively flat terrain, cultivation, mowing, and herbicides can be used.

Often a combination of these methods is used. Frequently a herbicide band is used down the tree row and cultivation or mowing is used between rows. This permits mowing or cultivation and minimizes the amount of herbicide used

TABLE 19-1. Herbicides Used on Orchard Crops[1]

	Ametryn	AMS	Atrazine	Bromacil	Cacodylic acid	2,4-D amine	Dalapon	Devine®	Dichlobenil	Diuron	DSMA and/ or MSMA	EPTC	Fluazifop	Glyphosate	Metolachlor	Napropamide	Norflurazon	Oryzalin	Oxyfluorfen	Paraquat	Pronamide	Sethoxydim	Simazine	Terbacil	Trifluralin
Almond									X		X	X	X	X	X	X	X	X	X	X		X	X		X
Apple		X				X	X		X	X	X		X	X	X	X	X	X	X	X	X	X	X	X	
Apricot							X				X		X	X	X	X	X	X	X	X		X			X
Avocado									X				X	X		X		X		X		X	X		
Cherry									X		X		X	X	X	X	X	X	X	X	X	X	X		
Chestnuts																X			X						
Date													X						X			X			
Fig									X				X			X		X	X	X		X			
Filbert									X				X	X		X	X	X	X	X		X	X		
Grapefruit	X			X	X		X	X	X	X	X	X	X	X	X	X	X	X		X		X	X	X	X
Hazelnut																X									
Hickory nut																X			X						
Kumquat													X	X	X										

Lemon				X	X				X	X	X		X	X	X	X	X	X	X	X		X	X	X	X
Lime					X			X	X		X	X	X	X	X		X			X		X		X	
Macadamia			X				X			X			X	X	X			X	X	X		X	X		
Mango														X											
Nectarine									X	X			X	X	X	X	X	X	X	X	X	X			X
Olive											X		X			X		X	X	X		X	X		
Orange	X			X	X		X	X	X	X	X	X	X	X	X	X	X	X	X	X		X	X	X	X
Papaya														X						X					
Peach						X	X		X	X	X		X	X	X	X	X	X	X	X	X	X	X	X	X
Pear		X				X	X		X	X	X		X	X		X	X	X	X	X	X	X	X		
Pecan									X	X			X	X	X	X	X	X	X	X		X	X	X	X
Persimmon																X									
Pistachio													X	X		X		X	X	X		X			
Plum and/or							X		X		X		X	X	X	X	X	X	X	X	X	X	X		X
prune															X										
Tangelo															X				X	X					X
Tangerine					X	X				X	X	X	X	X	X	X	X			X		X			X
Walnut, English									X	X	X	X	X	X	X	X	X	X	X	X		X	X		X
Walnut, black									X			X		X		X		X	X	X		X	X		X

[1]Most of these control annual weeds. However, specific ones also control certain perennial weeds. Several also have restrictions as to soil type, geographic location, application techniques, tree age, use on bearing or nonbearing crops, and minimum time between application and harvest. *Read labels carefully* and follow directions.

Figure 19-1. Strip application of a herbicide in an almond orchard. The untreated areas between the treated strip will be mowed, cultivated, or both.

(Figure 19-1). Cultivation close to the trees may cause mechanical damage to the trunks and to the root systems.

Numerous herbicides can be used in orchards to control annual and perennial weeds. Table 19-1 lists those registered for use in orchard crops. Some of these herbicides control only annual weeds or germinating seeds of perennial weeds. However, bromacil, 2,4-D, Devine® ("biological herbicide"), dichlobenil, fluazifop, EPTC, glyphosate, metolachlor, oxyfluorfen, sethoxydim, and terbacil are also effective on certain perennial weeds. As for perennial weeds, (1) glyphosate is effective on most species, (2) 2,4-D controls only broadleaf species, (3) fluazifop and sethoxydim are grass herbicides, (4) metolachlor is useful for yellow nutsedge control while EPTC suppresses both yellow and purple nutsedge, and (5) Devine is a *mycoherbicide* for control of perennial strangler or milkweed vine in citrus groves. This biological herbicide formulation consists of live *chlamydospores* of the root rot fungus *Phytophthora palmivora.* The mycoherbicide is applied to the soil surface, which must be moist at time of application to achieve root infection. The fungus will not infect citrus roots.

Dichlobenil, bromacil, oxyfluorfen, and terbacil are capable of controlling certain perennial weeds. Labels should be consulted and followed. Understanding how herbicides work should help each grower develop a comprehensive program giving best results from these practices (Chapters 8–14).

Resistant Weed Species The key to successful chemical weed control in orchards is closely related to the weed species present. Once a herbicide program has been selected, it will usually have to be revised periodically to prevent an increase of resistant weed species; for example, Baron and Monaco (1986) reported that continuous use of terbacil for weed control in blueberries resulted in buildup of tolerant goldenrod. Skroch et al. (1974) observed a buildup of brambles and Virginia clematis with 5 years of terbacil usage in apple orchards. Schubert (1972) reported that after six annual applications of 2,4-D or amitrole the vegetation under apple trees was nearly a pure stand of grasses. Herbicide rotation on sequential years or using a combination of two or more herbicides usually prevents resistant weeds from becoming a problem.

Grapes

As with orchards, annual weeds in the crop row are often controlled with herbicides, and the interrow area is cultivated or mowed. Cultivation close to the grapevine often injures the plant. In some soils, repeated cultivations for weed control can hasten the development of a hard pan, which can impede water penetration. A mowed cover crop between rows with a herbicide-treated strip in the row is used in some areas (Figure 19-2).

Annual weeds in vineyards can be controlled by soil applications of diuron, metolachlor, napropamide, oryzalin, oxyfluorfen, pronamide, simazine, or

Figure 19-2. Strip application of a herbicide in a vineyard. Untreated areas between treated strips will be mowed, cultivated, or both.

TABLE 19-2. Herbicides Used in Grapes[1]

Herbicide	Table Grapes	Wine Grapes
Dalapon	Post-1	Post-1
Dichlobenil	Pre[2]	
Diuron	Pre	
Fluazifop	Post	Post
Glyphosate	Post-1	Post-1
Metolachlor	Pre	Pre
Napropamide	Pre	Pre
Oryzalin	Pre	Pre
Oxyfluorfen	Pre	Pre
Paraquat	Post-1	Post-1
Pronamide	Pre	Pre
Sethoxydim	Post	Post
Simazine	Pre	Pre
Trifluralin		INC

[1] Most of these control annual weeds. However, specific ones also control perennial weeds. Several also have restrictions as to soil type, geographic location, application techniques, vine age, use on bearing or nonbearing grapes, and minimum time between application and harvest. *Read labels carefully* and follow directions.

[2] Pre, preemergence to weeds; Post, postemergence to weeds; Post-1, postemergence to weeds but directed or shielded to eliminate contact with the crop; INC, mechanically soil-incorporated.

trifluralin (Table 19-2). Herbicide selection is dependent on type of grape, weeds to be controlled, age of planting, and geographic location. Paraquat applied postemergence controls a wide array of annual weeds whereas postemergence applications of fluazifop or sethoxydim control many annual grasses and certain perennials such as rhizome johnsongrass, quackgrass, and bermudagrass. Glyphosate controls many weeds but like paraquat contact with green tissue must be avoided to prevent crop injury. Therefore, postemergence applications of paraquat or glyphosate must be directed to the base of the vine or shielded. When applied properly with regard to stage of growth, glyphosate controls many troublesome perennial weeds such as field bindweed, rhizome johnsongrass, bermudagrass, artemisia, Canada thistle, quackgrass, and Russian knapweed. Perennial grasses in vineyards can also be controlled with directed sprays of dalapon. Repeat applications of dalapon, fluazifop, glyphosate, and sethoxydim to weed regrowth are required for acceptable levels of perennial weed control.

Blueberries

Weed problems and general methods of control of weeds in blueberries depends on the species of blueberry. Lowbush blueberries (*Vaccinium angustifolium*) grown in the northeastern region of the United States are managed in the wild. They have unique weed problems and rely heavily upon herbicides because tillage is not possible. However, in highbush (*V. corymbosum*) and rabbiteye (*V. ashei*) blueberry plantings, weeds are managed by cultivation, mowing, and herbicides. Highbush blueberries are cultivated in the mid-Atlantic states through Florida and states such as Michigan. Rabbiteye blueberries are grown primarily in the southeastern states.

Preemergence herbicides that are registered for blueberries include dichlobenil, diuron, hexazinone, napropamide, norflurazon, oryzalin, oxyfluorfen, pronamide, simazine, and terbacil. These herbicides control primarily annual weeds; however, dichlobenil, hexazinone, and terbacil control certain perennial weeds. Postemergence applications of fluazifop and sethoxydim can be made to control annual grasses and certain perennial grasses. Carefully directed or shielded postemergence applications of paraquat or glyphosate control all annual weeds and in addition glyphosate controls many perennial weeds. Selection of a herbicide program is dependent on type of blueberry, weed(s) to be controlled, geographic location, and age of crop. Labels must be consulted for specifics.

Caneberries

Caneberries or brambles include mainly blackberries, boysenberries, loganberries, and raspberries. The herbicides registered for use in these crops are given in Table 19-3. Paraquat and glyphosate are applied to the foliage of weeds. Both must be directed or shielded to avoid crop injury. Paraquat is a contact

TABLE 19-3. Herbicides Used on Caneberries[1]

Herbicide	Blackberries	Boysenberries	Loganberries	Raspberries
Dichlobenil	Pre[2]			Pre
Diuron	Pre	Pre	Pre	Pre
Fluazifop	Post-1	Post-1	Post-1	Post-1
Glyphosate	Post-2			Post-2
Napropamide	Pre	Pre	Pre	Pre
Norflurazon	Pre			Pre
Oryzalin	Pre	Pre	Pre	Pre
Paraquat	Post-2	Post-2		Post-2
Pronamide	Pre			Pre
Sethoxydim	Post-1	Post-1	Post-1	Post-1
Simazine	Pre	Pre	Pre	Pre
Terbacil	Pre	Pre	Pre	Pre

[1]Check labels for limitations such as geographic areas, plant age, stage of growth, and methods of application.

[2]Pre, preemergence to weeds; post-1, postemergence to weeds, broadcast or directed; post-2, directed or shielded postemergence to avoid contact with the crop.

spray for the control of annual weeds while glyphosate is systemic and will control both annual and perennial weeds.

Annual grasses and certain perennial grasses such as johnsongrass, quackgrass, and bermudagrass can be controlled by postemergence applications of fluazifop and sethoxydim. The other herbicides listed are applied preemergence to the soil to control annual weeds as they germinate. However, diclobenil, norflurazon, and terbacil also control certain perennials.

Strawberries

Weed control in strawberries may cost up to several hundred dollars per acre per year. By using proper herbicides and other management practices, this cost can be reduced drastically. Lack of labor for hand weeding has stimulated the use of herbicides on this crop. Black polyethylene plastic or straw also control weeds in strawberries, as well as conserving moisture and keeping berries clean (Figure 19-3). Weeds that emerge through holes in the plastic are usually removed by hand.

Strawberry beds are usually abandoned after 1 to 3 years as a result of either disease or severe weed infestation. Appropriate use of herbicides can considerably extend the productive life of many plantings, especially if more effective disease controls become available.

In some areas, fields intended for strawberries are fumigated with a mixture of methyl bromide and chloropicrin before planting. This treatment not only kills many weed seeds but also soil-borne disease organisms, for example *Verticillium*, and nermatodes. Methyl bromide is the primary weed killer of this mixture.

Figure 19-3. Black polyethylene plastic used as a mulch for weed control and clean strawberry production.

Malva, burclover, filaree, field bindweed, and morningglory seeds are often not killed by this treatment.

Methyl bromide and chloropicrin are gases under normal temperatures and pressures. They are sold in pressurized containers as a liquid. Both are injected into the soil about 6 inches deep through chisels spaced about 12 inches apart on a tool bar. Immediately after treatment, the treated area is sealed with a gas-tight tarpaulin (polyethylene) for at least 48 hours. Injection of the fumigant and covering with the tarpaulin are completed in one operation, with all equipment mounted on a single tractor.

Strawberries may be planted 3 days after removing the tarpaulin. If weeds resistant to this treatment emerge later, they will have to be controlled by cultivation and hand weeding.

Preemergence herbicides that can be used in strawberries are DCPA, napropamide, and terbacil. DCPA and napropamide can be applied to actively growing strawberries; however, terbacil must be applied to dormant plantings in the winter or immediately following renovation procedures after harvest provided leaves have been removed by mowing. Growers need to recognize that varietal responses of strawberries vary considerably and the degree of dormancy differs in various locations.

Sethoxydim can be applied overtop of strawberries for control of emerged annual grasses and certain perennial grasses. Sethoxydin cannot be applied within 30 days preceding harvest. Winter-annual broadleaf weeds can be controlled with postemergence applications of 2,4-D when strawberries are dormant. The crop must be dormant to avoid injury and only the amine formulation of 2,4-D can be used. After harvest and prior to renovation, 2,4-D can be applied to control emerged broadleaf weeds.

Paraquat can be applied as a directed spray between strawberry rows to control annual weeds. Shields are generally used to eliminate contact with the crop. Paraquat cannot be applied more than three times per year or within 21 days preceding harvest.

SUGGESTED ADDITIONAL READING

Baron, J. J., and T. J. Monaco, 1986, *Weed Sci.* **34**, 824.

Crop Protection Chemicals Reference, 1989, Wiley, New York (updated annually).

Genez, A. L., and T. J. Monaco, 1983, *Weed Sci.* **31**, 56.

Horowitz, M., 1973, *J. Am. Soc. Hort. Sci.* **48**, 135.

Jordan, L. S., and B. E. Day, 1970, *FAO International Conference on Weed Control*, Champaign, IL, p. 128.

Lange, A. H., 1970, *FAO International Conference on Weed Control*, WSSA, Champaign, IL, p. 143.

Schubert, O. E., 1972, *Weed Sci.* **21**, 124.

Shribbs, J. M., and W. A. Skroch, 1986, *J. Am. Soc. Horti. Sci.* **111**, 525.

Skroch, W. A., T. J. Sheets, and T. J. Monaco, 1974, *Weed Sci.* **23**, 52.

Skroch, W. A., and J. M. Shribbs, 1986, *Hort Science* **21**, 390.

USDA, SEA, 1980, *Suggested Guidelines for Weed Control*, U.S. Government Printing Office, Washington, D. C. (621-220/SEA 3619).

USDA, SEA, 1981, *Compilation of Registered Uses of Herbicides.*

Weed Control Manual and Herbicide Guide, 1989, Willoughby, OH (updated annually).

For herbicide use, see the manufacturer's or supplier's label and follow these directions. Also see Preface.

20 Lawn, Turf, and Ornamentals

Lawn- and turf-weed problems have been largely underestimated. Nearly every home and apartment house has a weed problem. Other turf areas that have weed problems are golf courses, public and private parks, athletic fields, other recreation areas, grounds surrounding many commercial and governmental buildings, and roadsides.

No other type of weed control directly affects so many people. In the United States over 5 million acres are in home lawns, and an additional 10 million acres are in other types of turf.

Because there are millions of consumers, numerous turf species, and a multitude of ornamental plants, the job of educating users is more complex than in other areas of weed control.

The importance of management practices that produce a strong and vigorous turf cannot be overemphasized. In general, these practices are well understood for each area of the United States. They are specific for each geographic area. They include choice of an adapted lawn grass, proper grading and seedbed preparation, fertilization, mowing, and watering, as well as control of insects, diseases, and weeds.

LAWN

Weed control alone cannot guarantee a beautiful lawn. Other recommended practices must also be followed. For example, ridding a lawn of crabgrass may leave a bare area unless plans are made to encourage desirable turf grasses to become established. However, desirable turf grasses are usually present in areas in which crabgrass predominates. With elimination of crabgrass and other weeds, proper fertilization, mowing, and watering, lawn grasses such as Kentucky bluegrass or bermudagrass soon cover the area.

Topsoil Added to New Lawns

After new homes or buildings are finished, only subsoil may remain for starting a lawn. Often topsoil is hauled in to cover the area 2 to 4 inches deep. Most topsoil contains weed seeds and weedy plants that soon infest the area. The owner can (1) attempt to get topsoil from fields known to be free of serious perennial lawn weeds, (2) use a soil fumigant such as methyl bromide to rid the soil of weeds, or (3) not use the topsoil before planting.

Proper fertilization, including thorough mixing into the upper 3 to 4 inches of soil, makes it possible to grow many turf plants on subsoil. Turf plants are favored by adding peat moss at the rate of 1 bale (7 ft^3) per 200 ft^2 of surface plus liberal fertilization, with both worked into the surface 3 to 4 inches of subsoil. Well-rotted sawdust is as good as peat; however, the fertilization program depends on the degree of sawdust decomposition. Well-decomposed sawdust will require less fertilizer than fresh sawdust. Also, *well-rotted* manure can be used in place of the peat moss. Fresh manure will contain weed seeds. A surface mulch of peat moss, wheat straw, or other materials at seeding time may reduce soil erosion and keep the surface from drying rapidly.

Before Planting

Serious weeds that cannot be controlled after turf is established should be eliminated before planting. Such methods may delay planting, but will reduce the work required after the turf is planted.

Weed control before seeding may include shallow cultivation after emergence of most annual weeds, or the use of herbicides. To be effective, the herbicide needs to control all weedy growth and to have no residual toxicity to the turf to be seeded or sodded later. Two herbicides especially useful for this purpose are glyphosate and paraquat.

Glyphosate is a nonselective, broad-spectrum, foliar-applied herbicide that kills annual and most perennial plants. It is absorbed principally through foliage, requiring 7 to 10 days for translocation to the roots in some species. Application to foliage of desirable plants should be avoided. Glyphosate has no residual soil activity. It does not control dormant seeds in the soil. Glyphosate reacts with galvanized steel to form hydrogen gas, which is explosive. Stainless steel, aluminum, fiberglass, and plastic are suitable sprayer components.

Paraquat is a contact, nonselective herbicide that kills the tops of plants but seldom kills the roots of perennial plants. At usual rates of application it has no residual herbicidal activity through the soil.

Following the above herbicide treatment and after seeding or sodding, appropriate herbicide treatment should be made to continue the weed control program.

At Planting

Siduron can be applied at seeding of cool-season turf (Kentucky bluegrass, ryegrass, fescue, and some bent grasses) to control annual grasses. Siduron selectively controls annual grasses such as crabgrass, foxtail, and barnyardgrass for about 1 month.

After Planting

Postplanting control is also often needed after turfgrass species have emerged, but before they become well-established. This can be done by hand weeding, mowing, use of herbicides, or a combination of these practices. Mowing of newly planted turf will control some erect broadleaf weeds, but is not effective on grasses or prostrate broadleaf weeds. While the lawn is young, mowing height should be kept high so that a minimum of foliage of the turf species is removed.

Herbicides must be used with care on young turfgrass species to avoid injury. Uniform distribution and proper rate of application are essential. Many emerged broadleaf weeds can be controlled by low rates of bromoxynil, 2,4-D, mecoprop, or dicamba once the grass seedlings have reached the three- to four-leaf stage. At this stage, dosage rates should be reduced to about one-fourth to one-half of those recommended for established turf. As the grass becomes established these rates can be gradually increased.

ESTABLISHED TURF

A turf is established when the grasses have developed an extensive root system and are well tillered or when the rhizome (runner) system is well developed. With good management practices, weeds in established turf often can be controlled by hand pulling or cutting the occasional weed out of a home lawn. Since an established turf tolerates herbicides much better than a new planting, many herbicides may be used.

Herbicides used on established turf are either soil-active or foliar-active. Generally, soil-active compounds are applied about 2 to 3 weeks before

TABLE 20-1. Herbicides for Control of Grassy Weeds in Turf[1,2]

Large Crabgrass, Foxtails, Goosegrass		Annual Bluegrass		Nutsedge and Annual Sedges
Pre	Post	Pre[3]	Post	Post
Siduron	Fenoxaprop	Atrazine	Pronamide	CMA
Benefin	Sethoxydin	Simazine	Ethofumesate	DSMA
Bensulide	CMA[4]	Fenarimol	Metribuzin	MSMA
DCPA	DSMA[4]	Pronamide	Glyphosate	Bentazon
Napropamide	MSMA[4]	Ethofumesate		Imazaquim
Oryzalin				
Oxadiazon				
Pendimethalin				

[1] Labels must be consulted for turfgrass tolerance, application timing, type of planting, rates, formulations, and all other instructions and precautions.
[2] Adapted from North Carolina State University Agricultural Extension Service Bulletin AG-408, 1989.
[3] Herbicides listed for pre control of large crabgrass, foxtails, and goosegrass also control annual bluegrass.
[4] These herbicides also control dallisgrass and sandbur.

germination of weed seeds to be controlled. They are preemergence-type herbicides. Some require immediate sprinkle irrigation to minimize foliar injury to desirable turf species. All must be leached into the soil within a few days after application by sprinkle irrigation or rainfall to be effective.

Foliar-active herbicides are applied to emerged weeds. Because it usually takes several hours or sometimes a few days for maximum absorption, they should not be applied if rain is expected or if sprinkle irrigation is to be applied during that time.

TABLE 20-2. Tolerance of Established Cool-Season Turfgrass to Preemergence Herbicides for Control of Annual Weedy Grasses[1,2]

Herbicide	Kentucky Bluegrass	Tall Fescue	Fine Fescue	Perennial Ryegrass	Bentgrass (Golf Greens)
Benefin	T	T	M	T	NR
Bensulide	T	T	T	T	T
DCPA	T	T	M	T	NR
Napropamide	NR	T	T	NR	NR
Oryzalin	NR	T	NR	NR	NR
Oxadiazon	T	T	NR	T	NR
Pendimethalin	T	T	T	T	NR
Siduron	T	T	T	T	M

[1]T, tolerant; M, marginally tolerant; NR, not registered for use on this turfgrass. Consult labels for specifics and limitations.
[2]Adapted from North Carolina State University Agricultural Extension Service Bulletin AG-408, 1989.

TABLE 20-3. Tolerance of Established Warm-Season Turfgrass to Preemergence Herbicides for Control of Annual Weedy Grasses[1,2]

Herbicide	Bahia-grass	Bermuda grass	Bermuda-grass Golf Greens	Centipede-grass	St. Augustine-grass	Zoysia-grass
Benefin	T	T	NR	T	T	T
Bensulide	T	T	T	T	T	T
DCPA	T	T	NR	T	T	T
Napropamide	T	T	T	T	T	NR
Oryzalin	T	T	NR	T	T	T
Oxadiazon	NR	T	NR	NR	T	T
Pendimethalin	T	T	NR	T	T	T
Siduron	NR	NR	NR	NR	NR	T
Simazine	NR	T	NR	T	T	T
Atrazine	NR	T	NR	T	T	T

[1]T, tolerant; NR, not registered for use on this turfgrass. Consult labels for specifics and limitations.
[2]Adapted from North Carolina State University Agricultural Extension Service Bulletin AG-408, 1989.

TABLE 20-4. Tolerance of Turfgrasses to Postemergence Herbicides for Grass and/or Broadleaf Weed Control[1,2]

	DSMA, MSMA, CMA	Asulam	Atrazine	Bentazon	Etho-fumesate	Fenoxa-prop	Glypho-sate	Imaza-quin	Metri-buzin	Prona-mide	Sethoxy-dim
Cool-season grass											
Bentgrass	I	S	S	T	S	S	S	S	S	S	S
Kentucky bluegrass	I	S	S	T	S	T	S	S	S	S	S
Tall fescue	I	S	S	T	S	T	S	S	S	S	S
Fine fescue	I	S	S	T	S	T	S	S	S	S	S
Perennial ryegrass	T	S	S	T	T	T	S	S	S	S	S
Warm-season grass											
Bahiagrass	S	S	S	T	S	S	D	S	S	S	S
Bermudagrass	T	T[3]	T	T	D	S	D	T	T	T	S
Centipedegrass	S	S	T	T	S	S	S	T	S	S	T
St. Augustinegrass	S	T	T	T	S	S	S	T	S	S	S
Zoysiagrass	I	S	T	T	S	S	S	T	S	S	S

[1] D, apply only during dormant season; I, intermediately tolerant; S, sensitive; T, tolerant. Consult labels for specifics and limitations.
[2] Reproduced from North Carolina State University Agricultural Extension Service Bulletin AG-408, 1989.
[3] Tifton 419 bermudagrass.

TABLE 20-5. Herbicides for Control of Broadleaf Weeds in Turf[1,2]

Weed	2,4-D	Mecoprop	Dicamba	2,4-D + Mecoprop	2,4-D + Dichlorprop	2,4-D + Mecoprop + Dicamba
Bittercress, hairy	S	I	S	S	S	S
Black medic	R	I	S	I	S	S
Buttercups	S-I	I	I-R	S	S	S
Buttonweed, Va.	S-I	I	I	I	S-I	S-I
Carolina geranium	S	S-I	S	S	S	S
Carpetweed	S	I	S	S	S	S
Catsear	S-I	I	S	S	S	S
Chickweed, common	R	S-I	S	S	S	S
Chickweed, mousear	I-R	S-I	S	S	S	S
Chicory	S	S	S	S	S	S
Clover, hop	I	S	S	S	S	S
Clover, white	I	S	S	S	S	S
Dandelion	S	S	S	S	S	S
Dichondra	S	I	S-I	S	S	S
Dock, broadleaf and curly	I	I-R	S	I	I	S-I
Garlic, wild	S-I	R	S-I	S-I	S-I	S-I
Ground ivy	I-R	I	S-I	I	I	S-I
Hawkweed	S-I	R	S-I	S-I	S-I	S-I
Healall	S	R	S-I	S	S	S

Henbit	I-R	I	S	I	S-I	S
Knawel	R	I	S	S-I	S-I	S
Knotweed, prostrate	R	I	S	S-I	S-I	S
Lespedeza	I-R	S	S	S-I	I	S
Mallow	I-R	I	S-I	S-I	S-I	S-I
Mugwort	I	I-R	S-I	I	I	I
Parsley-piert	R	S-I	S-I	S-I	R	S-I
Pennywort, lawn	S-I	S-I	S-I	S-I	S-I	S-I
Plantains	S	I-R	R	S	S	S
Purslane, common	I	R	S	I	I	S-I
Red sorrel	R	S	S	S-I	I	S
Speedwell, corn	I-R	I-R	I-R	I-R	I-R	I-R
Spurge, prostrate	I	I	S	I	S-I	S-I
Spurge, spotted	I-R	S-I	S-I	S-I	S-I	S
Spurweed (lawn burweed)	I	S-I	S	S-I	I	S
Strawberry, India mock	R	I	S-I	I	R	S-I
Violet, johnnyjumpup	I-R	I-R	S-I	I-R	I	I-R
Violet, wild	I-R	I-R	S-I	I-R	I	I-R
Woodsorrel, common yellow	R	R	I	I-R	I-R	I-R
Yarrow	I	I-R	S	I-R	I	S-I
Yellow rocket	S-I	I	S-I	S-I	S-I	S

[1] S, susceptible; I, intermediately susceptible; R, resistant in most cases. Labels must be consulted for specifics.
[2] Adapted from North Carolina State University Agricultural Extension Service Bulletin AG-408, 1989.

Grass Weeds

Herbicides that are commonly used to control annual grasses, some perennial grasses and sedges in turf are listed in Table 20-1. Some are used preemergence while others are foliar active and can be applied to control emerged weed grasses in turf.

Tolerance of established cool-season and warm-season turfgrasses to preemergence herbicides used for annual grass control is summarized in Table 20-2 and 20-3. Tolerance of both types of turfgrasses to postemergence herbicides for grass and/or broadleaf weed control is listed in Table 20-4. Information provided in this table indicates that only a few of the turfgrasses are tolerant to herbicides such as fenoxaprop, sethoxydim, CMA, DSMA, MSMA, pronamide, and ethofumesate, that are used for control of grassy weeds. Fenarimol is a fungicide used in turf that is capable of controlling annual bluegrass from seed.

Broadleaf Weeds

Many types of broadleaf weeds may be found in turf including winter annuals, summer annuals, biennials, and perennials. Many of the growth regulator type herbicides can be used for postemergence control of broadleaf weeds in turf. Table 20-5 provides an extensive list of broadleaf weeds and their susceptibility to 2,4-D, mecoprop, dicamba, and various combinations of these products. Many commercial formulations contain two way and three mixes of these herbicides to provide control of a wide spectrum of these weeds.

The tolerance of cool-season and warm-season turfgrasses to herbicides applied postemergence for control of broadleaf weeds is presented in Table 20-6. Bromoxynil and triclopyr are listed in Table 20-6 but not in Table 20-5. These two herbicides control several annual and perennial or broadleaf weeds.

ORNAMENTALS

Weeds in ornamentals present serious problems for nurserymen as well as professional and home gardeners. Weed control methods include *hand weeding*, *mulches*, *cultivation*, and *herbicides*. Selecting the most suitable method or methods of controlling weeds in ornamentals is a difficult task.

Hand weeding, although tedious, is often very useful for the home gardener; however, for the nurseryman or professional gardener it is usually too expensive. *Mulches* are commonly used quite effectively. To control most annual weeds, a mulch 2 or 3 inches thick is sufficient. Good mulching material includes wood or bark chips, sawdust, peatmoss, small grain straw free of weed seeds, pine needles, and gravel or stones. An ideal mulch allows free passage of moisture and air, but smothers growth of young weeds and prevents germination of seeds that require light. *Cultivation* is used widely by the nurseryman, somewhat by the professional gardener, but little by the home gardener.

TABLE 20-6. Tolerance of Turfgrasses to Postemergence Herbicides for Broadleaf Weed Control[1,2]

	2,4-D	Mecoprop	Dicamba	Dichlorprop	Bromoxynil	Triclopyr
Cool-season grass						
Bentgrass	S-I	T	I	I	T	S-I
Kentucky bluegrass	T	T	T	T	T	T
Tall fescue	T	T	T	T	T	T
Fine fescue	T	T	T	T	T	I
Perennial ryegrass	T	T	T	T	T	T
Warm-season grass						
Bahiagrass	S-I	T	T	I	T	S
Bermudagrass	T	T	T	T	T	S
Centipedegrass	S-I	T	I-T	I	I	S
St. Augustinegrass	S-I	S-I	S-I	I	I	S
Zoysiagrass	T	T	T	T	T	S

[1] T, tolerant; I, intermediately tolerant; S, sensitive. Consult label for specifics. Consult labels for specifics and limitations.
[2] Reproduced from North Carolina State University Agricultural Extension Service Bulletin AG-408, 1989.

Woody ornamentals around the home, in nurseries, forest plantings, Christmas trees, and cemeteries can be safely treated with herbicides. It is possible to spray the soil and small weeds around the stem and below the foliage of woody ornamentals with postemergence herbicides such as amitrole, glyphosate, and paraquat. However great care must be exercised to avoid contact with crop foliage and stem.

Since woody ornamentals are perennials, it is possible to apply a soil-residual type of herbicide preemergence to the germination of weed seeds. Selectivity to the crop is often achieved by formulation. Many herbicides used for preemergence weed control in woody ornamentals are formulated as granules. Foliar contact is avoided if application of the granular herbicide is made when foliage is dry.

Herbicides that can be applied to ornamentals prior to weed emergence include alachlor, EPTC, and trifluralin. EPTC must be soil incorporated prior to planting. Alachlor or trifluralin can be applied either preplant soil-incorporated or postplant to the soil surface. Other preemergence herbicides that can be applied are bensulide, chloramben, dichlobenil, DCPA, metolachlor, naptalam, napropamide, oryzalin, oxadiazon, oxyfluorfen, pendimethalin, and simazine.

Postemergence herbicides registered in various woody ornamental crops include amitrole, asulam, fluazifop, glyphosate, paraquat, and sethoxydim. Selectivity with asulam, glyphosate, and paraquat is achieved by carefully directing or shielding the application to the base of woody ornamentals. Fluazifop and sethoxydim control emerged annual and perennial grasses. These two herbicides can contact foliage of most ornamentals without causing damage. Oxyflurofen and pronamide when applied preemergence can also control certain emerged weeds.

Nonselective sprays such as glyphosate and paraquat are especially effective for killing all vegetation in brick walks along borders or under woody ornamentals. For best results, weeds should be treated when 1 to 2 inches tall. One should be careful not to spray the foliage of valued ornamentals.

To help select proper herbicides for weed control in ornamentals, local county extension agents or other public agencies and local landscape architects should be consulted. One should always read the herbicide manufacturer's literature and labels.

SUGGESTED ADDITIONAL READING

Anonymous, 1989, *Pest Control Recommendations for Turfgrass Managers*, North Carolina State Univ., Agricultural Extension Service, AG-408.

Anonymous, 1988, *Pest Management Guide for Turfgrass*, Virginia Cooperative Extension Service, Publication 456-009.

Coats, G. E., 1986, *Weed Control in Turfgrass*, Mississippi Agricultural and Forestry Experiment Station, Information Bulletin 95.

Crop Protection Chemicals Reference, 1989, Wiley, New York, (updated annually).

Skroch, W. A., and J. C. Neal, 1986, *Weed Control in Field and Container-Grown Ornamentals*, North Carolina Agricultural Extension Service Nursery Crops Production Manual No. 17.

USDA, ARSES, 1984, *Weed Control in Lawns and Other Turf*, Home and Garden Bulletin Number 239.

Weed Control Manual and Herbicide Guide, 1989, Meister, Willoughby, OH (updated annually).

For herbicide use, see the manufacturer's or supplier's label and follow these directions. Also see Preface.

21 Pastures and Rangelands

Hundreds of kinds of weeds infest pastures and ranges. These include trees, brush, broadleaf herbaceous weeds, poisonous plants, and undesirable grasses. Control of trees and brush in pastures and range will be discussed in Chapter 22. This chapter is primarily limited to control of broadleaf weeds and grasses.

Almost half of the total land area of the United States is used for pasture and grazing (Klingman, 1970). Nearly all of this forage land is infested with weeds, some of it seriously. Weeds interfere with grazing, lower yield and quality of forage, increase costs of managing and producing livestock, slow livestock gains, and reduce quality of meat, milk, wool, and hides. Some weeds are poisonous to livestock (see Figure 1-4). The total cost of these losses is hard to estimate. Losses from undesirable woody plants are discussed in Chapter 22. Controlling heavy infestations of some woody weeds has increased forage yields from two to eight times.

Grass yields increased 400% after the removal of sagebrush in Wyoming. Forage consumed by cattle increased 318% on native Nebraska pasture after perennial broadleaved weeds were controlled by improved agronomic practices and use of 2,4-D. This 318% increase was a result of better pasture species, deferred and rotational grazing, and effective weed control (Klingman and McCarty, 1958). Spraying with 2,4-D on high-level-fertility weedy pastures increased consumption of forage by cattle 1000 lb/acre over the no-weed-control, high-fertility treatment (Peters and Stritzke, 1971).

These results emphasize the usual need for improved management practices along with a better weed-control program. Both must be used together.

Weed-control programs for pastures and ranges often include a combination of management, mechanical, fire, and chemical methods. Biological control has been effective on certain species (see Figure 3-6).

Chemical Composition of Grassland Weeds

Nutrient or chemical composition of grassland weeds is important to livestock farmers for two reasons. First, weeds contribute to the livestock ration. Second, weeds compete for nutrients and water needed by more palatable and more desirable species and thus cut down yields of desirable forage.

Scientists collected forage and weed samples before mowing in an intensive agricultural area of the Connecticut River Valley in Massachusetts. Table 21-1 shows nutrient content of various plants as determined by chemical analysis. In

TABLE 21-1. Chemical Composition of Grassland Weeds Compared to Timothy and Red Clover (Sampling Date June 5–10)

Plant	Growth Stage	Number of Samples	Mean Percentage Composition (Air-Dry Basis)				
			N	P	K	Ca	Mg
Timothy	Early heading	19	1.55	0.26	2.17	0.34	0.10
Red clover	In buds, before bloom	19	2.84	0.25	1.09	1.88	0.42
Tufted vetch	Early bloom	3	3.58	0.30	1.52	1.52	0.30
Yarrow	In buds, before bloom	8	1.56	0.31	2.35	0.82	0.18
Oxeye daisy	50% heads in bloom	7	1.63	0.34	2.48	0.94	0.21
Daisy fleabane	In buds, before bloom	11	1.47	0.38	2.12	1.12	0.20
Common dandelion	Mostly leaves	14	2.25	0.44	3.39	1.21	0.43
Yellow rocket	After bloom	7	1.44	0.24	1.55	1.23	0.17
Plaintain	Mostly leaves	11	1.48	0.30	2.10	2.55	0.46
Narrowleaf plantain	Mostly leaves	5	1.85	0.37	1.90	1.90	0.33
Yellow dock	50% heads in bloom	13	1.84	0.30	2.29	1.11	0.42
Tall buttercup	In bloom	7	1.45	0.31	1.98	0.94	0.25
Wild carrot	Vegetative growth leaves	5	2.52	0.54	2.37	1.92	0.44
Mouseear chickweed	In bloom	11	1.73	0.41	3.14	0.70	0.26
Cinquefoil	Early bud stage	2	1.49	0.28	1.31	2.08	0.33
Common milkweed	Vegetative growth	2	3.02	0.47	3.08	0.80	0.45
Sensitive fern	Vegetative growth	6	2.27	0.48	2.50	0.65	0.39
Quackgrass	Before heading	11	1.82	0.28	2.14	0.36	0.10

From Vengris et al. (1953).

addition to chemical composition, one must evaluate freedom from toxins, palatability, yield, and persistence in determining the worth of a species for forage. Also see Table 3-5.

Management

Choice of the most desirable forage species for an area is perhaps the first step in pasture improvement.

Proper management will control some weeds by itself. Proper management favors heavy plant growth, and many annual weeds are crowded out. For example, broomsedge disappears from pastures of the southeastern United States when the pastures are properly fertilized and seeded to high-yielding species such as Ladino clover, orchardgrass, or fescue.

Some weeds are favored by the recommended agronomic program. For example, dock responds to high soil fertility and favorable moisture. It is favored as much or more than the desired species.

In some cases weedy pastures should be plowed, fertilized if needed, and reseeded. These are usual steps in areas favored with adequate moisture. In many other areas, desirable species will flourish if grazing pressure is temporarily removed or time of use adjusted, and weedy species brought under control. This latter program is often most practical where rainfall is limited. Reseeding may also hasten establishment of desirable species in dryland areas. A weed-control program is usually necessary to allow new seedlings to become established.

Mowing

Mowing in the past was often recommended to control pasture weeds. Now, mowing is of little importance for rangeland weed control, and it is steadily becoming less important or more intensively grazed areas.

Mowing has an effect on some kinds of weeds, but not on others; it is more effective on upright-growing annuals, but is ineffective on those with leaves and seed heads close to the ground. Mowing during early flowering usually slows down or stops seed production of the upright types. Repeated mowing will kill most tall-growing annual and some tall-growing perennial weeds. With most perennials, mowing is most effective during early flowering and repeated as needed (see Figure 3-3).

Mowing is often disappointing. It tends to improve the appearance of the area, but kills few perennial weeds. In North Carolina, 5 years of monthly mowing was required to control horsenettle. Weekly mowing for three years (about 18 times during each growing season) reduced wild garlic plants in bermudagrass turf by only 52%.

In Nebraska, mowing a native grass pasture in either June or early July for 3 years left 65% of the perennial broadleaved weeds still living at the end of the experiment. After 20 years of mowing, 24–38% of the ironweed plants still persisted.

Mowing may be used to good advantage in new grass-legume seedlings to lessen weed competition. Clipping the tops off of broadleaved weeds may sufficiently reduce weed competition to permit survival of seedling grasses and legumes. However, mowing also clips the tops off the forage species, setting them back to some extent.

Herbicide Control

Pasture and grazing areas are well suited to chemical weed control. It is often possible to control a weed with little or no injury to desirable forage species. These forage species then respond with increased ground cover and larger yields (Figures 21-1 and 21-2).

Most broadleaf weeds can be controlled with phenoxy-type sprays. Control of undesirable grasses, especially annual grasses such as downy brome, has become an increasingly serious problem. Use of a preemergence herbicide, such as atrazine or diuron, before weed–seed germination may provide control.

Control of brush and woody weeds is covered in Chapter 22 and weed control in small-seeded legumes is discussed in Chapter 16. Control of these weed problems in pastures is essentially the same.

Figure 21-1. Herbicide use for weed control in pasture. *Left*: Before herbicide treatment. *Right*: After herbicide treatment. (Fisons Pest Control Ltd., Chestford Park, England.)

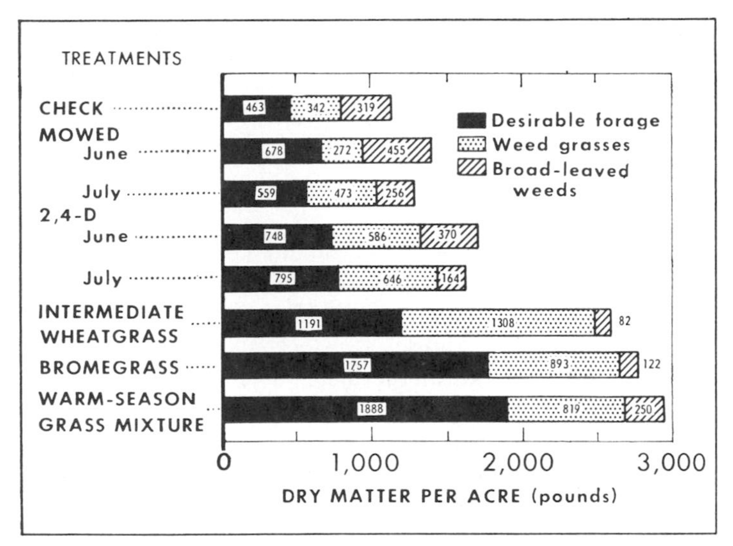

Figure 21-2. Effect of weed-control treatments on average pounds of dry matter per acre eaten be cattle. Seedbed for wheatgrass, bromegrass and warm season grass mixture was initially plowed and sprayed annually with 2,4-D. (D. L. Klingman and M. K. McCarty, USDA and University of Nebraska.)

Table 21-2 lists the herbicides and methods of treatment for weed control, including woody plants, in permanent pastures and rangeland.

Livestock Poisoning and Abortion

Most herbicides in common use in grazing lands are relatively nonpoisonous. If recommendations on the label are followed, no poisoning should occur.

TABLE 21-2. Weed Control Including Woody Plants, on Permanent Pastures and Rangeland by Various Herbicide Applications[1]

Herbicide	Foliage Spray	Basal Bark Spray	Stump Spray	Frill, Notching	Dormant Stem Spray	Tree Injection	Soil Treatment
Atrazine	X						X
Amitrole	X						
AMS	X						
2,4-D	X	X	X	X	X	X	
Dichlorprop	X						
Dicamba	X					X	X
Diuron	X						X
Fosamine	X						
Glyphosate	X						
Hexazinone	X						X
Karbutilate	X						X
MCPA	X						
Metsulfuron	X						
Paraquat	X						
Picloram	X						X
Tebuthiuron							X
Triclopyr	X	X	X		X		

[1]Weeds controlled by a number of herbicides are given on pages 439–442. Also, commercial herbicide labels list different weeds controlled, give instructions for use, and precautions. *Follow the labels instructions.*

Some herbicides, such as 2,4-D and similar compounds, are thought to temporarily increase the palatability of some poisonous plants. Livestock may eat these plants after treatment, whereas before spraying animals would have avoided them. Under such conditions livestock may be poisoned after plants have been sprayed, even though the herbicide is nonpoisonous.

Hydrocyanic acid (HCN) or prussic acid may be produced from glucosides that are found in various species of the sorghum family and in wild cherry. If cattle, sheep, and other animals with ruminant stomachs eat these plants, HCN is produced, and animals may be poisoned. Studies of wild cherry indicated no increase in HCN content following treatment with 2,4-D.

"Tift" Sudan grass was treated when 8 inches tall with 1 lb/acre of 2,4-D and MCPA. Chemical analysis showed that HCN increased after both chemical treatments (Swanson and Shaw, 1954). The use of phenoxy herbicides evidently does not change the usual precautions necessary to prevent HCN poisoning. Ruminant animals should not be permitted access to plants containing HCN, whether treated or not.

Livestock may also be poisoned by eating plants having a high nitrate content. Nitrate is reduced to nitrite by microorganisms in the animal's intestinal tract. Nitrite in the bloodstream interferes with effective transport and use of oxygen, and the animal dies of asphyxia (suffocation). The lethal dose of potassium nitrate is 25 g/100 lb of animal weight.

In Mississippi a number of pasture weeds were analyzed for chemical content. The scientists reported that redroot pigweed and horsenettle contain enough nitrate nitrogen to be toxic to livestock. Even though many weeds were found to have sufficient crude protein to meet animal needs, the authors concluded that most weeds should be considered detriments and should be removed. Factors cited as detrimental were bitterness, spines, toxic mineral levels, other toxic components of weeds, and the fact that weeds caused lower yields of forage (Carlisle et al., 1980).

The effect of herbicides on nitrate content of 14 weed species was studied in Michigan. 2,4-D, MCPA, dinoseb, chloropropham, and MH were tested. Before treatment, 10 of the 14 weeds contained enough nitrate to cause poisoning if consumed in considerable quantities. Following herbicide treatment, four species showed no change in nitrate content, five species showed increases, and one a definite decrease. Variation in the other species prevented drawing definite conclusions. Scientists concluded that many weeds contain enough nitrate to cause poisoning if eaten by livestock, whether sprayed or not sprayed with herbicides (Frank and Grigsby, 1957).

A high nitrate content in plants has been associated with abortion in cattle in Wisconsin. In Portage County, 400 abortions in cattle were reported in 1954. Reproductive diseases and pathogens accounted for only a very small number of abortions. "Poisonous weeds" were considered a possible explanation. Further study revealed that the muck soils were high in nitrogen but lacking in phosphorus and potassium; this condition is conducive to nitrate storage in plants. Weed species were analyzed for nitrate nitrogen and classified according to nitrate content (Table 21-3). Pastures were treated with 2,4-D to eliminate weeds thought to contribute to the high abortion rate. Pasture areas were divided for experimental purposes. On one pasture, 2,4-D was applied in both 1956 and 1957. The area was weed free in 1957. Ten heifers that grazed on this area calved normally in 1957, but all 11 heifers that grazed on nontreated and weedy pastures aborted in the same year.

A feeding trial was conducted to test the effectiveness of dosing pregnant cattle with nitrate to induce abortion. Three 700-pound heifers that were given 3.56

TABLE 21-3. Nitrate Nitrogen Content of Plants

High NO_3 Content (above 1000 ppm)	Medium NO_3 Content (300–1000 ppm)	Little or No NO_3 Content (below 300 ppm)
Elderberry	Goldenrod	Linaria
Canada thistle	Cinquefoil	Meadow rue
Stinging nettle	Boneset	Yarrow
Lambsquarters	Mints	Vervain
Redroot pigweed	Foxtail	Dandelion
White cockle	Aster	Milkweed
Burdock	Groundcherry	Willow
Smartweed	Toadflax	Dogwood
		Spirea

From Sund and Weight (1959).

ounces of potassium nitrate each day aborted after 3 to 5 days. The aborted fetuses and placentas were similar to those aborted on the weedy pastures (Simon et al., 1958).

In summary, herbicides are generally nonpoisonous to livestock if used as directed on the label. Poisoning may occur if palatability is increased following spraying so that livestock consume larger-than-usual quantities of poisonous weeds. Killing poisonous weeds with herbicides may reducc thc poisoning hazard.

Preventing Livestock Poisoning by Weeds

Livestock should be immediately isolated from poisonous plants to prevent livestock poisoning. It may be necessary to fence the infested area or remove livestock from the area. A small number of poisonous plants may be cut and removed from the pasture or killed by herbicide treatment.

Hundreds of plants cause livestock poisoning. Usually eradication, or at least very effective control, is needed. Cutting and removal of plants, or treatment with an effective herbicide may be most desirable. If the area is small, a soil sterilant may be the best answer, as will be discussed in Chapter 24. Small, isolated areas can probably be treated best with hand equipment or with granular materials. Large areas may be treated best by broadcast-type equipment, by either ground or aerial application.

SUGGESTING ADDITIONAL READING

Bovey, R. W., 1977, *Agriculture Handbook No. 493*, USDA, ARS.

Carlisle, R. J., V. H. Watson, and A. W. Cole, 1980, *Weed Sci.* **28**(2), 139.

Crop Protection Chemicals Reference, 1989, Wiley, New York (updated annually).

Frank, P. A., and B. H. Grigsby, 1957, *Weeds* **5**(3), 206.

Klingman, D. L., 1970, *Int. Conf. on Weed Contr.*, WSSA, Urbana, II, p. 401.

Klingman, D. L., and M. K. McCarty, 1958, *USDA Bulletin 1180.*

Peters, E. J., and J. F. Strizke, 1971, *USDA Technical Bulletin 1430.*

Simon, J., J. M. Sund, M. J. Wright, and A. Winter, 1958, *J. Am. Vet. Med. Assoc.* **132**, 164.

Sund, J. M., and M. J. Weight, 1959, *Down to Earth* (Dow Chem) **15**(1), 10.

Swanson, C. R., and W. C. Shaw, 1954, *Argon. J.* **46**(9), 418.

Vengris, J., M. Drake, W. G. Colby, and J. Bart, 1953, *Argon J.* **45**(5), 213.

Weed Control Manual and Herbicide Guide, 1989, Meister, Willoughby, OH (updated annually).

For herbicide use, see the manufacturer's or supplier's label and follow these directions. Also see Preface.

22 Brush and Undesirable Tree Control

Control of woody-plant growth is a problem affecting most types of property. This includes grazing and recreational areas; telephone, telegraph, highway, and railroad rights-of-way; and industrial plants and home sites.

There are about 1 billion acres of pasture, pasturelands, and grazing lands in the United States. On parts of nearly all of this area, woody plants present some problem. In range and pasture areas it is often desirable to eliminate all or most of the woody plants, leaving only grasses and legumes for livestock grazing.

On many Western dryland ranges, native grasses increase rapidly where brush is controlled. In Wyoming, range forage yields doubled the first year after sagebrush was treated and increased fourfold during a 5-year control program. In Oklahoma, control of brush with chemicals plus proper management (principally keeping livestock off during the first summer) increased the growth of grass four to eight times during the first 2 years after treatment. Figure 22-1 shows the relationship between mesquite control and increase of perennial grass forage on Southwestern rangeland. With 150 mesquite trees/acre, production of grass forage was reduced by about 85%, and total production including that produced by the mesquite was reduced by nearly 60%.

In the southeastern United States, over 100 million acres of land well suited to loblolly and short-leaf pine are being invaded by less-desirable hardwoods and heavy brush undergrowth. The same is true in northwestern United States and Canada, where dominant, but inferior, hardwoods may invade stands of Douglas fir, balsam fir, and spruce.

Some brush species act as alternate hosts for disease organisms affecting other plants. Common barberry harbors the organism causing stem rust in wheat and some other grasses. Other brush plants such as poison ivy and poison sumac are poisonous to man. Wild cherry and locoweed are poisonous to ruminant animals.

Control methods for woody plants vary with nature of the plant and size of the area infested. Some plants can be easily killed by one cutting. For example, many conifer trees have no adventitious buds on the lower parts from which new shoots or sprouts may develop. However, on large acreages even these species are controlled more efficiently by herbicidal sprays than by mechanical means. In Oregon and Washington, studies compared manual brush control with herbicide control in forests. Brush cut close to the ground by hand reached an average height of over 4 feet in 6 months, and this brush overtopped more than 40% of

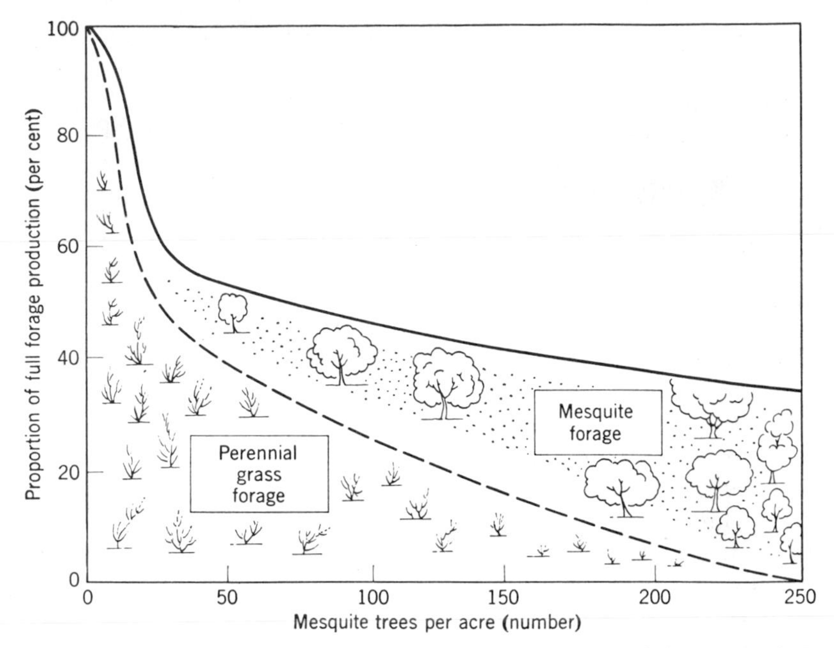

Figure 22-1. Effect of mesquite on forage production. (Adapted from USDA Leaflet No. 421, 1957.)

young timber (crop) trees. Furthermore, 22% of small Douglas firs were damaged by this handwork. Costs for the manual control averaged $230/acre compared with $15 to $30/acre for aerial herbicide sprays.

Plants hardest to control have underground buds from which new shoots or sprouts may develop. Some woody plants such as velvet and honey mesquite have these buds on the lower part of the trunk; other plants have buds on underground horizontal roots or stems (Figure 22-2). Most of the serious woody weeds are in one of these categories. As long as such plants have enough stored food and are not inhibited in some other way, new shoots will develop for as long as 3 years or more, even when all top growth is repeatedly removed. Higher plants produce their own food. This food-building process starts with photosynthesis. If photosynthetic plant organs, mainly the leaves, are repeatedly destroyed, the plant will eventually die of starvation.

Control methods for woody species are similar to those for other weeds, except that they must be adapted to woody and heavy type of growth. Methods include repeated cutting or defoliation, digging or grubbing, chaining and crushing, burning, girdling, and herbicide treatments.

Figure 22-2. Section of mesquite stump. The wartlike structures are buds that porduce sprouts. (Texas A&M University, Department of Range and Forestry.)

REPEATED CUTTING OR DEFOLIATION

The object of repeated cutting or repeated defoliation is to reduce food reserves within the plant until it dies. Plants commonly die during winter months if reserve foods have been reduced enough to make them more susceptible to winter injury (Figure 3-3).

Most woody species bear their leaves well aboveground, making repeated cutting or defoliation effective. Most species must be defoliated several times each season if they are to be controlled. Plants may be cut by hand, by mowers, by special roller-type cutters, or by saws. They may be defoliated with special equipment or by chemicals. Species having palatable leaves may be sufficiently defoliated by livestock. Sheep and goats have eradicated undesirable shrubs and tree sprouts, and insects may also defoliate a plant, gaining control as a biological predator.

Plants have well-established, annual patterns of food usage and food storage. This knowledge of root reserves, as they vary during the year, is valuable in reducing the number of treatments needed. With most deciduous plants, root reserves are highest during the fall. These reserves gradually diminish through the winter months as the plant uses this energy to survive. In the spring, the plant needs considerable energy to send up new stems and leaves. Thus food reserves in roots are usually lowest just before full-leaf development. Many plants reach this stage at about spring flowering or slightly before; therefore, mowing or defoliation is best started at that time (Figure 3-3). Repeat treatments are ordinarily needed for one or more seasons. Usually treatments are repeated when leaves reach full development. For more precise timing, root-reserve studies must be made on each plant species. As leaves become full grown, they start synthesizing more food than they need. As the excess accumulates, at least part of it is moved back to replenish the depleted root reserves.

Apical Dominance

The terminal bud of a twig or a shoot may hold back development of buds lower on the stem or roots. When the apical bud is cut off, lateral buds are no longer inhibited and may develop shoots. This mechanism is controlled by hormones within the plant. When apical dominance is destroyed, more stems may develop than were present originally. Following cutting, the stand may appear thickened rather than thinned. However, if cuttings are persistently continued so as to keep root reserves low, the stand will start to thin as the food supply is depleted. Under such conditions, the increase in number of shoots may actually speed reduction of root reserves (see "Mowing," page 41).

DIGGING OR GRUBBING

Plant species that sprout from an individual stem or root and remain as a "bunch" or "clump" can be effectively controlled by digging or grubbing. Whether this method is practical depends on the ease of removing the "clump," and density of stand. Mesquite is an example of this type of plant. Young plants can easily be removed by grubbing.

Giant bulldozers have been equipped with heavy steel blades or fingers (root plows) that run under the mesquite, lifting out the bud-forming crown along with the rest of the tree. The operation is expensive and hard on desirable forage species. Other weeds may quickly invade the disturbed area. Reseeding the disturbed area with desirable grasses will help reduce invasion by weeds. Disadvantages of this method include the following: (1) remaining roots may resprout, (2) the soil surface is severely disturbed, (3) the method is difficult in rocky soils, and (4) dormant weed seeds are brought to the surface to germinate and invade the area.

CHAINING AND CRUSHING

Chaining is a technique of pulling out trees without digging. A heavy chain is dragged between two tractors. This action tends to uproot most of the trees if the soil is moist. Then the chain may be pulled in the opposite direction to finish tearing trees loose from the soil. The method cannot be used on large, solidly rooted trees, and it is not effective on small, flexible brush, or on any species that sprouts from underground roots or rhizomes.

Crushing mashes and breaks solid stands of brush and small trees by driving over them with a bulldozer and/or with rolling choppers. Rolling choppers have been developed with drums 3 to 8 feet in diameter and weighing at least 2000 lb/ft of length. Chopper blades, about 8 inches wide, are attached across the entire length of the drum. This treatment crushes and breaks most of the branches and main stems. The chopper is usually pulled by a large, track-type tractor.

Crushing has disadvantages: it disturbs the soil, releasing buried-dormant weed seed to germinate and infest the area, and it requires expensive equipment. In experiments in Texas, roller chopping before, during, or after herbicide application did not increase herbicide effectiveness in control of live oak. Roller chopping was followed by no increase in livestock grazing per acre (Meyer and Bovey, 1980). Similar results have been observed with other species including white brush, McCartney rose, and mesquite. The main advantage to chopping was a temporary increase in visibility.

BURNING

Prior to modern weed science (about 1950), many U.S. grasslands and woodlands were burned regularly. These fires were started by lightning, Indians, ranchers, and farmers. Some fires were accidental; others were intentional.

Effective chemical methods are available to control many brush species on rangeland, yet burning is still used in many places. There are two reasons: cost is low and certain species are resistant to most herbicides but controlled by burning (e.g., red cedar). In Kansas experiments, late-spring burning controlled brush best and injured desirable forage species least.

Many people consider forest fires only as destructive. This is usually true of large, uncontrolled fires that occasionally ravage forests. However, fire in forest areas can produce benefits. Since about 1950, "prescribed burning" or "controlled burning" has become a method of periodic burning. This method leaves desirable forest trees uninjured when done properly. Such burning removes some undesirable trees, brush, leaves, branches, and other debris, and serves other useful purposes. These include (1) less competition from "weed" species, (2) the seedbed is improved for desirable tree seedlings and made less desirable for others, (3) less fuel remains for wild fires, and (4) some disease organisms may be controlled.

Thickness of bark is important in determining susceptibility to fire. If the phloem and cambium of the trunk are killed by heat, the tree may show an effect similar to that of gridling.

Most pine seedlings develop best with less than $1\frac{1}{2}$ inches of debris on the forest floor; oak seedlings need more debris. One reason for prescribed burning is to prepare seedbeds favorable to pine and unfavorable to oak. All tree seedlings, when small, are susceptible to fire. Most pine seedlings become resistant after the first year and will withstand considerable fire after the trunk has reached a 2-inch diameter at breast height (DBH). Oak and most other hardwoods the same age are still susceptible. Therefore, where pine forests are desired, controlled fire can effectively and selectively kill the deciduous hardwoods. Frequency of prescribed burning will depend on local conditions; usually every 2 to 3 years is best.

Local authorities should be consulted before setting a fire for regulations and precautions; firebreaks, fire-fighting equipment, wind direction and possibility of a shift, and other fire hazards must be considered. A large share of the cost of burning is for personnel and equipment on standby for emergency use. Obviously an escaped fire can be extremely costly. During several years of protection from fire, considerable debris and undergrowth may accumulate. If so, the first burning may be hot enough to damage even large trees. Burning when woods are only partially dry will reduce the intensity of the fire.

GIRDLING

Girdling consists of completely removing a band of bark around the woody stem. The method is effective on most woody, dicotyledonous plants, but is too laborious to be efficient on thick stands of smaller stemmed species. In most species, sprouting near the base of the tree is eliminated or at least considerably reduced on large, mature trees. Young trees may sprout profusely following girdling.

Plant foods move down to the roots through the phloem, and water and nutrients move upward through the xylem (see Chapter 5). When bark is removed, phloem tissue is destroyed, and plant foods can no longer move to the roots. Since roots depend on tops for food, the roots eventually die of starvation. This process usually takes from 1 to 3 years. The length of time depends on the rate of food usage and amount of food stored in the roots at the time of girdling. The top of the tree shows little injury until the root is near death if girdling is properly done. The top also succumbs as soon as the root dies.

The girdle must be wide enough to prevent healing and deep enough to prevent the cambium from developing a new phloem. The girdled area should dry in a short time. Chemicals similar to those used for frill treatment or heat may help kill an active cambium. The xylem should not be injured. If the top is killed too rapidly through injury to the xylem (sapwood), root reserves may not be depleted enough to prevent shoot or sprout growth. If sprouts or lateral branches develop below the girdle, these will soon provide enough food for the roots to keep the tree alive. Thus any sprouts that develop may be cut or treated with suitable chemical sprays to prevent food from returning to the roots.

HERBICIDES

Use of herbicides to control brush and undesirable trees has expanded rapidly since 1950. Proper use of chemicals has generally proven to be more effective and less costly than most other methods.

Herbicides can be applied in many different ways to accomplish different purposes. These can be classified as foliage sprays, bark treatment, trunk injections, stump treatment, and soil treatment. The herbicides used for these treatments are given in Table 22-1. The effectiveness of herbicides should not be evaluated too quickly, wait at least until the next growing season. In some cases, trees may die 2 to 3 years after herbicide treatment, after repeatedly developing new leaves followed by defoliation.

Foliage Sprays

Foliage sprays require uniform coverage to be most effective. The need for thorough wetting will depend on the chemical used and the species treated. In most cases, uniform coverage is more important than thorough wetting; the latter is more often simply a means of obtaining uniform coverage. Effectiveness of foliage sprays varies considerably among species and with size of woody plants. Several chemicals kill aboveground parts, but sprays have been less effective in killing roots. A resurgence of sprouts is common with some species. Repeat treatments are often needed as new sprouts develop.

Absorption by leaves is the first major problem of herbicide effectiveness. Many woody plants have coarse, thick leaves with a heavy cuticle. On such plants a nonpolar substance is absorbed more effectively than a polar substance. Thus esters of 2,4-D, dichlorprop, and other oil–soluble herbicides are absorbed better than salt formulations. An effective wetting agent added to the spray solution often increases absorption. Once absorbed, water-soluble or polar forms (especially of 2,4-D) are translocated more readily through the phloem than the esters. Esters appear to be hydrolyzed to their respective acids on entering the plant. A second important problem is acute versus chronic toxicity. An excessively high application rate may kill or injure the phloem and inhibit symplastic translocation. The chemical will not be translocated to the roots and roots will not be killed. Absorption is also discussed in Chapters 5 and 7.

For foliar application this suggests three practical steps: (1) low-acute-toxicity herbicides and wetting agents should be used for maximum absorption and translocation of growth substances through phloem tissues, (2) herbicides should be applied at rates that cause chronic toxicity, but not acute toxicity, and (3) low rates of treatment and repeated treatment should be used for some species as they often give superior final results.

Foliage sprays of growth substances (such as 2,4-D and triclopyr) have usually given best root kills when large quantities of food are being translocated to the roots. With species that become somewhat dormant in summer, this peak translocation period usually occurs about the time of full-leaf development in the

TABLE 22-1. Herbicides Used to Control Some Common Woody plants[1,2]

	2,4-D	Hexazinone	Imazapyr	Dicamba[3]	Picloram[3,4]	Fosamine	Amitrole[4]	Glyphosate	Tebuthiuron[4]	Triclopyr[3]
Alder	F, B			C, S		F			S	
Ash		S	B, C	C, S	S	F	F	F	S	F, S
Blackberry		S		F, S		F	F	F	S	F, S
Buckbrush	F, B								S	
Elderberry	F, B	S								F, S
Elm	C	S			F	F		F	S	F, S
Locust	F, B	S			S	F	F		S	F, S
Manzanita	F, B				F					
Maple		S	B, C	C, S	S	F	F	F		F, S
Mesquite				F, S	F				S	
Oak	F, C	S	B, C	C, S	S	F		F	S	
Poison ivy or oak	F, B	S		F, S	F		F		S	
Rabbitbrush	F, B								S	
Rose			B, C		S	F		F	S	
Sagebrush	F, B								S	
Saltcedar	F, B, C			S	S				S	
Sassafras			B, C			F				
Sumac	F, B	S	B, C	F, S		F	F		S	F, S
Willow	F, B	S	B, C	F,S		F		F	S	F, S

[1] F, foliar; B, stems; C, cutsurface (frill, injection or stump treatment); S, soil.

[2] All formulations of these herbicides are not suitable for all the uses indicated. *Check manufacturer's label* for specific species controlled with various formulations and for use precautions. *Follow label instructions.*

[3] Dicamba, picloram, dichlorprop, and triclopyr are often combined with 2,4-D.

[4] Noncropland only.

spring; a second peak period may occur in the fall. Plants that continue rapid growth and translocation throughout the summer can be treated anytime during the summer.

Chemicals are applied by hand sprayers, mist nozzles, power-mist applicators, truck-, trailer-, or tractor-mounted power-spray equipment, and airplanes. Water is normally used as the carrier with ground equipment. Many plants develop a thickened, waxy cuticle. On such plants oil–water mixes are more effective. Airplane application may use water, oil, or oil–water emulsions depending on the type of foliage to be treated and the relative humidity.

Herbicides such as 2,4-D and dichlorprop are applied at rates of 1 to 4 lb/acre. Amitrole, 2 to 4 lb/100 gal of water, is effective on poison ivy, poison oak, kudzu, ash, locust, blackberry, dewberry, and sumac.

Picloram, 0.06 to 0.5 lb/acre, and dicamba, 0.5 to 8 lb/acre, are effective as a foliar or soil application for certain woody species. They are used on species resistant to phenoxy-type herbicides. They are often combined with 2,4-D to control a broad range of woody species.

Ammonium sulfamate is an effective foliage treatment. It is usually suggested where phenoxy compounds are too hazardous to use because of spray drift. The application rate is 60 lb/100 gal and the foliage should be completely wetted.

Fosamine is foliar applied in late summer or early fall. Response is not normally observed until the next spring. Susceptible treated plants fail to refoliate and later die. It is applied at the rate of 6 to 12 lb/acre.

Glyphosate is a nonselective, foliar-applied herbicide. It will translocate from foliage to roots, resulting in kill of many species. It is applied at rates of 3 to 3.75 lb/acre when plants are actively growing and when most plants are at or beyond the full-bloom stage of growth. It is particularly effective against blackberries, honeysuckle, kudzu, maple, multiflora rose, trumpetcreeper, and willow.

Triclopyr is an analog of picloram and is effective on a number of woody species. The ester formulation is applied at rates of 4 to 6 lb/acre is much more active than the amine formulation which is applied at rates of 6 to 9 lb/acre as a foliar spray. Triclopyr is often tank mixed with picloram for foliar applications.

Bark Treatment

In bark treatments the chemical is applied to the stem near the ground and is absorbed through the bark. Two methods are used: broadcast bark treatment and basal bark treatment.

Broadcast Bark Treatment In broadcast bark treatment, the herbicide is applied to the entire stem area of the plant. This treatment is especially adapted to thick stands of small-stemmed woody plants. Spray is usually applied between fall leaf drop and midwinter. Phenoxy herbicides are effective on several species: from 4 to 12 lb/100 gal are applied, mixed in oil. The spray is applied so that the greatest amount of it will cover the stems evenly. For example, treating upright stems with a horizontally directed spray exposes the greatest stem surface to the spray.

Basal Bark Treatment This method and most of the following methods have the advantage that the operator can selectively treat only those trees he wishes to kill, without injury to other trees. With this method the herbicide is applied so as to wet the stem base, 10 to 15 inches aboveground, until rundown drenches the stem at groundline. The chemical can be applied from a sprayer or from a container that lets the chemical "trickle" on the stem base. A small number of trees can easily be treated with a small container to pour the chemical on the base of the tree trunk. Brush or large trees are left standing. This method is especially effective on brush and trees less than 6 inches in diameter. It controls some plants that sprout from horizontal underground roots and stems, especially if applied during early summer.

Herbicides used most often are 2,4-D, picloram, dicamba, triclopyr, imazapyr and various combinations of 2,4-D with dichlorprop, picloram, dicamba, or triclopyr. For 2,4-D, 8 to 16 lb active ingredient of herbicide is mixed in enough diesel oil to make 100 gal of mixture. The lower rate is suggested when trees are growing actively and the higher rate for when they are dormant. The herbicide mixture is usually applied at the rate of 1 gal/100 inches of tree diameter; for example, 50 trees 2 inches in diameter or 33 trees 3 inches in diameter.

Scientists do not fully understand the way basal bark treatment kills plants. It is possible that treatment either kills the phloem or immobilizes it to the point that the tree is chemically girdled. In addition, the herbicide may inhibit bud formation and sprout development. If the tree is chemically girdled, the tree root dies from starvation, as it does with girdling.

Trunk Injections

Trunk injections often help the herbicide penetrate through the bark. In some cases, injections may serve as a girdle, with the herbicide acting as a chemical girdle. Techniques used are frilling or notching with an ax or tree injector.

Frill Treatment Frill treatment consists of a single line of overlapping, downward ax cuts around the base of the tree (Figure 22-3). The herbicide is then sprayed or squirted into the cut entirely around the tree. The method is effective on trees too large in diameter for basal bark treatment.

Phenoxy compounds such as 2,4-D were the principal herbicides applied in this manner. Presently a wide range of materials can be used. These include dicamba, glyphosate, imazapyr, picloram, triclopyr, and various combinations such as 2,4-D + picloram, 2,4-D + triclopyr, and 2,4-D + dichlorprop. Ammonium sulfamate is applied by saturating the frill with water containing 3 to 5 lb of chemical per gallon.

Notching The tree trunk may be notched with an axlike frill treatment, except that one notch is cut for every 6 inches of trunk circumference. Notches filled with one teaspoonful of ammonium sulfamate crystals will kill many kinds of broadleaved trees (Figure 22-4). This method is less effective than frill treatment, which uses a continuous notch around the tree.

The tree-injector tool speeds notch treatment, and when properly used does a satisfactory job (Figure 22-5). The oil-soluble amine or ester form of the phenoxy compounds plus diesel oil (2:9 ratio) can be used; one cut for every 2 inches in trunk diameter. This tool is especially effective against elm, postoak, white oak, live oak, and willow. More resistant trees are treated by spacing injections closer together. Trees that are more resistant include ash, cedar, hackberry, hickory, blackjack oak, red oak, persimmon, and sycamore.

Figure 22-3. Frill application is used to apply various herbicides to trees too large for effective basal bark applications. (Dow Chemical Company, Midland, MI.)

Figure 22-4. Application of ammonium sulfamate in notches. (F. A. Peevy, ARS, USDA.)

Figure 22-5. The tree injector speeds notch application of herbicides. (W. C. Elder, Oklahoma State University.)

Other herbicides that can be injected include dicamba, glyphosate, hexazinone, imazapyr, triclopyr, 2,4-D, and 2,4-D + picloram. Dicamba is injected as the undiluted commercial liquid concentrate added to water at a rate of 1:1 for resistant trees and 1:4 for susceptible trees, with 0.5 to 0.1 ml applied per injection cut.

Stump Treatment

Stumps of many species may quickly sprout after trees or brush are cut. Most trees can be prevented from sprouting by proper stump treatment (Figure 22-6). However, weedy trees that can develop sprouts from underground roots or stems are difficult to control by stump treatment alone. With most such species, stump treatment is effective if followed by a midsummer basal bark treatment of the sprouts when they develop.

Herbicides such as 2,4-D or 2,4-D + triclopyr are usually used in stump treatment at the rate of 8 to 12 lb of acid in enough diesel oil to make 100 gal of spray solution. Enough spray should be applied to "wet" the tops and sides of the stump so that rundown drenches the stem at groundline. Ammonium sulfamate crystals are also highly effective on many species. Crystals are usually applied to the cut top of the stump at $\frac{1}{4}$ oz (about $1\frac{1}{2}$ teaspoonfuls)/inch of diameter or as a spray solution containing 3.5 to 5 lb/gal of water.

Stump treatments are most effective if applied immediately after the tree is cut. Sprouting from older cuts can usually be controlled by increasing the amount of chemical.

Figure 22-6. Hand spray application of a herbicide to prevent stumps from sprouting. (Dow Chemical Company, Midland, MI.)

Soil Treatments

Some herbicides control woody plants when applied to the soil. Most of these herbicides are used as dry pellets. They require rainfall to leach them into the soil as deep as the feeder roots. Therefore, they are usually applied just before or early in the rainy season. Herbicides used this way usually persist in the soil for more than 1 year for maximum efficacy. Effects may develop slowly and not be apparent for 1 to 2 years after treatment (Figure 22-7).

Research has shown that bromacil, dicamba, hexazinone, picloram, and tebuthiuron can be used effectively by this method. The species to be controlled determines which of these herbicides is used. Sometimes combinations of these have been more effective than either one alone.

Since the early 1970s, soil-applied herbicides have been used increasingly as a method to control undesirable brush and trees to improve grazing capacity on rangelands. Some of these herbicides are formulated as pellets or granules. They can be applied by hand, ground equipment, or airplane.

Many woody plants have extensive finely branched root systems, whereas many desirable grasses and forbs have root systems restricted to a small area. Some soil-applied herbicides used for woody plant control are also phytotoxic to desirable grasses and forbs. Therefore, studies have been conducted to determine the most efficient spacing of pellets for good woody plant control with minimal injury to desirable grasses and forbs. In Oklahoma, a grid application of tebuthiuron spaced 6 × 6 feet was equally or more effective than a broadcast

Figure 22-7. Brush and tree control with a soil treatment of tobuthiuron. *Right*: Treated. *Left*: Not treated. Photograph taken 2 years after application.

application for control of oak (Stritzke, 1976). Doses were placed in a small area 6 feet apart in each direction. In Texas, tebuthiuron placed in rows from $4\frac{1}{2}$ to 18 feet apart was equally effective as broadcast treatment for control of several woody species. Researchers concluded that this spacing principle makes application of herbicides from airplanes in more or less grid patterns commercially possible (Meyer et al., 1978).

With highly concentrated herbicide application in a 6 × 6 grid, a bare area 8 to 24 feet in diameter developed around the dosage sites about 90 days after application. Bare spots covered from 1.5 to 4.1% of the total area. After 1 year, in most cases, the bare areas had disappeared. Desirable grasses and forbs in the interspaces were not injured (Scifres et al., 1978). Obviously for successful use of this method for each herbicide, the percent concentration of active ingredient, size of pellet, and area covered by each pellet must be determined through research.

Hexazinone is formulated as a Gridball® to be distributed in a random grid pattern for control of woody plants. To apply 10 lb/acre, pellets are spaced about 6 feet apart in each direction on the soil surface; for 20 lb/acre, $4\frac{1}{2}$ feet apart; and for 40 lb/acre, 3 feet apart. The herbicide controls elm, hawthorn, maple, red oak, post oak, white oak, sweetgum, and wild plum.

SUGGESTED ADDITIONAL READING

Baldwin, F. L., and J. W. Boyd, 1989, *Recommended Chemicals for Weed and Brush Control*, Coop. Ext. Serv., Univ. of Arkansas, MP-44.

Crop Protection Chemicals Reference, 1989, Wiley, New York (updated annually).

Meyer, R. W., and R. W. Bovey, 1980, *Weed Sci.* **28**(1), 51.

Meyer, R. W., R. W. Bovey, and J. R. Baur, 1978, *Weed Sci.* **26**(5), 444.

Scifres, C. J., J. L. Mutz, and C. H. Meadors, 1978, *Weed Sci.* **26**(2), 139.

Stritzke, J. F., 1976, *Abstract, Proc. South. Weed Sci. Soc.* **29**, 255.

Weed Control Manual and Herbicide Guide, 1989, Meister, Willoughby, OH (updated annually).

For herbicide use, see the manufacturer's or supplier's label and follow these directions. Also, see Preface.

23 Aquatic Weed Control

Aquatic plants, as the term is used here, include those plants that normally start in water and complete at least part of their life cycle in water. They have both positive and negative aspects in regard to the welfare of man.

On the positive side, aquatic weeds may reduce erosion along shorelines, and some plant species provide food and protection for aquatic invertebrates, fish, fowl, and game. Algae are the original source of food for nearly all fish and marine animals; and swamp smartweed, wildrice, wild millet, and bulrush provide food and protection for waterfowl, especially ducks.

On the negative side, excessive growth of aquatic weeds causes many serious problems for people who use ponds, lakes, streams, and irrigation and drainage systems. Excess weeds (1) obstruct water flow and increase water losses, (2) interfere with navigation, fishing, and other recreational activities, (3) destroy wildlife habitats, (4) cause undesirable odors and flavors, (5) lower real-estate values, (6) create health hazards, and (7) speed up the rate of silting by increasing the accumulation of silt and debris.

Controlling aquatic weeds sometimes causes problems other than those of the chemical itself. For example, the rapid killing of dense, weedy growth may kill fish, which happens even though the chemical is nontoxic to the fish. During photosynthesis, living plants release oxygen, and fish depend on this oxygen for respiration. When plants are killed, they produce no more oxygen. In addition, dead plants are decomposed by microorganisms that require oxygen for respiration. These two actions may reduce oxygen content in the water, causing the fish to suffocate. The answer is to treat only a part of very heavily infested areas at one time; fish will move to the untreated part.

A recreation area suitable for both swimming and fishing presents management problems. For example, the right fertilization favors microscopic plants, which through a food chain are ultimately used as food by fish. This heavy growth of microscopic plants makes the water appear cloudy or dirty and may give it an undesirable odor; hence it is less desirable for swimming. Therefore, it is difficult to manage a body of water so that it is optimal for both swimming and fishing.

METHODS OF CONTROLLING AQUATIC WEEDS

Methods of controlling aquatic weeds include the following: (1) proper construction of pond, (2) competition for light, (3) pasturing, (4) drying,

(5) mowing, (6) hand cleaning, (7) chaining, (8) dredging, (9) burning, (10) biological control, and (11) chemical control.

Proper Construction of Pond

Proper pond construction is very important in controlling pond weeds. Many rooted aquatic plants are not easily established in deep water. The pond should be built so that as much water as possible is at least 3 feet deep. You can have water 3 feet deep only 9 feet from shoreline if all the edges of the pond have a slope of 3 to 1. Such a slope greatly reduces the area where cattails, rushes, and sedges first start growing. However, steep banks are hazardous for swimming. Gentle slopes should be provided for swimming areas.

Competition for Light and Fertilization

Ponds adequately fertilized develop millions of tiny plants and animals that give the water a cloudy appearance (bloom). If the water has a bloom and is at least 3 feet deep, submerged aquatic weeds have almost no chance to grow because of inadequate light. The benefits of fertilization include the following: (1) increased growth of beneficial microscopic life, including phytoplankton and zooplankton, (2) increased food supply for fish from the food chain that develops from the above, and (3) effective weed control by shading. Plants that do reach the surface should be cut off; otherwise they will be stimulated by the fertilizer.

A 16-20-4 or similar fertilizer is suggested at about 50 lb/acre. The first application is applied in early spring and is repeated as needed to maintain the cloudy appearance. A light-colored object should not be visible 1.5 feet below the surface. Fertilization of ponds is practical only where there is little loss of water from the pond, because fertility is lost with the overflow.

Pasturing

Pasturing is economical and effective in controlling marginal aquatic grasses, weeds, and some weedy species. A good legume–grass pasture mixture, if properly managed and grazed, will give the banks and dam a lawnlike appearance. A good sod also protects the banks against erosion and helps to control undesirable species. Excessive trampling may destroy the banks and muddy the water. Also, leeches in the water may attack animals, and some diseases are spread in the water to livestock, especially to dairy animals.

Drying and/or Freezing

Drying is a simple way to control many submerged aquatics. If the water can be withdrawn from the pond or ditch, leaf and stem growth of submerged weeds may be killed after 7 to 10 days of exposure to sun and air. Drying usually must be repeated to control regrowth from roots or propagules in the bottom mud or sand. In ditches this operation may be repeated several times per season. Especially in cold climates, if the water is drawn down in late fall and the lake not allowed to refill until early spring, many aquatic weeds will be killed. As the lake refills, reinfestation may occur from weed propagules from the deeper part of the lake.

Mowing

Mowing effectively controls some ditch-bank weeds. Power equipment can be most easily used where the banks are relatively smooth and not too steep. Underwater power-driven weed saws and weed cutters are also available. The effects usually are only for a short interval. Mowing is usually required at rather frequent intervals, and disposal of mowed weeds is often difficult.

Hand Cleaning

In lightly infested areas, hand cleaning may be the most practical method of control. A few hours spent in pulling out an early infestation may prevent the weed from spreading. The method is particularly effective on new infestations of emergent weeds such as cattail, arrowhead, and willow.

Chaining

Chaining aquatic weeds resembles chaining woody plants (Chapter 22). A heavy chain, attached between two tractors, is dragged in the ditch. The chain tears loose the rooted weeds from the bottom. The method is effective against both submerged and emergent aquatics.

Chaining should be started whenever new shoots of emersed weeds rise about 1 foot above the water or when submersed weeds reach the water surface. It should be repeated at regular intervals. Dragging the chain both ways may be effective in tearing loose most of the weeds. The method is limited primarily to ditches of uniform width, accessible from both sides with tractors, and free of trees and other obstructions. After chaining it is usually necessary to remove plant debris from the ditch to keep it from accumulating and stopping the flow of water.

Dredging

Dredging is a common method of cleaning ditches that are accessible from at least one side. The dredge may be equipped with the usual bucket, or a special weed fork may be used. Dredging may solve two problems: removal of weeds and removal of silt and debris. Dredging has been tried in ponds from specially built pontoons, but in general the pontoon dredge has not proven practical. Dredging is an expensive operation, because of high equipment costs and the large amount of labor involved.

Burning

Burning may control ditch-bank weeds such as cottonwoods, willows, perennial grasses, and many annual weeds. Green plants are usually given a preliminary searing. After 10 to 14 days, vegetation may be dry enough to burn from its own heat. Burning can also be combined with chemical- or mechanical-control programs. Burning the previous year's debris allows better spray coverage of regrowth. It may be desirable to burn the dead debris after chemical treatment. Mowing followed by burning the dried weeds may increase the effectiveness of the mowing.

Biological Control

Aquatic weeds have been controlled by fish, snails, insects, microorganisms, and higher plants. Ducks often effectively control duckweed in small ponds. Biological control has appeal because of the continuing control potential and the nonuse of chemicals in the water. However, as with other biological-control methods, care must be taken not to introduce a biological-control organism that will have undesirable side effects; for example, a fish that reduces the population of game fish.

Some freshwater fish will eat aquatic vegetation. The white amur (Chinese grass carp), tilapia, and silver dollar fish are used to control aquatic weeds in certain areas of the world. The white amur has been used in the People's Republic of China, Czechoslovakia, Poland, and the Soviet Union. It is now being used in Arkansas (see Figure 3-7).

For alligatorweed control a fleabeetle (*Agasicles hybrophila*), and a moth (*Vogtia malloi*) are providing control in the southeastern United States.

Although biological control holds great potential, its actual use has been limited. The federal government and state governments now support considerable research on this method. Hydrilla and eurasian watermilfoil are two aquatic weeds receiving considerable funding from the federal and state agencies.

Chemical Control

Chemicals effectively control many aquatic and ditch-bank weeds. The following information is needed to use this method: (a) the name or names of the weed species, (2) the appropriate chemical, recommended rate, and time of treatment, and (3) the amount of water or size of area to be treated.

Water surface areas are usually measured in acres, like field areas. One acre is 43,560 ft^2, or an area 208.7 feet square (on each side). One acre-ft of water means 1 acre of water 1 foot deep, or 43,560 ft^3 of water (325,828 gallons, or 2,719,450 pounds). Thus, a chemical concentration of 1 ppm would require 2.7 pounds of the chemical (active ingredient) per acre-ft of water. One ppm to an average depth of 3 feet would require 8.11 pounds. A closely related technique involves treating the "bottom acre-foot." Two methods of application are used. Using formulations that are heavier than water, the chemical can be applied to the water surface as either a spray or granule. The chemical sinks to the bottom. The second method involves drop hoses or weighted hoses dragged behind a boat, releasing the chemical at the lake bottom. Rates of application are usually based on a bottom acre (43,560 ft^2), much the same as with field-crop application. For conversion factors to other measurements see pages 448–451.

Running water is measured by several methods. Rates are usually given as *cubic feet per second*; 1 ft^3/sec is equal to 450 gal/min. Usually the rate of water flow is determined by the use of a weir and gauge.

CHEMICALS USED IN AQUATIC-WEED CONTROL

The more important chemicals used in aquatic-weed control are discussed below. The use of specific chemicals for control of certain aquatic weeds is given in Table 23-1. For more complete discussions of chemicals see Chapters 8 through 14. Trade names are on pages 443–445.

Restrictions on the use of water that has been treated with an aquatic herbicide are *extremely important*. Detailed instructions and restrictions are printed on the label. *Those instructions must be followed.*

Acrolein

Acrolein is useful for treating weed-infested drainage ditches and irrigation canals. It controls most submersed water weeds and many snails (Table 23-1). In small irrigation canals in western states at rates of 1 to 2.5 gal/ft^3/sec flow of water, this chemical controls weeds from 6 to 20 miles downstream. However, the principal use of acrolein is in large canals where the usual treatment is 0.1–0.6 ppm on a volume basis (ppmv) for 8 to 48 hours. Weed control with these treatments may extend 20 to 50 miles downstream.

Treated water does not harm crops when used for irrigation at low concentrations. Higher concentrations may cause injury to susceptible crops, such as cotton. Acrolein is not effective in the control of waterplantain. Acrolein is a potent irritant and lacrimator. It is also toxic to fish and should not be used where a fish kill cannot be tolerated.

Amitrole

Amitrole is safe to fish at normal rates. It is especially effective on cattails and bulrushes. A special form of amitrole and sodium thiocyanate known as amitrole-T is very effective on waterhyacinth and certain other emergent and floating species. Amitrole is applied at 7 to 9 lb/acre active ingredient as a foliage spray. It should be applied on cattails between the time of flowering and seed maturity.

Copper Sulfate and Complexes

Copper sulfate (bluestone, blue copperas, blue vitriol) and chelated copper (such as ethylenediamine, triethanolamine, and alkanolamine) are very effective against

TABLE 23-1. Herbicides[1] Used to Control Several Common Aquatic Weeds

	Class[2]	Acrolein	Amitrole	Copper Complexes	2,4-D	Dicamba	Dichlobenil	Diquat	Endothall	Fluridone	Glyphosate	Simazine
Algae												
Unicellular	S			X					X			X
Filamentous	S			X				X	X			X
Chara	S			X			X					X
Alligatorweed	E				X	X					X	
Arrowhead	E				X					X		
Bladderwort	S	X			X		X	X	X	X		
Bulrush	E		X		X						X	
Cattail	E		X		X	X					X	
Coontail	S	X			X	X	X	X	X	X		X
Duckweed	F				X			X		X		X
Elodea	S	X			X	X	X	X		X		
Hydrilla	S			X				X	X	X		
Naiad	S	X			X		X	X	X	X		X
Pickerelweed	E				X					X	X	
Pondweed	S	X			X		X	X	X	X		X
Rush	E				X							
Spikerush	S, E				X							
Waterhyacinth	F		X		X			X			X	
Waterlily	E				X		X			X		
Watermilfoil	S, E	X			X		X	X	X	X		X
Waterprimrose	S, F				X		X		X	X	X	

[1] See the manufacturer's label for rates, method of usage, species susceptibility, and precautions. Follow the label—regardless of statements in this book.
[2] F, Floating (unattaced, tops above water); E, emersed (rooted underwater or growing on wet soil, tops above water); S, submersed (usually rooted in soil, tops mostly underwater).

most kinds of algae, including chara. Soon after treatment the algae's color changes to a grayish white. Several days after treatment, the water should be nearly free of algal growth.

2,4-D

Many aquatic plants are susceptible to 2,4-D dissolved in water at rates of 1 to 5 ppm by weight (Table 23-1). To provide a concentration of 1.0 ppm in 1 acre-ft of water, 2.7 lb of 2,4-D (acid equivalent) are required. Water-soluble liquid forms of 2,4-D or granular forms may be used.

2,4-D should be exposed in water to susceptible weeds at full concentration for at least 10 hours. In *nonmoving* water, 2,4-D in water-soluble form will tend to *distribute itself equally* over several days for distances up to 40 feet. *In contrast*, the granular form falls to the pond bottom, where a relatively high concentration may develop at the soil–water interface.

Ester formulations of 2,4-D are 50 to 200 times more toxic to fish than amine formulations, but toxic effects have rarely been experienced under field conditions (Westerdahl and Getsinger, 1988). Esters are oil-like and oil-soluble materials, and oils are known to be toxic to most fish. Esters may be acting similar to oils, or the oil solvents, emulsifying agents, or other additives may be killing the fish. When pure 2,4-D ester is used in granular form, without surface-active agents, there is considerably less hazard to fish.

For irrigation water, use of treated water should be delayed for 3 weeks after treatment or until the water does not contain more than 0.1 ppm of 2,4-D.

Dalapon

Dalapon is harmless to fish at normal rates. Lake emerald shiners apparently suffered no ill effects from 3 days in 3000 ppm of dalapon, but 5000 ppm was fatal. Dalapon is especially effective against grasses and cattails. It is best applied as a foliage spray. Most species are susceptible if not rooted in standing water. Draining is advisable, if possible, several weeks before treatment with dalapon.

Dicamba

Dicamba is useful for the control of several submersed, floating, and emersed aquatic weeds. It is relatively nontoxic to fish, other aquatic organisms, and wildlife. However, it is generally not applied in water where human contact is likely. Dicamba is a powerful growth regulator; great care must be exercised when applications are made near sensitive crops or desirable vegetation.

Dichlobenil

Granular dichlobenil controls many aquatic weeds in static water such as lakes and ponds. It is particularly effective on elodea, watermilfoil, coontail, chara, and pondweeds (*Potamogeton* spp.). It is applied in early spring at rates of 7 to 10 lb/acre before weeds start to grow. Although it does not harm fish at these rates, humans may not eat these fish until 90 days after treatment. It cannot be used in commercial fish or shellfish waters. It should not be applied to water that will be used for irrigation, livestock, or human consumption.

Diquat

Diquat controls many submersed aquatic weeds and algae in static water. Bladderwort, coontail, elodea, naiad, pondweeds, watermilfoil, spirogyra, and pithophora are among the species controlled. It is applied at rates of 1 to 4 lb/surface acre.

Diquat can be applied by pouring directly from the container into the water while moving slowly in a boat. Early in the season strips 40 feet apart are suggested, later in the season 20 feet. For best results, it should be applied before weeds reach the surface of the water. Because diquat is a contact spray, repeated applications may be necessary to give seasonlong control.

It should not be used in muddy water because diquat is rendered ineffective by its tight adsorption on soil particles. Treated water should not be used for animal or human consumption, irrigation, agricultural sprays, swimming, or domestic purposes within 14 days after treatment. It is not harmful to most fish at the recommended rate. In heavy weed infestations only one-third to one-half of the area should be treated to avoid fish kill through oxygen depletion. Wait 10 to 14 days between treatments.

Endothall

Endothall is available as a number of derivatives and in several liquid and granular formulations. It controls most algae and submersed aquatic weeds in static and in some flowing waters. However, elodea is not controlled. It is applied at a broad range of rates—from 0.3 to 14 lb/acre-ft of water. Lower rates are for algae control and higher rates for coontail, watermilfoil, pondweeds, naiad, bass weed, and burreed control. Treated water should not be used for irrigation, agricultural sprays, animal consumption, or domestic purposes within 7 days. Fish may be used as food three days after treatment.

Fluridone

Fluridone controls at least 14 submersed and emersed plants. After application, chlorosis and a pinkish color show up in 7 to 10 days, but full herbicidal effects may take up to 60 days. About 4 weeks after treatment, susceptible weeds begin to sink to the bottom. This slow herbicidal response reduces the potential of fish kill from oxygen depletion.

Fluridone is adsorbed by the hydrosoil (lake-bottom soil), reaching a maximum 1 to 4 weeks after treatment. These hydrosoil residues declined to a nondetectable level after 16 to 52 weeks.

Glyphosate

Glyphosate is primarily effective for control of emersed aquatic weeds such as alligatorweed, cattail, maidencane, paragrass, spatterdock, waterhyacinth, giant cutgrass, and torpedograss. Glyphosate is applied at the rate of 2 to 3 lb/acre; a surfactant is required for optimal results. Glyphosate must be retained on foliage of actively growing plants and is not effective for control of submersed or mostly submersed vegetation.

Simazine

Simazine may be used in ponds for control of several submerged and floating aquatic weeds and algae. Control of weeds is usually seasonlong. Algae may be controlled for 1 to 3 months or may require retreatment later in the season. Simazine may be used in ponds containing fish. It should be used only in ponds that will have little or no outflow after treatment. Simazine should not be used as a spot treatment. Algae control requires 1.4 to 3.4 lb/acre-ft; submerged weeds, 2.7 to 6.8 lb; and floating weeds, 2.7 to 5.4 lb.

Ponds are treated after seasonal flow has ceased early in the weed and algae growth period. They should be treated when 5 to 10% of the pond surface is covered with scum (algae mats) or floating weeds, and/or while submerged aquatic weeds are actively growing and before they reach the surface of the water. As a general rule, ponds in northern areas may be treated between May 1 and

Figure 23-1. Several submersed aquatic weeds. *Left to right*: Watermilfoil, coontail, eelweed, pondweed, and largeleaf pondweed. (King and Penfound, 1946.)

Figure 23-2. Emersed and floating aquatic weeds. *Left to right*: Burreed, duckweed, white waterlily, floating pondweed. (King and Penfound, 1946.)

TABLE 23-2. Common Aquatic Weeds

Type	Name
Emersed plants	Alligatorweed (*Alternanthera philoxeroides*), arrowhead (*Sagittaria* spp.), bulrush (*Scirpus* spp.), cattail (*Typha* spp.); common reed (*Phragmites australis*), cutgrass (*Leersia hexandra*), fragrant waterlily (*Nymphaea odorata*), maidencane (*Panicum hemitomon*), pickerelweeds (*Pontederia* spp.), sawgrass (*Cladium jamaicense*), smartweeds (*Polygonum* spp.), southern watergrass (*Hydrochloa caroliniensis*), spatterdock (*Nuphar luteum*), torpedograss (*Panicum repens*), water pennywort (*Hydrocotyle umbellata*), water primrose (*Ludwigia uruguayensis*), watershield (*Brasenia schreberi*), waterwillow (*Justicia americana*)
Floating plants	Duckweed (*Lemna minor*), giant duckweed (*Spirodela polyrhiza*), mosquito fern (*Azolla caroliniana*), salvinia (*Salvinia rotundifolia*), slender duckweed (*Wolffiella floridana*), waterhyacinth (*Eichhornia crassipes*), waterlettuce (*Pistia stratiotes*), watermeal (*Wolffia columbiana*)
Submersed plants	Bladderworts (*Utricularia* spp.), coontail (*Ceratophyllum demersum*), egeria (*Egeria densa*), elodea (*Elodea canadensis*), eurasian watermilfoil (*Myriophyllum spicatum*), fanwort (*Cabomba caroliniana*), horned pondweed (*Zannichellia palustris*), hydrilla (*Hydrilla verticillata*), naiads (*Najas* spp.), parrotfeather (*Myriophyllum aquaticum*), pondweeds (*Potamogeton* spp.), water buttercup (*Ranunculus aquatilis*)
Algae	Planktonic—*Anabaena, Chlorella, Pediastrum, Scenedesmus, Oocystis* Filamentous—*Spirogyra, Cladophora, Rhizocionium, Zygnema, Hydrodictyon* Attached—erect—*Chara, Nitella*

June 15. In southern areas, where water warms up and weed growth is earlier, ponds may be treated between April 1 and May 15.

High water temperatures cause more rapid natural decay of dead weeds and algae that can cause fish distress; therefore, simazine should be applied before water temperatures exceed 75°F. Ponds having an extremely heavy infestation of weeds and algae such as occurs in mid and late summer should not be treated, since rapid decomposition of heavy growth greatly reduces the oxygen content of the water, and this can cause fish distress and/or death.

Use of Fish and Water following Use of Simazine The following regulations pertain to ponds following application of simazine: (1) fish taken from treated ponds may be used for human consumption; (2) treated ponds may be used for swimming; and (3) water from treated ponds may not be used for irrigation or spraying of agricultural crops, lawns, or ornamental plantings, or for watering cattle, goats, hogs, horses, poultry, or sheep, or for human consumption until 12 months following treatment. Ponds that have bordering trees with roots visibly extended into the water must not be treated, since injury to these trees may occur. Usually, trees 50 feet or more from the pond's edge will not be injured.

Aquatic Weed Identification

Identification of aquatic weeds is essential for selecting the appropriate control measure. Aquatic weeds are commonly classified by their growth habits: (1) *floating*, (2) *emersed*, and (3) *submersed* (Figures 23-1 and 23-2). Algae occupy a special category because of their growth form and undesirable characteristics. Algae may annoy bathers, causing a type of dermatitis and symptoms of hay fever. Blue-green algae have been known to cause poisoning of horses, cattle, sheep, hogs, dogs, and poultry. Odors and fishy tastes often result from decaying algae in water reservoirs. Extremely heavy algae growth may suffocate fish by depleting the supply of oxygen in the water at night.

A list of common aquatic weeds and their scientific names are listed in Table 23-2. Many excellent manuals are available to aid in the identification of aquatic weeds (Westerdahl and Getsinger, 1988; Schmidt, 1987; Tarver et al., 1978; Lewis amd Miller, 1984; Thayer et al., 1986).

SUGGESTED ADDITIONAL READING

Bruns, V. F., J. M. Hodgson, H. F. Arle, and F. L. Timmons, 1955, USDA Circular No. 971.

Burkhalter, A. P., L. M. Curtis, R. L. Lazor, M. L. Beach, and J. C. Hudson, 1974, *Aquatic Weed Identification and Control Manual*, Florida Department of Natural Resources, Tallahassee.

Crop Protection Chemicals Reference, 1989, Wiley, New York (updated annually).

Lewis, C. W., and J. F. Miller, 1984, *Identification and Control of Weeds in Southern Ponds*, Coop. Ext. Serv., Univ. of Georgia, B-839.

Muenscher, W. C., 1944, *Aquatic Plants of the United States*, Comstock, Ithaca, NY.

Schmidt, J. C., 1987, *How to Identify and Control Water Weeds and Algae*, Applied Biochemists, Inc., Mequon, WIS.

Tarver, D. P., J. A. Rodgers, M. J. Mahler, and R. L. Lazor, 1978, *Aquatic and Wetland Plants of Florida*, Florida Dept. of Natural Resources, Tallahassee, Florida.

Thayer, D. D., W. T. Haller, and J. C. Joyce, 1986, *Weed Control in Aquaculture and Farm Ponds*. Florida Coop. Extension Service.

Westerdahl, H. E., and K. D. Getsinger, 1988, *Aquatic Plant Identification and Herbicide Use Guide*. Volume I: *Aquatic Herbicides and Application Equipment*, U.S. Army Corp. of Engineers Technical Report A-88-9.

Westerdahl, H. E., and K. D. Getsinger, 1988, *Aquatic Plant Identification and Herbicide Use Guide*. Volume II: *Aquatic Plants and Susceptibility to Herbicides*, U.S. Army Corp. of Engineers Technical Report A-88-9.

For herbicide use, see the manufacturer's or supplier's label and follow these directions. Also see Preface.

24 Total Vegetation Control

Total control of vegetation is the removal of all higher green plants and maintenance of these areas vegetation free. Complete absence of vegetation is desirable on many sites such as railway roadbeds, industrial areas, highway brims, fencerows, and irrigation and drainage-ditch banks. Industrial sites include storage and work areas, lumberyards, utility transmission stations, and railroad right-of-ways (Figure 24-1).

Vegetation can be controlled totally by mechanical or chemical methods. Sometimes both methods are used. Brush and trees may be removed by mechanical methods at first clearance, although this method adds considerable cost. These areas are then maintained weed free with herbicides. Herbaceous species can be eliminated by either mechanical or chemical methods. Discing or other mechanical means may need to be repeated several times a season to keep the area weed free. But certain residual herbicides at high rates are needed only annually or less often.

FOLIAR HERBICIDES

To obtain vegetation-free areas, certain nonselective foliar-applied herbicides are used together with persistent, nonselective, soil-applied herbicides. These foliar herbicides include contact herbicides such as paraquat or diquat and translocated materials such as 2,4-D, picloram, dicamba, triclopyr, and glyphosate (Table 24-1). Combination products are available containing mixtures of foliar herbicides with persistent soil-applied herbicides.

SOIL HERBICIDES

Persistent nonselective herbicides used for total vegetation control are given in Table 24-1. Chemical and physical properties of these compounds, as well as other uses, have been discussed in previous chapters. Although some of these herbicides are also used as selective herbicides, the rate of application is usually lower for selective than for nonselective uses.

Soil-applied herbicides may remain toxic to plants for more than 1 year. Factors affecting the length of time that a herbicide remains toxic in the soil were discussed in Chapter 6. In general, dry weather with little or no leaching, cool or cold temperatures, and heavy soils tend to lengthen the time that a herbicide will

Figure 24-1. Total vegetation control on a railroad with tebuthiuron, also used on industrial areas and noncrop sites. (Elanco Products Company, a division of Eli Lilly and Company.)

TABLE 24-1. Herbicides Commonly Used for Total Vegetation Control of Weeds on Noncropland[1]

Soil-Applied Herbicides[2]	Rate (lb/acre)	Foliar-Applied Herbicides[2]	Rate (lb/acre)
Atrazine	4.8–40	Ametryn	1–2
Bromacil	2.4–40	Amitrole	1.8–10
Dicamba	2–8	AMS	95
Diuron	3.2–48	Asulam	3.3–6.6
Imazapyr	0.5–1.5	Bromoxynil	1.0
Metsulfuron	0.02–0.15	Cacodylic acid	1.25–5.0
Picloram	2–8.5	Dicamba	1–3
Prometon	10–15	2,4-D	1–3
Simazine	10–40	Dichlorprop	0.5–12
Sodium chlorate	190–760	Diquat	0.5
Sulfometuron	0.3–0.6	Fosamine	6–12
Tebuthiuron	1.2–16	Glyphosate	0.75–4.0
		Hexazinone	1–12
		Linuron	1–3
		MSMA	2.5
		Paraquat	0.5–1
		Triclopyr	1–9

[1] Various combinations of these herbicides are often used to give greater persistence and/or broader spectrum of weeds controlled. Trade names are given on pages 443–445.
[2] Some herbicides have considerable foliar activity as well as soil activity. See the manufacturer's label for rates, method of usage, species susceptibility, and precautions. Follow the label—regardless of statements in this book.

remain toxic. Under any given condition the length of time that a herbicide will remain toxic can be predicted with reasonable accuracy (Table 6-1). Annual applications of most of these herbicides at rates that persist somewhat longer than 1 year are usually more economical than massive rates that will persist for 2 years or more.

These persistent, nonselective soil-applied herbicides must be leached into the rooting of seed-germination zone of the weeds to be effective. Therefore, they are usually applied just before or during the rainy season. Persistent soil-applied herbicides gradually lose phytotoxicity. With loss of phytotoxicity weeds reinfest the area. Usually new plants are stunted and grow slowly at first. At that time, use of a broad-spectrum foliar herbicide will usually extend the period of total vegetation control.

PREVENTING INJURY TO TREES, SHRUBS, ORNAMENTALS, AND LAWNS

Nearly all nonselective herbicides will kill all types of plant growth. As one would expect, some plants are more susceptible to some herbicides than to others. At high rates they should all be considered very effective against trees, shrubs, ornamental flowers, and lawns. Soil sterilants should never be applied to the rooting zone of such plants or so that the chemical is washed into their rooting zone.

SUGGESTED ADDITIONAL READING

Baldwin, F. L., and J. W. Boyd, 1989, *Recommended Chemicals for Weed and Brush Control*, Coop. Ext. Serv., Univ. of Arkansas, MP-44.

Crop Protection Chemicals Reference, 1989, Wiley, New York (updated annually).

Weed Control Manual and Herbicide Guide, 1989, Meister, Willoughby, OH (updated annually).

For herbicide use, see the manufacturer's or supplier's label and follow these directions. Also see Preface.

Appendix

TABLE A-0. Herbicides that Control Sixty Common Herbeceous Weeds as Listed on the Manufacturer's Label[1,2,3]

Weed	Barnyardgrass	Bermudagrass	Bindweed, field	Bluegrass, annual	Brome, downy	Carpetweed	Cheat	Chickweed, common	Chickweed, mousear	Cocklebur, common	Crabgrass; large, smooth	Dallisgrass	Dandelion	Dock, curly	Dogfennel	Foxtail, yellow	Goosegrass	Groundcherry, cutleaf	Groundsel, spp.	Guineagrass	Henbit	Jimsonweed	Johnsongrass	Knotweed, prostrate	Kochia	Lambsquarters, common	Lettuce, prickly
1. Acifluorfen			×			×				×						×		×				×	×			×	
2. Alachlor	×					×					×					×	×	×				×			×	×	
3. Ametryn		×						×			×	×				×	×				×		×				
4. Amitrole		×			×	×					×						×		×			×			×		
5. Asulam	×										×						×						×				
6. Atrazine	×			×	×		×			×	×				×	×					×	×			×	×	
7. Benefin	×	×		×		×		×			×					×	×							×		×	
8. Bensulide	×			×		×					×						×									×	
9. Bentazon	×	×	×								×					×	×		×			×	×		×	×	
10. Bromacil	×	×		×	×		×				×	×	×		×		×				×		×		×		
11. Bromoxynil										×												×		×		×	
12. Butylate	×	×									×					×	×						×				
13. Chloramben	×					×		×			×					×	×		×						×	×	
14. Chlorimuron																						×					
15. Chlorsulfuron		×						×	×				×	×		×			×		×			×	×	×	×
16. Clomazone	×	×			×	×					×					×	×					×					×
17. Cyanazine	×			×	×	×	×	×		×	×			×	×	×	×		×		×	×		×	×	×	×
18. Cycloate	×			×							×					×					×					×	
19. 2,4-D			×			×		×		×			×	×							×	×			×	×	×
20. 2,4-DB										×				×					×			×			×	×	×
21. DCPA	×			×		×		×			×					×										×	
22. Desmedipham								×									×										
23. Dicamba			×			×		×	×	×			×	×	×						×	×			×	×	×
24. Dichlobenil				×		×					×			×	×				×		×					×	
25. Diclofop	×				×						×					×	×										
26. Diethatyl	×										×					×					×		×				
27. Difenzoquat																											
28. Endothall	×			×																	×				×		
29. EPTC	×	×		×		×		×			×					×	×				×	×			×	×	
30. Ethalfluralin	×			×		×		×			×					×	×				×	×	×		×	×	
31. Ethofumesate	×			×				×								×									×	×	
32. Fenoxaprop	×										×					×	×						×				
33. Fluazifop-P	×	×									×					×	×			×			×				
34. Fluometuron	×										×						×					×					
35. Fomesafen	×					×				×	×					×	×					×				×	
36. Glyphosate	×	×	×	×	×		×	×	×	×	×	×		×	×		×		×	×	×		×		×	×	×
37. Hexazinone	×	×		×	×		×	×		×	×		×	×	×				×	×	×						×
38. Imazamethabenz																											
39. Imazapyr		×	×	×	×	×	×	×		×	×	×	×		×		×			×			×		×	×	×
40. Imazaquin	×									×						×	×					×				×	
41. Isopropalin						×					×					×	×									×	
42. Lactofen						×				×								×			×						
43. Linurpn											×				×		×		×								
44. Methazole										×																×	
45. Metolachlor	×					×										×	×										
46. Metribuzin	×		×	×	×	×	×	×	×	×	×			×	×	×	×			×	×	×	×	×	×	×	×

TABLE A-0. (*Continued*)

Weed	Mallow, little	Morningglory, ivyleaf	Mustard, wild	Nightshade, black	Nightshade, hairy	Nutsedge, purple	Nutsedge, yellow	Oats, wild	Panicum, fall	Penneycress, field	Pigweed, redroot	Puncture vine	Purslane, common	Pusley, Florida	Quackgrass	Radish, wild	Ragweed, common	Rocket, London	Ryegrass, Italian	Sandbur, field	Shattercane	Shephardspurse	Sida, prickly	Smartweed, Penn.	Sowthisle, spp.	Speedwell, corn	Spurge, spp.	Sunflower, spp.	Thistle, bull	Thistle, Canada	Thistle, Russian	Velvetleaf	Wildbuckwheat
1.		×	×	×					×		×		×	×			×				×			×			×			×		×	×
2.				×	×		×		×					×						×	×		×									×	
3.			×						×				×	×							×			×	×			×				×	
4.								×							×							×			×		×			×			×
5.									×																								
6.			×	×			×	×		×		×	×		×		×							×	×	×	×	×	×	×	×	×	×
7.							×		×		×		×	×					×	×													
8.									×		×		×									×										×	
9.		×	×	×	×		×	×	×		×		×		×		×				×	×	×	×				×		×		×	×
10.								×				×		×	×																		
11.		×								×	×					×	×	×				×		×				×		×	×	×	×
12.						×	×		×											×	×												
13.			×	×	×				×		×	×	×				×					×	×	×	×		×				×	×	
14.		×					×										×											×					
15.			×							×	×	×	×			×		×				×		×	×			×		×	×		×
16.			×	×					×		×		×	×			×				×		×	×			×					×	×
17.			×	×		×	×	×	×		×	×	×	×	×	×	×	×	×		×	×	×	×	×		×	×			×	×	×
18.				×	×	×	×	×			×		×								×	×											
19.	×	×	×							×		×				×	×			×		×		×	×		×	×	×	×	×	×	
20.			×	×	×					×							×					×	×								×	×	
21.									×		×		×	×						×							×						
22.			×	×							×		×				×	×				×			×								×
23.		×	×	×						×	×	×	×	×		×	×					×	×	×	×		×	×	×	×	×	×	×
24.			×								×			×	×	×	×					×	×	×			×		×	×	×		
25.								×	×										×														
26.								×	×									×	×						×								
27.								×																									
28.											×																						×
29.				×	×	×	×	×	×		×	×	×	×	×				×	×	×	×	×									×	
30.			×	×	×	×	×	×	×		×		×	×			×		×		×	×	×	×							×	×	×
31.				×		×	×	×			×		×									×		×	×						×		×
32.								×	×												×												
33.								×	×						×				×	×	×												
34.									×					×									×	×									
35.		×	×	×			×		×		×		×	×			×						×	×								×	
36.						×	×	×	×	×	×		×	×	×		×	×	×	×	×	×	×	×	×	×		×		×	×	×	
37.								×		×	×				×		×	×				×		×			×			×			
38.			×					×		×																							×
39.	×							×	×				×		×		×		×						×			×	×	×			×
40.		×					×				×	×		×			×				×		×	×			×	×				×	
41.									×		×		×	×					×	×													
42.			×									×	×	×			×						×	×			×			×		×	
43.		×		×					×				×	×		×	×						×									×	×
44.																							×				×						
45.				×	×		×		×		×		×	×						×	×												
46.	×	×	×	×			×	×	×	×	×		×	×	×		×	×	×	×	×	×	×	×	×		×	×			×	×	×

TABLE A-0. (*Continued*)

Weed	Barnyardgrass	Bermudagrass	Bindweed, field	Bluegrass, annual	Brome, downy	Carpetweed	Cheat	Chickweed, common	Chickweed, mousear	Cocklebur, common	Crabgrass; large, smooth	Dallisgrass	Dandelion	Dock, curly	Dogfennel	Foxtail, yellow	Goosegrass	Groundcherry, cutleaf	Groundsel, spp.	Guineagrass	Henbit	Jimsonweed	Johnsongrass	Knotweed, prostrate	Kochia	Lambsquarters, common	Lettuce, prickly
47. Metsulfuron			×					×						×	×				×		×			×	×	×	×
48. MSMA	×							×		×	×	×					×						×				
49. Napropamide	×	×		×	×	×		×			×					×	×		×	×	×		×	×		×	×
50. Naptalam	×		×			×		×		×	×					×	×	×								×	
51. Norflurazon	×	×		×	×	×	×			×	×				×		×		×	×	×		×		×	×	
52. Oryzalin	×	×	×	×	×	×		×			×				×	×	×		×	×	×	×	×	×	×		×
53. Oxyfluorfen	×			×		×				×	×						×	×	×		×	×		×		×	
54. Pebulate		×									×					×	×				×						
55. Pendimethalin						×					×					×	×				×				×	×	
56. Prometryn	×										×						×										
57. Pronamide	×			×	×	×		×	×		×					×	×				×		×			×	
58. Propachlor	×					×					×					×	×					×			×	×	
59. Propanil	×										×					×	×		×						×	×	
60. Pryazon																					×						
61. Quizalofop	×															×	×										
62. Sethoxydim	×	×				×					×					×	×	×				×				×	
63. Siduron	×										×																
64. Simazine	×			×	×	×		×			×						×		×		×					×	×
65. Sulfometuron	×			×	×			×			×		×	×	×								×		×		×
66. Tebuthiuron	×				×	×	×				×	×			×				×		×				×		×
67. Terbacil	×	×			×			×			×				×				×	×	×	×					×
68. Thiameturon		×		×				×	×	×				×							×			×	×	×	×
69. Thiobencarb	×																										
70. Triallate																											
71. Triclopyr			×			×								×							×						
72. Tridiphane	×									×	×					×	×					×			×	×	
73. Trifluralin	×	×	×	×	×	×	×				×				×	×	×			×	×	×	×		×		
74. Vernolate	×	×		×		×	×				×					×	×						×			×	

[1] Prepared from data in *Crop Protection Chemicals Reference*, 5th Ed., 1989, Wiley, New York.

[2] Space limitations preclude the listing of all herbicides and all weeds; see Labels, Wiley's *Crop Protection Chemicals Reference*, and/or Meister's *Weed Control Manual* for more comprehensive coverage.

[3] Inclusion or omission of a herbicide or weed does not constitute a recomendation or nonrecomendation.

TABLE A-0. (*Continued*)

Weed	Mallow, little	Morningglory, ivyleaf	Mustard, wild	Nightshade, black	Nightshade, hairy	Nutsedge, purple	Nutsedge, yellow	Oats, wild	Panicum, fall	Penneycress, field	Pigweed, redroot	Puncture vine	Purslane, common	Pusley, Florida	Quackgrass	Radish, wild	Ragweed, common	Rocket, London	Ryegrass, Italian	Sandbur, field	Shattercane	Shephardspurse	Sida, prickly	Smartweed, Penn.	Sowthistle, spp.	Speedwell, corn	Spurge, spp.	Sunflower, spp.	Thistle, bull	Thistle, Canada	Thistle, Russian	Velvetleaf	Wildbuckwheat
47.	×		×							×	×		×									×			×			×	×	×	×		×
48.				×			×	×		×	×									×		×											
49.	×				×	×	×	×	×		×		×	×			×	×	×	×		×			×								
50.			×	×	×		×		×		×		×	×			×					×						×				×	
51.	×								×			×	×	×	×		×	×				×	×	×	×		×				×	×	×
52.	×		×	×	×	×		×	×		×	×	×	×	×	×	×	×	×	×		×	×	×	×	×	×		×		×	×	×
53.	×	×	×	×				×	×		×		×		×		×	×		×		×	×	×	×		×		×		×	×	×
54.					×	×	×	×			×		×	×								×											
55.									×			×	×	×						×	×						×					×	
56.				×				×	×					×									×	×									
57.	×		×	×	×			×	×				×		×	×		×	×			×											
58.				×	×				×								×			×		×	×	×								×	×
59.			×								×			×																			
60.										×												×										×	
61.								×	×						×						×												
62.			×	×			×	×	×		×		×	×	×		×			×	×	×	×				×			×		×	×
63.																																	
64.			×	×	×			×	×				×	×	×							×		×							×		
65.	×						×												×						×		×	×		×	×		
66.								×	×			×	×				×		×	×		×	×	×	×		×	×		×			
67.						×	×							×	×							×											
68.	×		×							×	×					×		×				×		×	×			×			×		×
69.																																	
70.								×																									
71.	×																								×	×	×		×	×			
72.				×				×					×				×							×				×				×	
73.			×	×	×	×	×	×	×		×	×		×			×		×	×	×	×	×	×			×				×	×	
74.				×		×	×	×	×		×	×		×	×		×			×	×											×	

TABLE A-1. Common Name of Herbicides in Alphabetical Order and Corresponding Trade Name and Manufacturer[1]

Common Name	Trade Name	Manufacturer
Acetochlor	Harness, Tophand	Monsanto
Acifluorfen	Blazer, Tackle	BASF, Rhone-Poulenc
Acrolein	Magnacide	Magna
Alachlor	Lasso	Monsanto
Ametryn	Evik	CIBA-GEIGY
Amitrole	Amitrol, Amizol	Rhone-Poulenc
AMA	AMA, Methar	Vineland, Cleary
AMS	Ammate	DuPont
Asulam	Asulox	Rhone-Poulenc
Atrazine	AAtrex, Atrazine	CIBA-GEIGY, DuPont
Benefin	Balan	Elanco
Bensulide	Prefar	ICI Americas
Bentazon	Basagran	BASF
Bensulfuron	Londax	DuPont
Bifenox	Modown	Rhone-Poulenc
Bromacil	Hyvar, Bromax	DuPont
Bromoxynil	Brominal, Buctril	Rhone-Poulenc
Buthidazole	Ravage	Sandoz
Butylate	Sutan, Genate	ICI Americas, Valent
Chloramben	Amiben	Rhone-Poulenc
Chlorimuron	Classic	DuPont
Chloroxuron	Tenoran	CIBA-Geigy
Chlorpropham	Furloe	Chevron, PPG
Chlorsulfuron	Glean, Telar	DuPont
Cinmethylin	Cinch	DuPont
Clethodim	Select	Chevron, Valent
Clomazone	Command	FMC
Cloproxydim	—	Chevron
Clopyralid	Lontrel, Reclaim, Stinger	Dow
CMA	Calar	Vineland
Cyanazine	Bladex	DuPont
Cycloate	Ro-Neet	ICI-Americas
2,4-D	Weedar, Weedone, Dacamine, Others	Fermenta, Rhone-Poulenc, Others
2,4-DB	Butyrac	Rhone-Poulenc
Dalapon	Dalapon 85, Others	Fermenta, Others
Dazomet	Basamid	Loveland/Hopkins
DCPA	Dacthal	Fermenta
Desmedipham	Betanex	Nor-Am
Dicamba	Banvel	Sandoz
Dichlobenil	Casoron	Uniroyal
Dichlorprop	Several	Several
Diclofop	Hoelon	Hoechst-Roussel
Diethatyl	Antor	Nor-Am

(*Continued*)

TABLE A-1. *(Continued)*

Common Name	Trade Name	Manufacturer
Difenzoquat	Avenge	American Cyanamid
Diphenamid	Enide	Nor-Am
Dipropetryn	Sancap	CIBA-GEIGY
Diquat	Diquat, Tag	Chevron, ICI Amer., Valent
Diuron	Karmex	DuPont
DSMA	Several	Several
Endothall	Aquathol, Endothal, Hydrothol, Herbicide 273	Pennwalt
EPTC	Eptam, Eradicane, Genep	ICI-Americas, Valent
Ethalfluralin	Sonalan	Elanco
Ethofumesate	Nortron	Nor-Am
Fenoxaprop	Acclaim, Horizon, Option, Tiller, Whip	FMC, Hoechst-Roussel
Fluazifop	Fusilade	ICI America
Fluazifop-P	Flusilade 2000	ICI America
Fluchloralin	Basalin	BASF
Fluometuron	Cotoran	CIBA-GEIGY
Fluridone	Sonar	Elanco
Fomesafen	Reflex	ICI Americas
Fosamine	Krenite	DuPont
Glyphosate	Roundup, Others	Mosanto
Haloxyfop	Verdict, Gallant	Dow
Hexazinone	Velpar	DuPont
Imazamethabenz	Assert	American Cyanamid
Imazapyr	Arsenal, Chopper, Contain	American Cyanamid
Imazaquin	Scepter	American Cyanamid
Imazethapyr	Pursuit	American Cyanamid
Isopropalin	Paarlan	Elanco
Isouron	Conserve	Elanco
Isoxaben	Gallery, Snapshot	Elanco
Lactofen	Cobra	Chevron, Valent
Linuron	Lorox	DuPont
MAA	Several	Several
MCPA	Several	Several
MCPB	Thistrol	Rhone-Poulenc
Mecoprop	Several	Several
Metham	Vapam	ICI Americas
Methazole	Probe	Sandoz
Metolachlor	Dual	CIBA-GEIGY
Metribuzin	Lexone, Sencor	DuPont, Mobay
Metsulfuron	Ally, Escort	DuPont
Molinate	Ordram	ICI Americas
MSMA	Ansar, Bueno, Daconate	Fermenta
Napropamide	Devrinol	ICI Americas

TABLE A-1. *(Continued)*

Common Name	Trade Name	Manufacturer
Naptalam	Alanap	Uniroyal
Norflurazon	Evital, Solicam, Zorial	Sandoz
Oryzalin	Surflan	Elanco
Oxadiazon	Ronstar	Rhone-Poulenc
Oxyfluorfen	Goal	Rhom and Haas
Paraquat	Cylone, Gramoxone	ICI Americas
Pebulate	Tillam	ICI Americas
Pendimethalin	Prowl, Stomp	American Cyanamid
Perfluidone	Destun	3M
Phenmedipham	Betanal, Spin-Aid	Nor-Am
Picloram	Tordon	Dow
Profluralin	Tolban	CIBA-GEIGY
Prometon	Pramitol	CIBA-GEIGY
Prometryn	Caparol	CIBA-GEIGY
Pronamide	Kerb	Rohm and Haas
Propachlor	Ramrod	Monsanto
Propanil	Stam, Stampede	Rohn and Haas
Propazine	Milogard	CIBA-GEIGY
Propham	Chem Hoe	Chervon, PPG
Pyrazon	Pyramin	BASF
Pyridate	Tough	Gilmore, Agrolinz
Quizalofop	Assure	DuPont
Sethoxydim	Poast	BASF
Siduron	Tupersan	DuPont
Simazine	Aquazine, Princep	CIBA-GEIGY
Sulfometuron	Oust	DuPont
Tebuthiuron	Graslan, Spike	Elanco
Terbacil	Sinbar	DuPont
Terbutryn	Igran	CIBA-GEIGY
Thiameturon	Harmony	DuPont
Thiobencarb	Bolero	Chevron, Valent
Triallate	Far-Go	Monsanto
Triclopyr	Garlon, Turflon, Rely	Dow, Hoechst-Roussel
Tridiphane	Tandem	Dow
Trifluralin	Treflan	Elanco
Vernolate	Vcrnam, Surpass	ICI Americas

[1] There may be other trade names and manufacturers for some of these herbicides. It is not intended that the ones listed are preferred over any others.

TABLE A-2. Trade Name of Herbicides in Alphabetical Order and Corresponding Common Name[1]

Trade Name	Common Name	Trade Name	Common Name
AAtrex	Atrazine	Contain	Imazapyr
Acclaim	Fenoxaprop	Cotoran	Fluometuron
Alanap	Naptalam	Cyclone	Paraquat
Ally	Metsulfuron	Dacamine	2,4-D
AMA	AMA	Daconate	MSMA
Amiben	Chloramben	Dacthal	DCPA
Amitrol	Amitrole	Dalapon 85	Dalapon
Amizol	Amitrole	Destun	Perfluidone
Ammate	AMS	Devrinol	Napropamide
Ansar	MSMA	Diquat	Diquat
Antor	Diethatyl	Dual	Metolachlor
Aquathol	Endothall	Endothal	Endothall
Aquazine	Simazine	Eptam	EPTC
Arsenal	Imazapyr	Enide	Diphenamid
Assert	Imazamethabenz	Eradicane	EPTC
Assure	Quizalofop	Escort	Metsulfuron
Asulox	Asulam	Evik	Ametryn
Atrazine	Atrazine	Evital	Norflurazon
Avenge	Difenzoquat	Far-Go	Triallate
Balan	Benefin	Flusilade	Fluazifop
Banvel	Dicamba	Flusilade 2000	Fluazifop-P
Basagran	Bentazon	Furloe	Chlorpropham
Basalin	Fluchloralin	Gallant	Haloxyfop
Basamid	Dazomet	Gallery	Isoxaben
Betanal	Phenmedipham	Garlon	Triclopyr
Betanex	Desmedipham	Genate	Butylate
Betasan	Bensulide	Genep	EPTC
Bladex	Cyanazine	Glean	Chlorsulfuron
Blazer	Acifluorfen	Goal	Oxyfluorfen
Bolero	Thiobencarb	Gramoxone	Paraquat
Bromax	Bromacil	Graslan	Tebuthiuron
Brominal	Bromoxynil	Harmony	Thiameturon
Buctril	Bromoxynil	Harness	Acetochlor
Bueno	MSMA	Herbicide 273	Endothall
Butyrac	2,4-DB	Hoelon	Diclofop
Calar	CMA	Horizon	Fenoxaprop
Caparol	Prometryn	Hydrothol	Endothall
Casoron	Dichlobenil	Hyvar	Bromacil
Chem Hoe	Propham	Igran	Terbutryn
Cinch	Cinmethylin	Karmex	Diuron
Chopper	Imazapyr	Kerb	Pronamide
Classic	Chlorimeturon	Krenite	Fosamine
Cobra	Lactofen	Lasso	Alachlor
Command	Clomazone	Lexone	Metribuzin
Conserve	Isouron	Londax	Bensulfuron

TABLE A-2. (*Continued*)

Trade Name	Common Name	Trade Name	Common Name
Lontrel	Clopyralid	Sonalan	Ethalfluralin
Lorox	Linuron	Sonar	Fluridone
Magnacide	Acrolein	Spike	Tebuthiuron
Methar	AMA	Spin-Aid	Phenmedipham
Milogard	Propazine	Stam	Propanil
Mowdown	Bifenox	Stampede	Propanil
Nortron	Ethofumesate	Stinger	Clopyralid
Option	Fenoxaprop	Stomp	Pendimethalin
Ordram	Molinate	Surflan	Oryzalin
Oust	Sulfometuron	Surpass	Vernolate
Paarlan	Isopropalin	Sutan+	Butylate
Poast	Sethoxydim	Tackle	Acifluorfen
Pramitol	Prometon	Tag	Diquat
Prefar	Bensulide	Tandem	Tridiphane
Princep	Simazine	Telar	Chlorsulfuron
Probe	Methazole	Tenoran	Chloroxuron
Prowl	Pendimethalin	Thistrol	MCPB
Pursuit	Imazethapyr	Tillam	Pebulate
Pyramin	Pyrazon	Tiller	Fenoxaprop
Ramrod	Propachlor	Tolban	Profluralin
Ravage	Buthidazole	Tophand	Acetochlor
Reclaim	Clopyralid	Tordon	Picloram
Reflex	Fomesafen	Tough	Pyridate
Rely	Triclopyr	Treflan	Trifluralin
Ro-Neet	Cycloate	Turflon	Triclopyr
Ronstar	Oxadiazon	Tupersan	Siduron
Roundup	Glyphosate	Vapam	Metham
Sancap	Dipropetryn	Velpar	Hexazinone
Scepter	Imazaquin	Verdict	Haloxyfop
Select	Clethodim	Vernam	Vernolate
Sencor	Metribuzin	Weedar	2,4-D
Sinbar	Terbacil	Weedone	2,4-D
Snapshot	Isoxaben	Whip	Fenoxaprop
Solicam	Norflurazon	Zorial	Norflurazon

[1]There may be other trade names for some of these herbicides. It is not intended that the ones listed are preferred over any others.

TABLE A-3. Conversion Factors

Liquid Measure

1 gallon (U.S.) = 3785.4 millilitere (ml); 256 tablespoons; 231 cubic inches; 128 fluid ounces; 16 cups; 8 pints; 4 quarts; 0.8333 imperial gallon; 0.1337 cubic foot; 8.337 pounds of water
1 liter = 1000 milliliters; 1.0567 liquid quarts (U.S.)
1 gill = 118.29 milliliters
1 fluid ounce = 29.57 milliliters; 2 tablespoons
3 teaspoons = 1 tablespoon; 14.79 milliliters; 0.5 fluid ounce
1 cubic foot of water = 62.43 pounds; 7.48 gallons

Weight

1 gamma = 0.001 milligram (mg)
1 grain (gr) = 64.799 milligrams
1 gram (g) = 1000 milligrams; 15.432 grains; 0.0353 ounce
1 pound = 16 ounces; 7000 grains; 453.59 grams; 0.45359 kilogram
1 short ton = 2000 pounds; 0.097 metric ton
1 long ton = 2240 pounds; 1.12 short ton
1 kilogram = 2.2046 pounds

Linear Measure

12 inches = 1 foot; 30.48 centimeters
36 inches = 3 feet; 1 yard; 0.914 meter
1 rod = 16.5 feet; 5.029 meters
1 mile = 5280 feet; 1760 yards; 160 rods; 80 chains; 1.6094 kilometers (km)
1 chain = 66 feet; 22 yards; 4 rods; 100 links
1 inch = 2.54 centimeters (cm)
1 meter = 39.37 inches; 10 decimeters (dm); 3.28 feet
1 micron (μm) = $\frac{1}{1000}$ millimeter (mm)
1 kilometer = 0.621 statute miles; 0.5396 nautical miles

Area

1 township = 36 sections; 23,040 acres
1 square mile = 1 section; 640 acres
1 acre = 43,560 square feet; 160 square rods; 4840 square yards; 208.7 feet square; an area $16\frac{1}{2}$ feet wide and $\frac{1}{2}$ mile long; 0.4047 hectare
1 hectare = 2.471 acres; 100 are

Capacity (Dry Measure)

1 bushel (U.S.) = 4 pecks; 32 quarts; 35.24 liters; 1.244 cubic feet; 2150.42 cubic inches

Pressure

1 foot lift of water = 0.433 pound pressure per square inch (psi)
1 pound pressure per square inch will lift water 2.31 feet
1 atmosphere = 760 millimeters of mercury; 14.7 pounds; 33.9 feet of water

Geometric Factors ($\pi = 3.1416$; r = radius; d = diameter; h = height)
Circumference of a circle = $2\pi r$ or πd

TABLE A-3. (*Continued*)

Diameter of a circle $= 2r$
Area of a circle $= \pi r^2$ or $\frac{1}{4}\pi d^2$ or $0.7854d^2$
Volume of a cylinder $= \pi r^2 h$
Volume of a sphere $= \frac{1}{6}\pi d^3$

Other Conversions

Multiply	by	To Obtain
Gallons per minute	2.228×10^{-3}	Cubic feet per second
Gallons per acre	9.354	Liters per hectare
Kilograms per hectare	0.892	Pounds per acre
Liters	1.05	U.S. quarts
Liters	0.2642	U.S. gallons
Liters per hectare	0.107	Gallons per acre
Miles per hour	88.0	Feet per minute
Miles per hour	1.61	Kilometers per hour
Pounds per gallon	0.12	Kilograms per liter
Pounds per square inch	70.31	Grams per square centimeter (Atm)
Pounds per 1000 square feet	0.489	Kilograms per acre
Pounds per acre	1.12	Kilograms per hectare
Square inch	6.452	Square centimeter
Parts per million	2.719	Pounds acid equivalent per acre foot of water

Temperature degrees

$F^\circ = C^\circ + 17.78 \times 1.8$
$C^\circ = F^\circ - 32.00 \times \frac{5}{9}$

°C	°F	°C	°F
100	212	30	86
90	194	20	68
80	176	10	50
70	158	0	32
60	140	−10	14
50	122	−20	−4
40	104	−30	−22

TABLE A-4. Weight of Dry Soil

Type	Pounds per Cubic Foot	Pounds per Acre, 7 inches Deep
Sand	100	2,500,000
Loam	80–95	2,000,000
Clay or silt	65–80	1,500,000
Muck	40	1,000,000
Peat	20	500,000

TABLE A-5. Length of Row Required for One Acre

Row Spacing (inches)	Length or Distance
24	7260 yards = 21,780 ft
30	5808 yards = 17,424 ft
36	4840 yards = 14,520 ft
42	4149 yards = 12,445 ft
48	3630 yards = 10,890 ft

TABLE A-6. Available Commercial Materials in Pounds Active Ingredients per Gallon Necessary to Make Various Percentage Concentration Solutions[1]

Pounds of Active Ingredient in 1 gal of Commercial Product	Pounds of Active Ingredient/Pint[1]	Liquid Ounces of Commercial Product to make one gallon of				
		$\frac{1}{2}$%	1%	2%	5%	10%
2.00	0.25	2.68	5.36	10.72	26.80	53.60
2.64	0.33	2.02	4.05	8.10	20.25	40.50
3.00	0.375	1.78	3.56	7.12	17.80	35.60
3.34	0.72	1.59	3.18	6.36	15.90	31.80
4.00	0.50	1.34	2.68	5.36	13.40	26.80
6.00	0.75	0.89	1.78	3.56	8.90	17.80

[1] Based on 8.4 lb/gal (weight of water) and 128 liquid oz = 1 gal, 16 liquid oz = 1 pint.

TABLE A-7. Equivalent Quantities of Liquid Materials When Mixed by Parts

Water	1–400	1–800[1]	1–1600
100 gal	1 qt	1 pt	1 cup
50 gal	1 pt	1 cup	$\frac{1}{2}$ cup
5 gal	3 tbs	5 tsp[1]	$2\frac{1}{2}$ tsp
1 gal	2 tsp	1 tsp	$\frac{1}{2}$ tsp

[1] Example: If a recommendation calls for 1 part of the chemical to 800 parts of water, it would take 5 tsp in 5 gal of water to give 5 gal of a mixture of 1–800.

TABLE A-8. Equivalent Quantities of Dry Materials (Wettable Powders) for Various Quantities of Water

Water	Quantity of Material					
100 gal[1]	1 lb	2 lb	3 lb	4 lb[1]	5 lb	6 lb
50 gal	8 oz	1 lb	24 oz	2 lb	$2\frac{1}{2}$ lb	3 lb
5 gal[1]	3 tbs[2]	$1\frac{1}{2}$ oz	$2\frac{1}{2}$ oz	$3\frac{1}{4}$ oz[1]	4 oz	5 oz
1 gal	2 tsp[2]	3 tsp	$1\frac{1}{2}$ tbs	2 tbs	3 tbs	3 tbs

[1] Example: If a recommendation calls for a mixture of 4 lb of a wettable powder to 100 gal of water, it would take $3\frac{1}{4}$ oz (approximately 12 tsp) to 5 gal of water to give 5 gal of spray mixture of the same strength.

[2] Wettable materials vary considerably in density. Therefore, the teaspoonful (tsp) and tablespoonful (tbs) measurements in this table are not exact dosages by weight but are within the bounds of safety and efficiency for mixing small amounts of spray.

TABLE A-9. Pressure Drop in Hose Due to Friction; Figures Are for 25 ft of Clean Hose with No Couplings

$\frac{1}{4}''$ i.d. Hose		$\frac{3}{8}''$ i.d. Hose		$\frac{1}{2}''$ i.d. Hose		$\frac{5}{8}''$ i.d. Hose		$\frac{3}{4}''$ i.d. Hose		$1''$ i.d. Hose		$1\frac{1}{4}''$ i.d. Hose	
Flow (gpm)	Pr. Drop (psi/25 ft)	Flow (gpm)	Pr. Drop (psi/25 ft)	Flow (gpm)	Pr. Drop (psi/25 ft)	Flow (gpm)	Pr. Drop (psi/25 ft)	Flow (gpm)	Pr. Drop (psi/25 ft)	Flow (gpm)	Pr. Drop (psi/25 ft)	Flow (gpm)	Pr. Drop (psi/25 ft)
0.2	0.8												
0.3	1.5												
0.4	2.5												
0.5	4.0	0.5	0.5										
0.6	5.0	0.6	0.8										
0.8	9.0	0.8	1.3										
		1.0	1.8	1.0	0.5								
		2.0	6.0	2.0	1.5								
		3.0	13.0	3.0	3.1	3.0	1.0						
				4.0	6.0	4.0	1.8						
				5.0	8.5	5.0	2.5	5.0	1.0				
				6.0	12.0	6.0	3.7	6.0	1.5				
						8.0	6.5	8.0	2.5	8.0	0.6		
						10.0	9.5	10.0	3.7	10.0	1.0		
								15.0	8.0	15.0	2.0	15.0	0.7
								20.0	14.0	20.0	3.4	20.0	1.2
										25.0	5.0	25.0	1.8
										30.0	6.5	30.0	2.5
										40.0	12.0	40.0	4.4
												50.0	6.0
												60.0	9.0
												70.0	13.0

INDEX

Major coverage in **bold type**, herbicides and crops in use tables in *italics*, and major use table for crops or herbicides in ***bold-italics***.